INTERNATIONAL ASTRONOMICAL UNION
UNION ASTRONOMIQUE INTERNATIONALE

ORGANIC MATTER IN SPACE

PROCEEDINGS OF THE 251st SYMPOSIUM OF THE
INTERNATIONAL ASTRONOMICAL UNION
HELD IN HONG KONG, CHINA
FEBRUARY 18–22, 2008

Edited by

SUN KWOK
Faculty of Science, The University of Hong Kong, Pokfulam Road, Hong Kong

and

SCOTT SANDFORD
Astrophysics Branch, NASA/Ames Research Center, USA

CAMBRIDGE
UNIVERSITY PRESS

CAMBRIDGE UNIVERSITY PRESS
The Edinburgh Building, Cambridge CB2 8RU, United Kingdom
32 Avenue of the Americas, New York, NY 10013-2473, USA
477 Williamstown Road, Port Melbourne, VIC 3207, Australia
Ruiz de Alarcón 13, 28014 Madrid, Spain
Dock House, The Waterfront, Cape Town 8001, South Africa

First published 2008

Printed in the United Kingdom at the University Press, Cambridge

Typeset in System LATEX 2_ε

A catalogue record for this book is available from the British Library

Library of Congress Cataloguing in Publication data

10 056 72001

ISBN 9780 521 889 827 hardback
ISSN 1743-9213

Table of Contents

Session I Observations of organic compounds beyond the Solar System

Session Chairs: William Irvine, Ewine van Dishoeck, Yvonne Pendleton & Hans Olofsson

Session II Organic compounds within the Solar System
Session Chairs: Scott Sandford, Ernst Zinner & Dale Cruikshank

Preface

The idea of proposing an IAU symposium on the topic of organic matter in space germinated in the summer of 2006, when we organized a special session on "Organic Compounds: From Stars to the Solar System" at the 208th general meeting of the American Astronomical Society in Calgary. The success of the half-day session convinced us that there are sufficient new developments in this field to warrant the organization of an international meeting on a larger scale. While the IAU has held symposia on the topic of astrochemistry, the coverage of this earlier series of meetings has largely focused on gas-phase molecules observed with millimeter-wave techniques. We planned to bring together observational astronomers, laboratory spectroscopists, and Solar System scientists to share their expertise in order to come up with new ideas for the solution to the many unsolved mysteries associated with the origin, evolution, and distribution of organic matter of all forms in space.

Since the 1980s, infrared ground-based, airborne, and space-based spectroscopic observations have found evidence of complex carbonaceous compounds with aromatic and aliphatic structures in circumstellar and interstellar media. Millimeter and submm observations have detected rotational transitions of over 140 molecules, including hydrocarbons, alcohols, acids, aldehydes, ketones, amines, ethers, and other organic molecules. Organic molecules and compounds are also believed to be the carriers of the diffuse interstellar bands, the 2175 Å extinction feature, and the extended red emission (ERE), and the family of "PAH" infrared emission bands seen in a wide variety of astrohysical environments. Laboratory isotopic analysis of meteorites and interplanetary dust collected in the upper atmosphere, and now cometary materials have revealed the presence of presolar grains similar to those formed in evolved stars. These Solar System materials also contain complex organics, at least some of which show isotopic evidence that they have an interstellar chemical heritage. Spectroscopic studies of comets and asteroids also show spectral signatures of organics. The direct link between star dust and the Solar System therefore suggests that the early Solar System was chemically enriched by both stellar ejecta and the produces of interstellar processing. The solution to these long-standing mysteries will depend on close working relationships between astronomers and laboratory spectroscopists.

The Cassini mission and the Huygens probe have returned new results on the chemical composition of planetary and satellite atmospheres. There is an increasing recognition that organic compounds are major constituents of the atmosphere and surface of Titan. The sample return from the STARDUST mission is currently providing us with a great opportunity to examine the content of stellar material in the Solar System. This, in conjunction with recent progress in laboratory simulations of circumstellar and interstellar dust, will open up a new area of research tying together astronomy, chemistry, geology, and even biology.

It is in this context that the IAU Symposium 251 on "Organic Matter in Space" was held on February 18-22, 2008 in Hong Kong. The meeting was attended by 162 participants from 22 countries and 2 regions. A total of 58 oral talks and 79 posters were presented.

The conference was opened by the President of the IAU, Dr. Catherine Cesarsky. After the opening review by Ewine van Dishoeck, the first day of the meeting was devoted to astronomical observations of gas-phase organic molecules in the interstellar medium, including emissions from molecules in star formation regions and absorptions in the diffuse interstellar medium. Also discussed was the role played by organic molecules on the formation of the diffuse interstellar bands. The second day of the meeting focused to the observations of organic molecules and solids in circumstellar environments, spectral line surveys, and the observation of organic species in external galaxies. On the third day, the analysis of organic materials in interplanetary dust particles, meteoroids, planetary surfaces, asteroids, and comets were discussed. Analysis of the organic content of Comet 81P/Wild 2 based on results from the Stardust Comet Sample Return Mission were also presented. On the fourth day, discussions on the Solar System continued, with emphasis on the planetary satellites Europa and Titan. Issues relating to presolar grains and their relations with AGB stars were also addressed. The sessions on Friday were devoted to laboratory studies, including simulations of molecular synthesis and the spectroscopic properties of possible laboratory analogs of interstellar organic compounds such as poly-cyclic aromatic hydrocarbons (PAH), quenched carbonaceous composites (QCC), and hydorgenated amorphous carbon (HAC). All these sessions were attended by virtually everyone in attendance at the meeting irrespective of expertise of academic background, and the result was an excellent interdisciplinary exchange.

Many participants expressed appreciation of the fact that they met new colleagues in other fields, and that the talks and posters in the meeting have stimulated them to explore new ideas and interdisciplinary studies. We hope that this symposium has helped to break down disciplinary boundaries and represents a beginning in our pursuit of a new understanding of stellar-Solar System connections.

In order to capture the spirit of the discussions, we have included some of the discussions transcribed from the audio recordings. Although this involved more work than the traditional written question and answer format, it reflects the more spontaneous atmosphere of the QA sessions. Finally, the editors would also like to express their gratitude to Ms. Anisia Tang who provided valuable assistance in the production of the proceedings with great enthusiasm and dedication.

Sun Kwok and Scott Sandford, co-chairs SOC,
Hong Kong, China, June 17, 2008

From the Local Organizing Committee

IAU Symposium 251 was the first IAU meeting ever held in Hong Kong, a city of strong international character with a mixed culture of the east and west. Although Hong Kong has been a popular destination of international conferences and trades shows, there has never been a large-scale astronomy conference in Hong Kong. With the growing astronomical research activities in Asia, it is appropriate that Hong Kong will play a role as a meeting place of new ideas from a broad geographical representation. Indeed, IAU symposium 251 was attended by participants from six continents (Africa, Asia, Australia, Europe, North America, and South America), 22 countries and 2 regions.

A welcome reception was held in the evening of February 17 at the Renaissance Harbour View Hotel next to the Hong Kong Convention Center, overlooking the harbor above the Wanchai Star Ferry terminal. The conference itself was held in the historical Loke Yew Hall of The University of Hong Kong, a university that has served Hong Kong for almost 100 years. The poster sessions were held in the Convocation Room, and the participants enjoyed the coffee breaks on the patio next to the poster room. During the Monday lunch break, the participants were welcomed with a traditional lion dance and a cutting of a roasted pig ceremony. Wednesday afternoon was free and many participants joined our tour of Hong Kong, where they visited the fishing village of Aberdeen, Victoria Peak, Repulse Bay, and Stanley on the south shore of the Hong Kong Island. The tour ended with views of the night lights of Hong Kong from the Avenue of the Stars and a night cruise of Victoria Habour. On Thursday evening, the conference banquet was held in the historical Repulse Bay Hotel, where the participants enjoyed a fascinating mask illusion show. The performer was able to change her face masks in a millisecond, right in front of the audience, including a change to the face of Scott Sandford while talking to him. Our after-dinner speaker was Prof. Clifford Matthews, who was a student of The University of Hong Kong and prisoner of war in Japan after his capture by the Japanese after the surrender of Hong Kong in 1941. Prof. Matthews is well-known for his theory on the role played by HCN polymers in the origin of life on Earth, and the recitation of his war-time experience has brought him a standing ovation after his speech at the banquet. After the conference, some participants joined the tour of the Lantau Island and the Po Lin Monastery. Additional details of the meeting and associated activities can be found in our web site www.hku.hk/science/iau251.

Significant local press coverage was given to the conference in Hong Kong, including two major stories in the South China Morning Post, a leading newspaper in Southeast Asia.

In addition to travel grants from the IAU, the Local Organizing Committee received financial assistance from the Lee Hysan Foundation, The Croucher Foundation, Fong Shu Fook Tong Foundation, K.C. Wong Education Foundation, and The University of Hong Kong. The support of these organizations are greatly appreciated.

The overall organization of the meeting was ably managed by Ms. Anisia Tang. She was assisted at the registration desk by a number of staff in the Faculty of Science office. Dr. Jason Pun took care of many of the local logistics and Dr. Steven Pointing helped to serve as expert guide to the Lantau Island. Many students (Henry Chan, Selina Chong, Gloria Cheung, Franky Wong, Daisy Mak, Xuemin Dai, Jay Tung, Xiaolei Tian, Vincent

Yu) also helped in various ways during the conference. The devotion and attention to details by Anisia was well beyond the call of duty and we all owe to her the smooth operation of conference and the enjoyable social events.

Sun Kwok, on behalf of the LOC
Hong Kong, China, May 20, 2008

THE ORGANIZING COMMITTEE

Scientific (SOC)

Peter Bernath (U.K.)
Thomas Geballe (U.S.A.)
Thomas Henning (Germany)
William Irvine (U.S.A.)
Sun Kwok (co-chair, China)
Karl Menten (Germany)

Tom Millar (U.K.)
Yvonne Pendleton (U.S.A.)
Scott Sandford (co-chair, U.S.A.)
Setsuko Wada (Japan)
Ernst Zinner (U.S.A.)

Local (LOC)

Kwing L. Chan (HKUST)
K.S. Cheng (HKU)
Albert C. Cheung (HK City U)
Allan S.C. Cheung (HKU)
Sun Kwok (co-chair, HKU)

Chun Ming Leung (HK Open U)
Junichi Nakashima (HKU)
Steve Pointing (HKU)
Jason C.S. Pun (co-chair, HKU)

Acknowledgements

The symposium is sponsored and supported by the IAU Divisions III (Planetary Systems Sciences), and VI (Interstellar Matter); and by the IAU Commissions No. 14 (Atomic & Molecular Data), No. 15 (Physical Studies of Comets & Minor Planets), No. 22 (Meteors, Meteorites & Interplanetary Dust) and No. 51 (Bio-Astronomy).

The Local Organizing Committee operated under the auspices of
The University of Hong Kong, Hong Kong, China.

Funding by the
International Astronomical Union,
and the institutions in Hong Kong
Lee Hysan Foundation,
The Croucher Foundation,
Fong Shu Fook Tong Foundation,
and
K.C. Wong Education Foundation,
are gratefully acknowledged.

Conference photograph

Participants

Yuri **Aikawa**, Kobe University, Japan   aikawa@kobe-u.ac.jp
Conel **Alexander**, Carnegie Institution of Washington, USA   alexande@dtm.ciw.edu
Daniel **Austin**, Brigham Young University, USA   austin@chem.byu.edu
Simon **Balm**, Santa Monica College, USA   balm_simon@smc.edu
Jeronimo **Bernard-Salas**, Cornell University, USA   jbs@isc.astro.cornell.edu
Max **Bernstein**, NASA Ames Research Center, USA   Max.P.Bernstein@nasa.gov
Samantha Kaj **Blair**, University of Georgia, USA   lugosam@physast.uga.edu
Sandrine **Bottinelli**, Leiden Observatory, The Netherlands   sandrine@strw.leidenuniv.nl
Philippe **Bréchignac**, University of Paris-Sud, France   philippe.brechignac@u-psud.fr
Rosario **Brunetto**, CNRS/Universite Paris Sud, France   rosario.brunetto@ias.u-psud.fr
Jorge Enrique **Bueno Prieto**, National university of Colombia, Colombia   jebuenop@unal.edu.co
Henner **Busemann**, The Open University, UK   h.busemann@open.ac.uk
Patrick **Cassam-Chenaï**, Université de Nice-Sophia Antipolis, France   cassam@unice.fr
Cecilia **Ceccarelli**, Observatoire de Grenoble, France   Cecilia.Ceccarelli@obs.ujf-grenoble.fr
Catherine **Cesarsky**, CEA-Saclay, France   catherine.cesarsky@cea.fr
Juhua **Chen**, Hunan Normal University, P.R. China   jhchen@hunnu.edu.cn
Isabelle **Cherchneff**, ETH Zuerich, Switzerland   isabelle.cherchneff@phys.ethz.ch
Allan S.C. **Cheung**, The University of Hong Kong, P.R. China   hrsccsc@hku.hk
Ching Chi, Fiona **Chu**, National Central University, Taiwan   cchu@phy.ncu.edu.tw
Svatopluk **Civiš**, J. Heyrovský Institute of Physical Chemistry ASCR, Czech Republic   civis@jh-inst.cas.cz
George **Cody**, Carnegie Institution of Washington, USA   g.cody@gl.ciw.edu
Martin **Cordiner**, Queens University Belfast, UK   m.cordiner@qub.ac.uk
Nick **Cox**, European Space Agency, Italy   Nick.Cox@sciops.esa.int
Nicolas **Crimier**, Laboratoire d'Astrophysique de Grenoble, France   ncrimier@obs.ujf-grenoble.fr
Dale **Cruikshank**, NASA Ames, USA   dcruikshank@mail.arc.nasa.gov
Maria **Cunningham**, University of New South Wales, Australia   maria.cunningham@unsw.edu.au
Herma **Cuppen**, Leiden Observatory, The Netherlands   cuppen@strw.leidenuniv.nl
Brad **Dalton**, Jet Propulsion Laboratory, USA   James.B.Dalton@jpl.nasa.gov
Ankan **Das**, Centre for Space Physics, India   ankan@csp.res.in,ankan.das@gmail.com
Walt **Duley**, University of Waterloo, Canada   wwduley@uwaterloo.ca
Alexandre **Faure**, Observatoire de Grenoble, France   afaure@obs.ujf-grenoble.fr
George **Flynn**, State University of New York at Plattsburgh, USA   flynngj@plattsburgh.edu
Hideaki **Fujiwara**, The University of Tokyo, Japan   fujiwara@astron.s.u-tokyo.ac.jp
Pedro **Garcia-Lario**, European Space Astronomy Centre, Spain   Pedro.Garcia-Lario@sciops.esa.int
Robin **Garrod**, Max-Planck-Institut fuer Radioastronomie, Germany   rgarrod@mpifr-bonn.mpg.de
Marie-Claire **Gazeau**, Université Paris XII-Val de Marne, France   gazeau@lisa.univ-paris12.fr
Wolf Dietrich **Geppert**, University of Stockholm, Sweden   wgeppert@hotmail.com
Yolanda **Gomez**, Centro de Radioastronomia y Astrofisica, Mexico   y.gomez@astrosmo.unam.mx
Arnold **Gucsik**, Max Planck Institute for Chemistry, Germany   gucsik@mpch-mainz.mpg.de
Ahmed Abdel **Hady**, Cairo University, Egypt   aahady@yahoo.com
Maria **Hajdukova Jr.**, Slovak Academy of Sciences, The Slovak Republic   astromia@savba.sk
DeWayne Terrence **Halfen**, University of Arizona, USA   halfendt@starfleet.as.arizona.edu
Mathias **Hamberg**, Stockholm University, Sweden   mathias.hamberg@gmail.com
Paul **Harvey**, University of Texas, Austin, USA   pmh@astro.as.utexas.edu
Thomas **Henning**, Max-Planck-Institute für Astronomie, Germany   henning@mpia.de
Kenneth **Hinkle**, National Optical Astronomy Observatory, USA   hinkle@noao.edu
Åke **Hjalmarson**, Chalmers University of Technology, Sweden   ake.hjalmarson@chalmers.se
Mitsuhiko **Honda**, Kanagawa University, Japan   hondamt@kanagawa-u.ac.jp
Bruce **Hrivnak**, Valparaiso University, USA   Bruce.Hrivnak@valpo.edu
Chih-Hao **Hsia**, National Central University, Taiwan   d929001@astro.ncu.edu.tw
Susana **Iglesias-Groth**, Instituto de Astrofísica de Canarias, Spain   sigroth@iac.es
William M. **Irvine**, University of Massachusetts, USA   irvine@astro.umass.edu
Thierry **Jacq**, OASU, France   jacq@obs.u-bordeaux1.fr
Cornelia **Jäger**, Max-Planck-Institut für Astronomie, Germany   conny@astro.uni-jena.de
Biwei **Jiang**, Beijing Normal University, P.R. China   bjiang@bnu.edu.cn
Paul **Jones**, The University of New South Wales, Australia   pjones@phys.unsw.edu.au
Kay **Justtanont**, Onsala Space Observatory, Sweden   kay.justtanont@chalmers.se
Claudine **Kahane**, Grenoble University, France   claudine.kahane@ujf-grenoble.fr
Chihiro **Kaito**, Ritsumeikan University, Japan   kaito@se.ritsumei.ac.jp
Hidehiro **Kaneda**, Japan Aerospace Exploration Agency, Japan   kaneda@ir.isas.jaxa.jp
Daniel Brett **Kaplan**, Stevens Institute of Technology, USA   dkaplan@stevens.edu
Maja **Kaźmierczak**, Nicolaus Copernicus University, Poland   maja.kazmierczak@gmail.com
Bishun N. **Khare**, Nasa Ames Research Center, USA   bkhare@mail.arc.nasa.gov
Rafael Kobata **Kimura**, Universidade de São Paulo, Brazil   kimura@astro.iag.usp.br
Yuki **Kimura**, Ritsumeikan University, Japan   ykimura@se.ritsumei.ac.jp
Claudia **Knez**, University of Maryland, USA   claudia@astro.umd.edu
Kensei **Kobayashi**, Yokohama National University, Japan   kkensei@ynu.ac.jp
Nico Adrian **Koning**, University of Calgary, Canada   nkoning@iras.ucalgary.ca
Yi-Jehng **Kuan**, National Taiwan Normal University, Taiwan   kuan@alioth.geos.ntnu.edu.tw
Akihito **Kumamoto**, Ritsumeikan University, Japan   rp008011@se.ritsumei.ac.jp
Sun **Kwok**, The University of Hong Kong, P.R. China   sunkwok@hku.hk
Chinshuang **Lee**, National Central University, Taiwan   cslee@phy.ncu.edu.tw
Man Hoi **Lee**, The University of Hong Kong, P.R. China   mhlee@hku.hk
Chun Ming **Leung**, The Open University of Hong Kong, P.R. China   cmleung@ouhk.edu.hk
Anny-Chantal **Levasseur-Regourd**, Université Pierre et Marie Curie, Paris, France   aclr@aerov.jussieu.fr
Chien-Hsien **Lin**, National Central University, Taiwan   m959007@astro.ncu.edu.tw
René **Liseau**, Onsala Observatory, Sweden   rene.liseau@chalmers.se
Xiang **Liu**, Urumqi Astronomical Observatory, P.R. China   liux@uao.ac.cn
Ralph **Lorenz**, Johns Hopkins University Applied Physics Laboratory, USA   ralph.lorenz@jhuapl.edu
Adriana Patricia **Lozano Medellin**, National University of Colombia, Colombia   aplozanom@unal.edu.co
Kam Biu **Luk**, UC Berkeley, USA   k_luk@lbl.gov
Matthias **Maercker**, Stockholm Observatory, Sweden   maercker@astro.su.se

John Paul **Maier**, University of Basel, Switzerland — j.p.maier@unibas.ch
Giuliano **Malloci**, INAF-Osservatorio Astronomico di Cagliari, Italy — gmalloci@ca.astro.it
Arturo **Manchado**, Instituto de Astrofísica de Canarias, Spain — amt@iac.es
Ingrid **Mann**, Kobe University, Japan — mann@diamond.kobe-u.ac.jp
Oscar **Martinez Jr.**, University of Colorado, USA — oscar.martinez@colorado.edu
Hiroko **Matsumoto**, University of Tokyo, Japan — matsumoto@astron.s.u-tokyo.ac.jp
Mikako **Matsuura**, National Astronomical Observatory of Japan — mkmatsuura@nifty.com,mikako@optik.mtk.nao.ac.jp
Clifford **Matthews**, USA — Hcnmatthews@cs.com
Henry **Matthews**, National Research Council of Canada, Canada — Henry.Matthews@nrc-cnrc.gc.ca
Vito **Mennella**, INAF Osservatorio Astronomico di Capodimonte, Italy — mennella@na.astro.it
Giacomo **Mulas**, INAF - Osservatorio Astronomico di Cagliari, Italy — gmulas@ca.astro.it
Michael **Mumma**, NASA Goddard Space Flight Center, USA — mmumma@ssedmail.gsfc.nasa.gov
Chinnathambi **Muthumariappan**, Indian Institute of Astrophysics, India — muthu@iiap.res.in
Jun-ichi **Nakashima**, The University of Hong Kong, P.R. China — junichi@hku.hk
Alicia **Negron-Mendoza**, Instituto de Ciencias Nucleares, Unam, Mexico — negron@nucleares.unam.mx
Larry **Nittler**, Carnegie Institution of Washington, USA — lrn@dtm.ciw.edu
Delphine **Nna-Mvondo**, Centro de Astrobiologia (CAB) / CSIC-INTA, Spain, Spain — nnamvondod@inta.es
Michel **Nuevo**, NASA Ames Research Center, USA — michel.nuevo-1@nasa.gov
Joseph **Nuth**, NASA Goddard Space Flight Center, USA — Joseph.A.Nuth@NASA.gov
Karin **Öberg**, Leiden University, The Netherlands — oberg@strw.leidenuniv.nl
Masatoshi **Ohishi**, National Astronomical Observatory of Japan, Japan — masatoshi.ohishi@nao.ac.jp
Hans **Olofsson**, Chalmers Inst. of Technology, Sweden — hans.olofsson@chalmers.se
Kasandra **O'Malia**, University of Colorado, Boulder, USA — jorgenkk@colorado.edu
Takashi **Onaka**, University of Tokyo, Japan — onaka@astron.s.u-tokyo.ac.jp
Renaud **Papoular**, France — papoular@wanadoo.fr
Amit **Pathak**, Indian Institute of Science, India — amitpathak1234@rediffmail.com
Yvonne **Pendleton**, NASA headquarters, USA — yvonne.pendleton@nasa.gov
Carina Margareta **Persson**, Onsala Space Observatory, Sweden — carina.persson@chalmers.se
Sergio **Pilling**, Pontifical Catholic University of Rio de Janeiro, Brazil — sergiopilling@gmail.com
Frank **Postberg**, MPI für Kernphysik, Germany — Frank.Postberg@mpi-hd.mpg.de
Thomas **Prevenslik**, Consultant, Discovery Bay, Hong Kong, P.R. China — thomas.prevenslik@gmail.com
Jason **Pun**, The University of Hong Kong, P.R. China — jcspun@hkucc.hku.hk
Eric **Quirico**, Université Joseph Fourier, France — eric.quirico@obs.ujf-grenoble.fr
Sofia **Ramstedt**, Stockholm Observatory, Sweden — sofia@astro.su.se
Mark **Rawlings**, Joint Astronomy Centre, Hawaii, USA — m.rawlings@jach.hawaii.edu
Helen **Roberts**, The Queen's University of Belfast, UK — h.roberts@qub.ac.uk
Vera Kalenikovna **Rosenbush**, National Academy of Sciences of Ukraine, Ukraine — rosevera@MAO.Kiev.UA
Paul **Ruffle**, National Radio Astronomy Observatory, USA — paulruffle@paulruffle.com
Midori **Saito**, Ritsumeikan University, Japan — rp010027@se.ritsumei.ac.jp
Itsuki **Sakon**, The University of Tokyo, Japan — isakon@astron.s.u-tokyo.ac.jp
Farid **Salama**, NASA/Ames Research Center, USA — Farid.Salama@nasa.gov
Scott **Sandford**, NASA/Ames Research Center, USA — ssandford@mail.arc.nasa.gov
Peter **Sarre**, The University of Nottingham, UK — peter.sarre@nottingham.ac.uk
Fredrik **Schöier**, Onsala Space Observatory, Sweden — schoier@chalmers.se
Yuriy **Serozhkin**, Institute of Semiconductor Physics, Ukraine — yuriy.serozhkin@zeos.net
Bhalamurugan **Sivaraman**, The Open University, UK — B.Sivaraman@open.ac.uk
Gregory C. **Sloan**, Cornell University, USA — sloan@isc.astro.cornell.edu
Erin **Smith**, University of California, Los Angeles, USA — erincds@astro.ucla.edu
Theodore P. **Snow**, University of Colorado, USA — tsnow@casa.colorado.edu
In-Ok **Song**, Kyung Hee University, Korea — songio@khu.ac.kr
Angela **Speck**, University of Missouri- Columbia, USA — speckan@missouri.edu
Ryszard **Szczerba**, N. Copernicus Astronomical Center, Poland — szczerba@ncac.torun.pl
Dahbia **Talbi**, CNRS, France — talbi@graal.univ-montp2.fr
Emily Dale **Tenenbaum**, University of Arizona, USA — emilyt@as.arizona.edu
Johanna **Teske**, American University, USA — jt8344a@american.edu
Alan **Tokunaga**, University of Hawaii, USA — alantoku@hawaii.edu,tokunaga@ifa.hawaii.edu
Gian Paolo **Tozzi**, INAF - Osservatorio Astrofisico di Arcetri, Italy — tozzi@arcetri.astro.it
Wei-Ling **Tseng**, National Central University, Taiwan — d939006@astro.ncu.edu.tw
Deepak Bhadramukh **Vaidya**, Gujarat Arts & Science College, India — dbv@satyam.net.in
Ewine **van Dishoeck**, Leiden Observatory, The Netherlands — ewine@strw.leidenuniv.nl
Erik **Vigren**, Stockholm University, Sweden — erik.vigren@physto.se
Ruud **Visser**, Leiden Observatory, The Netherlands — ruvisser@strw.leidenuniv.nl
Setsuko **Wada**, The University of Electro-Communications, Japan — wada@pc.uec.ac.jp
Jack **Waite**, Southwest Research Institute, USA — hwaite@swri.edu
Junxian **Wang**, University of Science and Technology of China, P.R. China — jxw@ustc.edu.cn
Ronald **Weinberger**, University of Innsbruck, Austria — Ronald.Weinberger@uibk.ac.at
Laurent **Wiesenfeld**, Université Joseph Fourier Grenoble, France — laurent.wiesenfeld@obs.ujf-grenoble.fr
Eva Sofia **Wirström**, Onsala Space Observatory, Sweden — eva.wirstrom@chalmers.se
Mau **Wong**, California Institute of Technology, USA — mau.c.wong@jpl.nasa.gov
Alwyn **Wootten**, National Radio Astronomy Observatory, USA — awootten@nrao.edu
Yanling **Wu**, Cornell University, USA — wyl@astro.cornell.edu
Hikaru **Yabuta**, Carnegie Institution of Washington, USA — hyabuta@ciw.edu
Bin **Yang**, University of Hawaii, USA — yangbin@ifa.hawaii.edu
Tai-Sone **Yih**, National Central University, Taiwan — tsyih@phy.ncu.edu.tw
David T **Young**, Southwest Research Institute, USA — dyoung@swri.org
Hanyu **Zhang**, Stevens Institute of Technology, USA — zhanghanyu1999@yahoo.com
Ke **Zhang**, Beijing Normal University, P.R. China — coco-tree@163.com
Yong **Zhang**, The University of Hong Kong, P.R. China — zhangy96@hku.hk
Ernst **Zinner**, Washington University, USA — ekz@wuphys.wustl.edu
Lucy **Ziurys**, University of Arizona, USA — lziurys@as.arizona.edu

Opening Address of Symposium 251

Catherine Cesarsky
European Southern Observatory
Garching bei Muenchen, Germany

The International Astronomical Union (IAU) was founded in 1919, with the mission to promote and safeguard the science of astronomy in all its aspects through international cooperation. Now, close to 90 years later, it is stronger than ever, with almost 10000 individual members, professional astronomers from 87 countries on the globe. One of the key activities is to organize scientific meetings, nine Symposia per year on carefully selected topics. Every three years, the IAU holds a General Assembly, which brings six of the Symposia of the year to one single location. These are accompanied by a flurry of other activities: high level invited discourses, joint discussions over interdisciplinary topics, special meetings on topical subjects, lunches for young astronomers or for discussions about women in astronomy, business meetings of the various scientific Divisions and Commissions, award ceremony for the Gruber prize in cosmology, forum on future facilities in astronomy. Recent IAU General Assemblies (GA) took place in Sydney (2003) and Prague (2006). The 2009 GA will take place in Rio, and in 2012 Beijing will be hosting the gathering. IAU also organizes, every three years, large regional meetings in various areas of the world. In 2008, APRIM, the 10th Asian Pacific Regional IAU Meeting will take place also in China, in Kunming, Yunnan.

Another important activity of IAU is the promotion of educational activities in astronomy, mainly aimed at developing countries which, most often, are not IAU members and do not yet have an established astronomical community.

Next year, 2009, will be very special: following the request of IAU, endorsed by UNESCO, it has been proclaimed by the United Nations "International Year of Astronomy". It will be a global celebration of astronomy; 400 years after Galileo first put his eye through a telescope to look at the moon. With the motto: "The universe yours to discover", our aim is to foster a greater appreciation of the inspirational aspects of astronomy. The search for our cosmic origin, our common heritage, interests and connects all of mankind. Astronomers, thanks to taxpayers' money, can devote their lives to try to unravel the mysteries of the universe. This time, rather than remaining in our usual environment of professional astronomers, IAU members and Observatories, we want to reach out to science museums, planetariums, and amateur astronomers, and to countries which have little opportunities to hear about past and recent astronomical discoveries. We have much to gain by sharing our knowledge and our passion. Not only is it rewarding for us, but also when interesting the public, and particularly the young, we obtain additional support from politicians and decision makers to be able to realize some of our dreams.

We are meeting here for the start of a Symposium which elicited great enthusiasm in the decision bodies of IAU, not only because it is the first IAU Symposium taking place in magnificent Hong Kong, but also or mainly because of the appealing topic: "Organic matter in space". It is worthwhile to trace back the development of this subject through the history of IAU Symposia and Colloquia. For many years, the main related topic, which has been reconsidered periodically, is planetary nebulae (PN). The first IAU Symposium on Planetary Nebulae was in Czechoslovakia in 1967. There was then a ten year delay until a meeting in Ithaca, USA, but after that PN Symposia took place practically every five years: London in 1982, Mexico in 1987, Innsbruck in 1992, Groningen in 1996. In 2001, Sun Kwok was co-chair, in Canberra, and as you can see he did not loose his enthusiasm for this type of endeavor. The most recent IAU PN meeting was called "Planetary Nebulae in our Galaxy and beyond", in Hawaii in 2006, showing the great progress obtained owing to large telescopes. Another connected subject that has been addressed several times in IAU meetings is that of interstellar dust, starting with a Colloquium in Jena in 1969, then a Symposium in Albany in 1972. It came back as a Symposium in 1988 in Santa Clara, California.

In parallel, the Solar System community was discussing "Asteroids, comets, meteoric matter", starting with a Colloquium in Nice in 1972, This became a sparse series, at least as seen from the IAU vantage point, with meetings in Lyon in 1976, in Italy in 1993, in Rio de Janeiro in 2005. There were also Colloquia on comets which considered the physical state of the cometary matter, starting with Greenbelt in 1974, then "Comets, gases, ices, grains and plasmas" in Tucson in 1981, followed by Bamberg in 1989 and Tenerife in 2002. Interplanetary dust properties were also addressed, first in connection with zodiacal light, in Honolulu in 1967 and in Heidelberg in 1975, then on its own merits: "Solid particles in the Solar System", featured at a Symposium in Ottawa in 1979, followed by one in Marseilles "Origin and evolution of interplanetary dust", a Colloquium in Kyoto in 1990, and one in Gainesville in 1995. By 2000, this expanded into: "Dust in the Solar System and other planetary systems" in Canterbury.

The existence and importance of molecules in interstellar space was highlighted at a Symposium in Quebec in 1979, and thoroughly analyzed again in Leiden in 1996. The advances in all these topics required the knowledge of chemists, and thus astrochemistry became a new branch of astrophysics, with a special Working Group in Division VI, currently chaired by Ewine Van Dishoeck. There have been already four IAU Symposia in Astrochemistry, in Goa in 1985, in Sao Paolo in 1991, in South Korea in 1999, and in Monterey in 2005.

Here we will discuss Organic matter, and of course our thoughts are immediately drawn to the exciting concepts relating to the search for life in the universe. In IAU meetings, this first appeared in a specific context, intelligent extra terrestrial life, at a Symposium in Boston in 1984. But then the interest switched to the search for life in any form, and another new discipline was born, bio-astronomy, which a dedicated Commission in Division III, chaired by Alan Boss, and has rise to two Colloquia, one in Lake Balaton in 1987 and the other in Capri in 1996, and one Symposium, in Australia in 2002.

The present meeting aims at integrating results discussed in this vast array of Symposia and Colloquia, plus a large amount of completely new data, based on the now accepted fact that organic matter and its possible consequence, life on the universe, must not be considered as an isolated phenomenon on the Earth, or in comets or meteors in the Solar system, or in interstellar clouds and protostellar disks, but must take account of all the stages through which cosmic matter and stardust have gone. We are at the threshold of momentous discoveries in this particularly intriguing field, which so attracts the attention of human society at large.

The fascination with the topic of the Symposium is enhanced by the fact that it is being held in beautiful and vibrant Hong Kong. It is a great privilege to be here, in the historic part of the Hong Kong University. We have grown accustomed, in recent years,

2. The need for laboratory astrophysics

Several decades ago, Henk van de Hulst accused cosmologists of 'playing tennis without a net', when they were putting forward many models that could not be tested by any observations. Similarly, much of astrochemistry (and, in fact, much of astronomy as a whole) would be 'playing tennis without a net' if there were no laboratory data available to analyse and interpret the observational data of astronomical sources. The list of required data for organic compounds is extensive, and getting such information for even a single molecule often involves the building up of sophisticated laboratory equipment followed by years of painstaking data taking.

The most basic required information is spectroscopy of organic molecules from UV to millimeter wavelengths to identify the sharp lines and broad bands observed toward astronomical sources. One recent development in this area is the use of cavity ringdown spectroscopy to increase the sensitivity compared with classic absorption spectroscopy by orders of magnitude, which has allowed measurements of rare species that can be produced only in small amounts. Examples include gaseous polycyclic aromatic hydrocarbons (PAH) (e.g., Tan & Salama 2005, Rouille *et al.* 2007) or carbon chains (e.g., Dzhonson *et al.* 2007, Linnartz *et al.* 2000), in addition to matrix-isolation studies of large samples of PAHs (e.g., Hudgins & Allamandola 1999). Spectroscopy data bases of solids, including silicates (e.g., Jaeger *et al.* 1998, 2003), carbonates (e.g., Posch *et al.* 2007), ices (e.g., Bisschop *et al.* 2007a, Bernstein *et al.* 2005) and carbonaceous material (e.g., Mennella *et al.* 1997, Jaeger *et al.* 2006, Muñoz-Caro *et al.* 2006) continue to grow.

The next step in understanding organics is to obtain rates for the various reactions that are expected to form and destroy organics under space conditions. Here recent developments include measurements and theory of gaseous neutral-neutral rate coefficients at low temperatures (e.g., Chastaing *et al.* 2001, Smith *et al.* 2006), branching ratios for dissociative recombination (e.g., Geppert *et al.* 2004), and rates for photodissociation of molecules exposed to different radiation fields (van Hemert & van Dishoeck 2008). Surface science techniques at ultra high-vacuum conditions are now being applied to study thermal- and photo-desorption (e.g., Collings *et al.* 2004, Öberg *et al.* 2007) and formation of simple organic ices at temperatures down to 10 K (e.g., Watanabe *et al.* 2004, Bisschop *et al.* 2007b), while more traditional set-ups continue to provide useful information on the formation of complex organics in ices exposed to UV (e.g., Elsila *et al.* 2007, Muñoz-Caro & Schutte 2003) and to higher energy particle bombardment (e.g., Hudson & Moore 2000). There is also a wealth of new literature on the formation of carbonaceous material in discharges (e.g., Imanaka *et al.* 2004) and its processing at higher temperatures and when exposed to UV (e.g., Dartois *et al.* 2007).

Finally, the techniques to analyze meteoritic and cometary material in the laboratory have improved enormously in the last decade, and now allow studies of samples on submicrometer scale. Examples include ultra-L^{2}MS and nano-SIMS (e.g., Messenger *et al.* 2007), XANES (e.g., Flynn *et al.* 2006) and NMR (e.g., Cody *et al.* 2005). Their development was essential for analysis of the samples returned by *Stardust*.

3. Which organic compounds are found where?

In the following sections, the observational evidence for organic material is summarized, together with the identification of the type of material, where possible (Figure 1). For small gas-phase molecules, the identification is unambiguous, but for larger compounds often only the types of carbon bonds making up the material can be specified. Carbon can be bonded in several ways: a triple CC bond with single H on the side (e.g.,

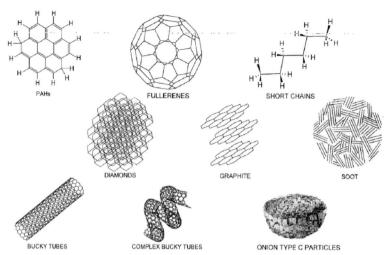

Figure 1. Examples of different types of carbonaceous material which are likely present in the ISM and Solar System (from: Ehrenfreund & Charnley 2000)

HC≡CH, denoted as *sp* hybridization); a double CC bond with two H's on each side (e.g, $H_2C=CH_2$, denoted as sp^2 hybridization) and a single CC bond with three H's on each side (e.g., H_3C-CH_3, denoted as sp^3 hybridization). In aromatic material, the electrons are delocalized over the entire molecule such as in a benzene ring or a polyethylene chain with alternating double and single CC bonds. The molecule thus contains sp^2 bonds. In aliphatic material, no double bonds occur, and only sp^3 hybridization is found. Spectral signatures of organic compounds include strong electronic transitions at optical and UV wavelengths, vibrational transitions at infrared wavelengths, and pure rotational transitions at millimeter wavelengths. In the latter case, the molecule needs to have a permanent dipole moment to be detected.

The majority of carbon in interstellar clouds (at least 50%) is in some form of carbonaceous solids with grain sizes large enough ($\sim 0.1\ \mu$m) not to have any clear spectroscopic signature, other than continuous opacity. Another fraction of the carbon (up to 30%) can be in gaseous C, C^+ and/or CO, or in CO and CO_2 ices. Most of the discussion in this paper concerns the remaining $\sim 20\%$ of carbon present in carbonaceous molecules, ices and small grains.

3.1. *Diffuse and translucent clouds*

Diffuse and translucent molecular clouds are concentrations of the interstellar gas with extinctions up to a few mag (for review, see Snow & McCall 2006). Typical temperatures range from 15–80 K and densities from a few 100–1,000 cm^{-3}. These are the only types of clouds for which high quality optical and UV spectra can be obtained by measuring the electronic transitions in absorption against bright background stars. In addition to the simplest organics CH, CH^+ and CN detected in 1937–1941, a series of more diffuse features called the 'diffuse interstellar bands' (DIBs) has been known since 1922. Nearly 300 DIBs are now known in the 4,000–10,000 Å range (e.g., Hobbs *et al.* 2008), but not a single one has yet been firmly identified despite numerous suggestions by the world's leading spectroscopists. Two bands at 9577 and 9632 Å are consistent with features of C_{60}^+, but laboratory spectroscopy of gaseous C_{60}^+ is needed for firm identifcation of the first fullerene in interstellar space (Foing & Ehrenfreund 1997). Long carbon chains with $n < 10$ and small PAHs have been excluded (Maier *et al.* 2004, Ruiterkamp *et al.* 2005),

3.3. *Dense star-forming regions*

Cold dense cores are the realm of the long unsaturated carbon chains such as HC_9N, discovered in the 1970's. Recent developments include the identification of negative ions such as C_6H^- and C_8H^- (McCarthy *et al.* 2006) as well as more saturated chains such as CH_2CHCH_3 (Marcelino *et al.* 2007). Taken together, these chains make up only a small fraction, $< 0.1\%$, of the total carbon budget.

Saturated complex organic molecules such as CH_3OH, CH_3OCH_3 and C_2H_5CN are commonly seen in high abundances toward warm star-forming regions such as Orion and Sgr B2, which have been surveyed at (sub)mm wavelengths for more than 30 years (e.g., Schilke *et al.* 2001). Such 'hot cores' have been detected around most massive protostars and are now commonly used as a signpost of the earliest stages of star formation. One recent development is that they are also found around low-mass protostars, with IRAS 16293–2422 as the prototypical example (Cazaux *et al.* 2003). Abundance ratios from source to source are remarkably constant (e.g., Bisschop *et al.* 2007c, Bottinelli *et al.* 2007) pointing to an origin in grain surface chemistry, although some variations between low- and high-mass sources are found. Also, a clear segregation of oxygen- and nitrogen-rich organics is seen (e.g., Wyrowski *et al.* 1999). One of the major questions is whether all observed complex organics are produced in the ice or whether some of them are formed in the hot gas following evaporation of ices (Charnley *et al.* 1992). Each organic molecule has an abundance of typically $10^{-9} - 10^{-7}$ with respect to H_2, but the total fraction of carbon locked up in these complex molecules can amount to a few %.

More complex organic molecules such as amino acids and bases, which are relevant for pre-biotic material, have not yet firmly been identified. Indeed, the spectra of hot cores are so crowded that line confusion is a serious issue. Ethylene glycol, CH_2OHCH_2OH, a complex organic found in comets, has been claimed in Sgr B2 (Hollis *et al.* 2002), but its detection is not yet fully secure.

The largest reservoir of volatile carbonaceous material is in the ices, whose strong mid-infrared absorption bands are seen not only toward most massive protostars (Gibb *et al.* 2004) but also toward a wide variety of low-mass YSOs (e.g., Boogert *et al.* 2008). Besides H_2O ice, CO, CO_2, OCN^-, CH_4, HCOOH, CH_3OH and NH_3 ice have been identified. The recent surveys of low-mass YSOs show that some molecules like CH_4 have relatively constant abundances of $\sim 5\%$ with respect to H_2O ice (Öberg *et al.* 2008), whereas those of CH_3OH vary from < 1 to 25% (see Bottinelli *et al.*, this volume). Altogether, the known organic molecules (excluding CO and CO_2) may lock up to 10% of the available carbon.

Most significant is the absence of PAH and 3.4 μm emission or absorption in the cold cores and deeply embedded stages of star formation. Indeed, a recent *Spitzer* and VLT survey of low-mass embedded YSOs shows no detections, indicating PAH abundances that are at least a factor of 10 lower than in the diffuse gas, perhaps due to freeze-out (Geers *et al.* 2008). No absorptions due to PAHs in ices have yet been found, but the lack of basic spectroscopy prevents quantitative limits.

3.4. *Protoplanetary disks*

Once the collapsing cloud has been dissipated, a young star emerges which can be seen at visible wavelengths but is still surrounded by a protoplanetary disk. PAH emission has been detected in roughly half of the disks surrounding Herbig Ae stars, i.e., intermediate mass young stars (Acke & van den Ancker 2004). More recently, PAHs have also been seen in a small fraction ($\sim 10\%$) of disks around solar mass T Tau stars (Geers *et al.* 2006). A quantitative analysis of the emission indicates PAH abundances that are typically factors of 10–100 lower than in the diffuse ISM, either due to freeze-out or caused by

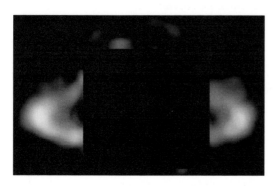
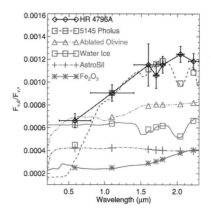

Figure 4. Left: False color HST image ($\sim 4'' \times 4''$) of the HR 4796A disk. Blue corresponds to 0.58 μm, green to 1.10 μm and red to 1.71 μm. Right: Disk / stellar flux ratio as function of wavelength. For comparison, grain models for candidate materials with a $n^{-3.5}$ size distribution with $a_{\min} = 3$ μm and $a_{\max} = 1,000$ μm are shown, normalized to the 1.10 μm data for HR 4796A and offset for clarity (from: Debes *et al.* 2008).

coagulation. The spatial extent of the PAH emission measured with adaptive optics on 8 m class telescopes is of order 100 AU, i.e., comparable with the size of the disk, but varying with feature (Habart *et al.* 2004). Modeling of the spatial extent as well as the destruction of PAHs by the intense UV or X-ray emission from the star indicates that the PAHs must be large, $N_C \approx 100$ (Geers *et al.* 2007b, Visser *et al.* 2007).

Smaller organics are present in high abundances in the inner disk (< 10 AU). Indeed, hot (400-700 K) C_2H_2 and HCN have been detected in absorption in edge-on disks with abundances factors of 1,000 larger than in cold clouds (Lahuis *et al.* 2006). Recently, they have also been seen in emission (Carr & Najita 2008). The observed abundances are consistent with models of hot dense gas close to LTE (e.g., Markwick *et al.* 2002).

A particularly intriguing class of disks is formed by the so-called transitional or 'cold' disks with large inner holes. An example is Oph IRS 48, in which a large (60 AU radius) hole is revealed in the large grain 19 μm image. Interestingly, PAHs are present inside the hole, indicating a clear separation of small and large grains in planet-forming zones (Geers *et al.* 2007a). Another intriguing case is formed by the more evolved disk around HR 4796A, which has likely lost most of its gas and is on its way to the debris-disk stage. Recent HST imaging shows colors with a steep red slope at 0.5–1.6 μm and subsequent flattening off (Debes *et al.* 2008) (Figure 4). These colors are reminiscent of those of minor planets in our Solar System, such as the Centaur Pholus, where the data are best fitted with tholins, i.e., complex organics produced in the laboratory in a CH_4/N_2 discharge with characteristics similar to Titan's haze (e.g., Cruikshank *et al.* 2005).

3.5. *Comets and minor planets*

Many volatile organics have been detected in bright comets like Hale-Bopp thanks to improved sensitivity at IR and mm wavelengths (see reviews by Bockelée-Morvan *et al.* 2000, 2006). Most of them are parent species evaporating directly from the ices. The list includes HCN, C_2H_2, C_2H_6, CH_3OH, ... , all of which except C_2H_6 have also been detected in star forming regions in the ice or gas. Typical abundances are 0.1–few % with respect to H_2O ice. A larger variety of comets originating from both the Oort cloud and Kuiper Belt have now been sampled, and variations in abundances between comets are emerging, with organics like CH_3OH and C_2H_2 depleted by a factor of 3 or more in some comets

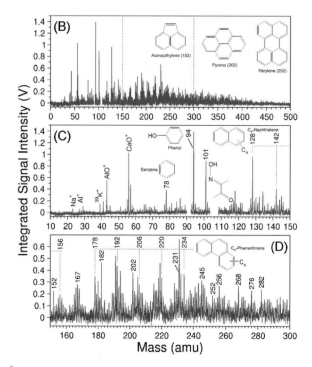

Figure 5. *ultra*-L^{2}MS analysis of one of the *Stardust* samples, showing the presence of small PAHs (from: Clemett *et al.* 2007).

(e.g., Kobayashi *et al.* 2007). PAHs have not yet been firmly identified by ground-based 3.3 μm spectra.

Fly-bys through the comae of Comets Halley, Borelly and Wild 2 have provided a much closer look at cometary material, including in-situ mass spectrometry of the gases. A major discovery of the *Giotto* mission to Halley was the detection of the so-called CHON particles: complex, mostly unsaturated, organics with only a small fraction of O and N atoms (Kissel & Krueger 1987).

A major question is whether the evaporating gases are representative pristine material unchanged since the comets were formed more than 4 billion yr ago, or whether they have been changed by 'weathering' (e.g., high-energy particle impact) during their long stay in the outer Solar System. The *Deep Impact* mission to Comet Tempel 1 was specifically designed to address this question, by liberating pristine ices from deep inside the comet following impact (A'Hearn *et al.* 2005). A sigificant increase in the IR emission around 3.5 μm, characteristic of CH$_3$CN and CH-X bands was seen immediate after impact, but no strong PAH bands were evident.

The *Stardust* mission has taken a major step forward in the study of primitive Solar System material by returning samples from Comet Wild 2 back to Earth, where they can be subjected to in-depth laboratory analysis using the most sophisticated experimental techniques (Brownlee *et al.* 2006). Wild 2 is a less evolved comet than others, having spent most of its lifetime in the Kuiper Belt and being captured into its currrent orbit only 30 years ago. Thus, it should not have suffered much thermal heating close to the Sun. Many complex organic molecules are found in the analysis of the *Stardust* particles to date, with a heterogeneous distribution in abundance and composition between particles. Many of the organics are PAHs, with typical sizes of just a few rings, i.e., generally smaller than the PAH size inferred in the ISM (Sandford *et al.* 2006, Clemett *et al.* 2007) (Figure 5).

Also, a new class of aromatic poor organic material is found compared with those seen in IDPs and meteorites, perhaps related to the fact that Wild 2 has had less thermal processing. The material appears richer in O and N than meteoritic organics. A major challenge for future studies will be to quantify the organics produced by the particle impacts inside the aerogel and isolate those from true cometary material.

Evidence for the presence of organics on other minor planets comes from their red colors at optical and near-IR wavelengths as seen in reflected sunlight. A particularly well studied case is the surface of the Centaur object Pholus (Cruikshank *et al.* 2005). Since discrete spectral features are lacking, identification of the material is not unique, but energy deposition in gas and ice mixtures containing CH_4 and N_2 produces tholins with colors similar to those observed (Imanaka *et al.* 2004).

Titan is particularly interesting because its atmosphere is thought to be similar to that of our (primitive) Earth, with the main difference being that it consists mostly of N_2 and CH_4 rather than N_2 and CO_2. The *Cassini* mission has studied Titan's haze in detail and the descent of the *Huygens* probe through the atmosphere has indeed revealed many nitrogen-rich organics (Niemann *et al.* 2005). Methane in the atmosphere must be continuously replenished by cryo-volcanism or other processes, since its lifetime due to photochemistry is short.

Besides tholins, HCN polymers have also been speculated to be part of the dark component present on outer Solar System bodies, including comets (Matthews & Minard 2006). It can also contribute to the orange haze in the stratosphere of Titan. Overall, it is clear that organics are a widespread component of Solar System material.

3.6. *Meteorites and IDPs*

The most primitive and least processed meteorites —the so-called carbonaceous chondrites — contain ample organic material. Well known examples are the Murchison, Orgueil and Tagish Lake meteorites, which contain up to 3% by weight in carbon-rich material. Most of the organics (60-80%) are in an insoluable macromolecular form, often described as 'kerogen-like'. The remaining 20% are in soluable form and have been found to contain corboxylic acids, PAHs, fullerenes, purines, amides, amides and other prebiotic molecules (e.g., Cronin & Chang 1993, Botta & Bada 2002). Amino acids –more than 80 different types– have also been found, but are likely formed from reactions of liquid water with HCN and H_2CO under the high pressure in the parent body rather than being primitive Solar System material.

Interplanetary dust particles (IDPs) have been collected through stratospheric flights over the past decade and analyzed in detail in the laboratory. Organic carbon, including aliphatic hydrocarbons and the carrier of the 2175 Å feature, are common (Flynn *et al.* 2000, Bradley *et al.* 2005).

4. Evolution of organic matter

As organic material evolves from the evolved stars to the diffuse and dense ISM, and subsequently from collapsing envelopes to disks, icy solar sytem bodies and meteorites, many processes can affect their composition and abundance (Ehrenfreund & Sephton 2006). From the AGB and PPN phase to the PN phase, UV processing changes aliphatics to aromatics material (Kwok 2007a). In the subsequent step, the organics can be shattered by shocks as they enter the diffuse ISM and are exposed to passing shocks from supernovae and winds (Jones *et al.* 1996). Destruction of graphite produces very small carbonaceous grains, including presumably the smaller PAHs. When the organics enter the dense cloud phase, freeze-out will affect all organics, coagulation can occur,

Gibb, E. L., Whittet, D. C. B., Boogert, A. C. A., & Tielens, A. G. G. M. 2004, *ApJS*, 151, 35

Goto, M., *et al.* 2007, *ApJ*, 662, 389

Habart, E., Natta, A., & Krügel, E. 2004, *A&A*, 427, 179

Hobbs, L. M., *et al.* 2008, *ApJ*, in press

Hollis, J. M., Lovas, F. J., Jewell, P. R., & Coudert, L. H. 2002, *ApJ* (Letters), 571, L59

Hudgins, D. M., & Allamandola, L. J. 1999, *ApJ* (Letters), 516, L41

Hudgins, D. M., Bauschlicher, C. W., & Allamandola, L. J. 2005, *ApJ*, 632, 316

Hudson, R. L., & Moore, M. H. 2000, *Icarus*, 145, 661

Imanaka, H., *et al.* 2004, *Icarus*, 168, 344

Jaeger, C., Molster, F. J., Dorschner, J., Henning, Th., Mutschke, H., & Waters, L. B. F. M. 1998, *A&A*, 339, 904

Jaeger, C., Dorschner, J., Mutschke, H., Posch, Th., & Henning, Th. 2003, *A&A*, 408, 193

Jaeger, C., *et al.* 2006, *ApJS*, 166, 557

Jones, A. P., Tielens, A. G. G. M., & Hollenbach, D. J. 1996, *ApJ*, 469, 740

Kissel, J. & Krueger, F. R. 1987, *Nature*, 326, 755

Kobayashi, H., Kawakita, H., Mumma, M. J., Bonev, B. P., Watanabe, J., & Fuse, T. 2007, *ApJ* (Letters), 668, L75

Kwok, S. 2007a, *Adv. Space Res.*, 40, 655

Kwok, S. 2007b, *Adv. Space Res.*, 40, 1613

Lahuis, F., *et al.* 2006, *ApJ* (Letters), 636, L145

Linnartz, H., *et al.* 2000, *J. Chem. Phys.*, 112, 9777

Lucas, R. & Liszt, H. S. 2000, *A&A*, 358, 1069

Maier, J. P., Walker, G. A. H., & Bohlender, D. A. 2004, *ApJ*, 602, 286

Marcelino, N., Cernicharo, J., *et al.* 2007, *ApJ* (Letters), 665, L127

Markwick, A. J., Ilgner, M., Millar, T. J., & Henning, Th. 2002, *A&A*, 385, 632

Matthews, C. N., & Minard, R. D. 2006, *Faraday Disc.*, 133, 393

McCarthy, M. C., Gottlieb, C. A., Gupta, H., & Thaddeus, P. 2006, *ApJ* (Letters), 652, L141

Mennella, V., Baratta, G. A., Colangeli, L., Palumbo, M., Rotundi, A., Bussoletti, E., & Strazzulla, G. 1997, *ApJ*, 481, 545

Messenger, S., Nakamura-Messenger, K., Keller, L., Matrajt, G., Clemett, S., & Ito, M. 2007, *Geochimica et Cosmochimica Acta*, 71, 15

Muñoz-Caro, G. M., Ruiterkamp, R., Schutte, W. A., Greenberg, J. M., & Mennella, V. 2001, *A&A*, 367, 347

Muñoz-Caro, G. M. & Schutte, W. A. 2003, *A&A*, 412, 121

Muñoz-Caro, G. M., *et al.* 2006, *A&A*, 459, 147

Niemann, H. B., *et al.* 2005, *Nature*, 438, 779

Öberg, K. I. *et al.* 2007, *ApJ* (Letters), 662, L23

Öberg, K. I. *et al.* 2008, *ApJ*, 678, 1032

Pendleton, Y. J. 2004, in: A. Witt *et al.* (eds.), *Astrophysics of Dust*, (ASP Vol. 304), p. 573

Pendleton, Y. J., & Allamandola, L. J. 2002, *ApJS*, 138, 75

Pety, J., *et al.* 2005, *A&A*, 435, 885

Posch, Th., Baier, A., Mutschke, H., & Henning, Th. 2007, *ApJ*, 668, 993

Rouille, G., *et al.* 2007, *J. Chem. Phys.*, 126, 174311

Ruiterkamp, R., *et al.* 2005, *A&A*, 432, 515

Sandford, S. A., *et al.* 2006, *Science*, 314, 1720

Schilke, P., Benford, D. J., Hunter, T. R., Lis, D. C., & Phillips, T. G. 2001, *ApJS*, 132, 281

Sloan, G.C., *et al.* 2005, *ApJ*, 632, 956

Smith, I. W. M., Sage, A. M., Donahue, N. M., Herbst, E., & Quan, D. 2006, *Faraday Discussions*, 133, 137

Snow, T. P., & McCall, B. J. 2006, *ARA&A*, 44, 367

Song, I.-O., Kerr, T. H., McCombie, J., & Sarre, P. J. 2003, *MNRAS* (Letters), 346, L1

Tan, K. F., & Salama, F. 2005, *J. Chem. Phys.*, 123, 14312

Tielens, A. G. G. M. 2008, *ARA&A*, in press

van Winckel, H., Cohen, M., & Gull, T. R. 2002, *A&A*, 390, 147 van Hemert, M. C. & van Dishoeck, E.F. 2008, *Chem. Phys.*, 343, 292

Visser, R., *et al.* 2007, *A&A*, 466, 229

Watanabe, N., Nagaoka, A., Shiraki, T., & Kouchi, A. 2004, *ApJ*, 616, 638

Wyrowski, F., Schilke, P., Walmsley, C. M., & Menten, K. M. 1999, *ApJ* (Letters), 514, L43

Yan, L., *et al.* 2005, *ApJ*, 628, 604

Discussion

MUMMA: Would you comment on the role of the X-ray processing around young stars, in particular through the enhanced ion-molecule processing such as H_3^+.

VAN DISHOECK: There is no doubt that young stars have X-rays and in fact you can even measure them. We know that the surfaces and interiors of a number of protoplanetary disks are exposed to X-rays. Any enhanced ionization rate will enhance H_3^+ and drives ion-molecule chemistry. However, to what extent that it will lead to complex organics is not certain. Of course too strong X-ray emission is not good because it will destroy the more volatile organic molecules. In the inner parts of the disks, it will be very difficult for small PAHs to survive there.

VAIDYA: What is against graphite as the carrier of the 2175 feature?

VAN DISHOECK: There is not much spectroscopic evidence because we just have one very broad feature. I have always felt that some form of graphite must be present, but exactly what the carrier is, is difficult to answer.

Michael Mumma, Ruud Visser, and Ewine Van Dishoeck having a lively discussion during the coffee break (photo by Dale Cruikshank).

The Loke Yew Hall.

Organic Matter in Space
Proceedings IAU Symposium No. 251, 2008
S. Kwok & S. Sandford, eds.

© 2008 International Astronomical Union
doi:10.1017/S174392130802108X

Molecular spectral line surveys and the organic molecules in the interstellar molecular clouds

Masatoshi Ohishi

Astronomy Data Center, National Astronomical Observatory of Japan, the Graduate
University for Advanced Studies (SOKENDAI), & the National Institute for Informatics,
2-21-1, Osawa, Mitaka, Tokyo, 181-8588, Japan
email: masatoshi.ohishi@nao.ac.jp

Abstract. It is known that more than 140 interstellar and circumstellar molecules have so far been detected, mainly by means of the radio astronomy observations. Many organic molecules are also detected, including alcohols, ketons, ethers, aldehydes, and others, that are distributed from dark clouds and hot cores in the giant molecular clouds. It is believed that most of the organic molecules in space are synthesized through the grain surface reactions, and are evaporated from the grain surface when they are heated up by the UV radiation from adjacent stars.

On the other hand the recent claim on the detection of glycine have raised an important issue how difficult it is to confirm secure detection of weak spectra from less abundant organic molecules in the interstellar molecular cloud.

I will review recent survey observations of organic molecules in the interstellar molecular clouds, including independent observations of glycine by the 45 m radio telescope in Japan, and will discuss the procedure to securely identify weak spectral lines from organic molecules and the importance of laboratory measurement of organic species.

Keywords. ISM: molecules, line: identification, molecular data

1. Introduction

In recent years highly sensitive observations on interstellar molecules have been conducted by, e. g., the Greenbank Telescope (GBT) and the Nobeyama 45 m telescope. Such observations provided lots of new insights on new species, molecular abundances, exsistence of organic species in a wide variety of objects, and so on. These will be the basis in understanding the interstellar chemistry.

Remarkable progress was made in observing large, complex organic species, e. g., glycolaldehyde, the simplest sugar, (Hollis *et al.* 2000) or glycine, the simplest α-amino acid, (Kuan *et al.* 2003), which may link to the understaning to the origin of life in the universe. However some observed spectra have weak signal intensities and these spectra are continated by other spectra, leading the debates that such observation results are plausible or not.

In this paper I will review several observations on organic molecules toward cold and dark clouds and star forming regions, and will discuss what are needed to verify a new identification in this era of crowded spectra.

2. Molecular Line Survey toward TMC-1

Cold dark interstellar clouds have been extensively studied as the formation sites of low-mass stars and planetary systems since their identification to the interstellar molecular clouds in 1970's. A variety of exotic chemical compounds found in molecular clouds, especially those containing carbon atoms, attracted strong interests in connection with the formation of planets and the origin of life in the universe. Recent radio and IR observations towards comets collected important evidence that comets, 4.6 billion year-old fossil bodies of the proto-solar-system nebula, keep molecular composition similar to that in cold dark clouds. Therefore, the chemical evolution in cold dark clouds is basically important as the initial process of interstellar matter evolution toward the planets, and, ultimately to life.

Chemical reactions in dark clouds are not yet fully understood, and many unknown molecules might be synthesized in dark clouds. It is essential for understanding the chemical reactions in dark clouds, therefore, to make an unbiased frequency survey that can detect all molecular lines, including unpredicted lines of unknown molecules.

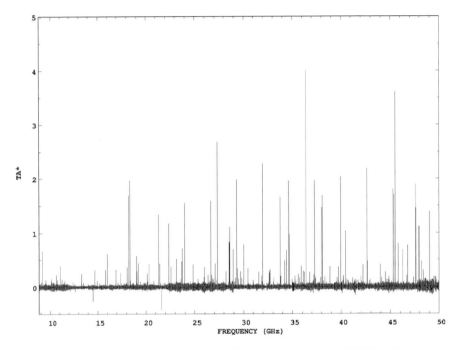

Figure 1. A compressed atlas of the 8.8–50 GHz spectrum toward TMC-1 (cyanopolyyne peak).

Kaifu *et al.* (2004) published a molecular spectral line survey data toward a dark cloud, the cyanoplyyne peak of TMC-1, in the frequency range between 8.8 and 50 GHz (see Figure 1), using the 45-m mm-wave telescope of the Nobeyama Radio Observatory†. They detected 414 lines from 38 molecules. Most of the molecules are linear carbon chain species and their derivatives, and there are only a few organic species such as CH_3OH, CH_3CHO, HCCCHO and CH_2CHCN. More saturated species, e. g., C_2H_5CN and $HCOOCH_3$, were not detected at all. According to their preliminary analysis (Ohishi & Kaifu 1998), these species generally have less abundances than major linear carbon chain molecules such as HC_3N and CCS, and it would be possible to conclude that the organic species are not the main constituent in the cold and dark clouds.

† Nobeyama Radio Observatory is an open-use facility for mm-wave astronomy, being operated under the National Astronomical Observatory of Japan (NAOJ).

3. Detection of Organic Species in the Early Stage of Protostellar Evolution

So far large organic molecules (e.g., $HCOOCH_3$, $(CH_3)_2O$, and C_2H_5CN) were observed with high abundances toward the hot cores such as Orion KL, Sgr B2(N) and W51 e1/e2 where O/B stars are formed. Such highly saturated molecules are difficult to be produced by gas-phase chemical reactions under low-temperature conditions, and hence the grain surface chemistry is thought to play an important role in their production. When star formation takes place, the grain mantles are heated up by various activities of newly born stars, supplying parent molecules, like CH_3OH and H_2CO, into the gas phase through evaporation processes. Subsequent gas-phase reactions under high-temperature and high-density conditions would produce large organic molecules.

Recently detections of large organic species were reported toward the hot corinos in IRAS 16293-2422 (Cazaux *et al.* 2003, Bottinelli *et al.* 2004, Kuan *et al.* 2004). Further detection of $HCOOCH_3$ toward NGC1333 IRAS4B (Sakai *et al.* 2006) and NGC2264 MMS3 (Sakai *et al.* 2007) were reported. Figure 2 shows the spectrum of $HCOOCH_3$ toward NGC1333 IRAS4B that is a class 0 low-mass protostar with an estimated age of a few 100 years. This suggests that the complex organic molecules appear from the very early stage of protostellar evolution.

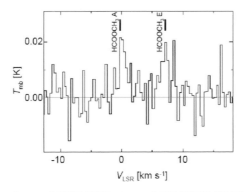

Figure 2. A Spectrum of $HCOOCH_3$ toward NGC1333 IRAS4B taken by the Nobeyama 45 m Telescope.

Sakai *et al.* (2007) found the $HCOOCH_3$ distribution, reveal by the Nobeyama Millimeter Array, seems to be similar to the case of Orion KL Compact Ridge where the molecular peak is shifted from the dust continuum peak, and suggests that such distribution similarity would give a hint to understand the formation of $HCOOCH_3$ in hot corinos and hot cores.

Since other highly saturated organic species, such as C_2H_5OH, $(CH_3)_2O$ and C_2H_5CN, are observed in hot cores, it would be necessary to conduct deep observations on these species toward low-mass protostars to further understand the formation mechanism and evaporation conditions of organic species.

4. Recent Reports of Large Organic Species toward High Mass Star Forming Regions

In the last several years many very large organic molecules were reported to be detected. This is primarily because of the powerful observation performance of the Greenbank Telescope (GBT) together with improvement of the receiver sensitivities.

These molecules include glycolaldehyde (CH_2OHCHO) (Hollis *et al.* 2000), ethyleneg-lycol (($CH_2OH)_2$) (Hollis *et al.* 2002), glycine (NH_2CH_2COOH) (Kuan *et al.* 2003), propenal (CH_2CHCHO) and propanal (CH_3CH_2CHO) (Hollis *et al.* 2004a), acetone (($CH_3)_2CO$) (Friedel *et al.* 2005), cyanoallene (CH_2CCHCN) (Lovas *et al.* 2006), ac-etamide (CH_3CONH_2) (Hollis *et al.* 2006), and cyanoformaldehyde ($CNCHO$) (Remijan *et al.* 2008).

The report on the detection of glycolaldehyde (Hollis *et al.* 2000) toward Sgr B2(N) by the Kitt peak 12 m telescope was an epoch making one, because glycolaldehyde is the simplest sugar and really a prebiotic molecule. However the reported spectra had low signal to noise ratios, and some of them appeared on the shoulders of other molecular lines. Such a situation had led to dabates if the detection was secure. In 2004 Hollis *et al.* (2004b) made observations of glycolaldehyde by the GBT in four frequency ranges below 22 GHz, showing data with sufficient signal to noise ratios. Three transitions out of four showed absorption features at 63 kms^{-1} where absorption features from other al-ready known species appear. The GBT observations made the detection of glycolaldehyde secure.

On the other hand the cases for propenal and ethyleneglycol are different. Only two propenal lines were reported toward Sgr B2(N), and the absorption features at 63 kms^{-1} is apparent for only one line. Five ethyleneglycol data were reported, however, they appeared on the shoulders of other molecular lines. This situation made difficult to in-vestigate if the intensity distribution was reasonable for species observed toward Sgr B2(N). Therefore it may be concluded that the detection of propenal and ethyleneglycol have not yet been confirmed.

The report on the detection of glycine, the simplest α-amino acid, by Kuan *et al.* (2003) had the similar situation in that the signal to noise ratios were not so high and many spectra seemed to be contaminated. They observed 27 frequency bands of glycine con-former I toward Orion KL, Sgr B2(N) and W51 e1/e2, however, only three of them were reported to be observed from three objects. Intensity analyses by means of the rotation diagram were made, and the distributions seemed plausible. However, the derived column densities of glycine for three sources had similar values, inconsistent with a fact that in most cases a column density of a molecule toward Sgr B2(N) is higher by about two order of magnitudes than those toward Orion KL and W51 e1/e2.

Therefore we made an independent observations by the 45 m telescope at Nobeyama. Because the glycine spectra had high possibility of contamination, we carefully examined our past observed data to find "clean spectrum windows" where no transitions from other known abundant species exist.

Figure 3 shows a spectrum observed toward Orion KL at around 90 GHz. There are transitions of glycine conformer I at 90043.13 MHz ($15_{1,15}-14_{1,14}$, upper energy = 24.63 cm^{-1}), 90049.71 MHz ($15_{0,15}-14_{0,14}$, upper energy = 24.62 cm^{-1}), and 90056.98 MHz ($15_{1,15}-14_{0,14}$, upper energy = 24.63 cm^{-1}). The expected brightness temperatures for these transitions are around 20 mK when we used the column density and the roration temperature derived by Kuan *et al.* (2003). One peak coincides with the glycine tran-sition at 90043.13 MHz with around the expected intensity, however, other two are not clearly seen. The line at 90043.13 MHz was not identified when we investigated available molecular line databases, leaving a possibility that the line could be identified to the glycine line. However we noticed that there is another line next to the 90043.13 MHz line with a similar line intensity. Such a "doublet line" could be due to the internal rotation, and it is well known that the molecular abundance of $HCOOCH_3$ toward Orion KL is so high. We contacted a laboratory molecular spectroscopist, Dr. Hitoshi Odashima at

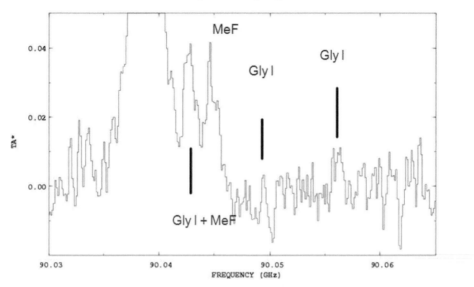

Figure 3. A Spectrum toward Orion KL at around 90 GHz where the Glycine (conformer I) transitions exist, that are shown by the vertical lines. MeF stands for $HCOOCH_3$.

Toyama University, Japan, asking if he had maesured weak lines of $HCOOCH_3$, and it was revealed that two unassigned lines of $HCOOCH_3$ matched the two lines above.

Therefore it can be concluded that the detection of interstellar glycine has not yet confirmed.

5. Toward Secure Line Identification

It was found that some reports on detection of a new interstellar molecule have not yet been confirmed.

Then what should we take into account toward secure line identifications?

Several issues may be listed toward secure line identifications, although not complete, as follows:

- Refer to reliable frequency data
- Many lines as possible, in "clean windows", with a consistent intensity distribution
- Expected lines should be observed with expected intensity
- Radial velocities should be consistent among observed transitions
- Observations in other sources with similar physical conditions (e.g. Orion KL, W51e1/e2) to see similar lines
- Less broad lines are desirable (e.g. NGC6334)

In the following subsections, I will disucuss some issues above.

5.1. *Reference to reliable frequency data*

It is crucial to refer to reliable frequency catalog data in order to make secure identifications.

There are several molecular line catalogs, such as the Cologne Database for Molecular Spectroscopy (CDMS) (Müller *et al.* 2001, Müller *et al.* 2005, http://www.ph1.uni-koeln.de/vorhersagen/), JPL Catalog (http://spec.jpl.nasa.gov/ftp/pub/catalog/catform.htm). These catalogues were compiled and maintained by microwave spectroscopisits, and the transition frequencies were calculated based on laboratory measurements. Most

Kaifu, N., Ohishi, M., Kawaguchi, K., Saito, S., Yamamoto, S., Miyaji, T., Miyazawa, K., Ishikawa, S., Noumaru, C., Harasawa, S., Okuda, M., & Suzuki, H. 2004, *PASJ*, 56, 69

Kuan, Y. J., Charnley, S. B., Huang, H. C., Tseng, W. L., & Kisiel, Z. 2003, *ApJ*, 593, 848

Kuan, Y. J., Huang, H. C., Charnley, S. B., Hirano, N., Takakuwa, S., Wilner, D. J., Liu, S. Y., Ohashi, N., Bourke, T. L., Qi, C. H., & Zhang, Q. Z. 2004, *ApJ* (Letter), 616, L27

Müller, H. S. P., Schlöder, F., Stutzki, J., & Winnewisser, G. 2005, *J. Mol. Struct.*, 742, 215

Müller, H. S. P., Thorwirth, S., Roth, D. A., & Winnewisser, G. 2001, *A&A* (Letter), 370, L49

Lovas, F. J., Remijan, A. J., Hollis, J. M., Jewell, P. R., & Snyder, L. E. 2006, *ApJ* (Letter), 637, L37

Odashima, H. 2006, private communication

Ohishi, M. & Kaifu, N. 1998, *J. Chem. Soc., Faraday Discussion*, 109, 205

Remijan, A. J., Hollis, J. M., Lovas, F. J., Stork, W. D., Jewell, P. R., & Meier, D. S. 2008, *ApJ* (Letter), 675, L85

Sakai, N., Sakai, T., & Yamamoto, S 2006, *PASJ* (Letter), 58, L15

Sakai, N., Sakai, T., & Yamamoto, S. 2007, *ApJ*, 660, 363

Snyder, L. E., Lovas, F. J., Hollis, J. M., Friedel, D. N., Jewell, P. R., Remijan, A. J., Ilyushin, V. V., Alekseev, E. A., & Dtubko, S. F. 2005, *ApJ*, 619, 914

Discussion

ZIURYS: It is a very big problem because basically it is wall-to-wall lines when you get down to more sensitive limits. What people really need to do is to have an energy level diagram and just go through and make sure all the transitions below a certain energy and above a certain line intensity are there. This requires probably observations of multiple bands. Otherwise, it's so easy to make a mistake. I think we are just beginning to realize that it's so treacherous.

OHISHI: I agree with that. I wouldn't say more.

ZIURYS: So I guess my comment is be careful on Sgr B2 and Orion and do lots of observations if you want to detect these things.

MATTHEWS: Thank you for your elegant non-detection of glycine. It is possible that amino acids will never be seen as free molecules in space, but that doesn't rule out that space maybe swarming with potential amino acids in the form of large molecules like HCN polymers, which will give rise to many amino acids once it gets hydrolyzed. They are very stable in space and we can receive them later from comets, meteorite, etc. So even if one makes the statement that there are no free amino acids in space, it has really very little to do with the whole origin of life question because space may be swarming with amino acid because like hydrogen cyanide polymers.

OHISHI: I use millimeter-wave observations and glycine is below our detection limit, but many people say if we go to the sub-millimeter wave region with very high spatial resolution, we can look for glycine in very dense regions. There remain a possibility that amino acids can be detected in the gas phase in the near future, e. g., with ALMA.

IRVINE: And of course we know there are lots of amino acids in meteorites.

MUMMA: I notice in the spectrum you show that there were four transitions which were marked to have these proper frequencies and this was regarded as a definition of the molecules being correctly identified. The problem that I saw on that slide was one that at least one of the lines is in absorption, the others were in emission. I want to point out that the issue of intensity is extremely critical, and particularly when radiative transfer is not being considered.

OHISHI: Yes, Sgr B2 is a very difficult source to understand. People say that there is hot area surrounded by cold cloud so the radiative transfer calculation needed to understand the intensity distribution correctly is very, very difficult.

MUMMA: so the question is how can one say that this is a detection unless you also explain the unusual intensity distribution?

OHISHI: We may need more high signal-to-noise ratio observations in different transitions, or we may need to observe other sources.

IRVINE: And also on what energy levels correspond to absorption and what energy levels correspond to emission.

The University of Hong Kong.

The delegates were welcomed by a traditional lion dance (photo by Dale Cruikshank).

Coffee break on the patio (by Mathias Hamberg).

Organic Matter in Space
Proceedings IAU Symposium No. 251, 2008
S. Kwok & S. Sandford, eds.

A confusion-limited spectral-line survey of Sgr B2(N) at 1, 2, and 3mm: Establishing the organic inventory in molecular clouds

DeWayne T. Halfen and Lucy M. Ziurys

Depts. of Chemistry and Astronomy, Arizona Radio Observatory, Laplace Center for
Astrobiology, and Steward Observatory, University of Arizona
933 N. Cherry Ave. Tucson, AZ 85721 USA
email: `halfendt@as.arizona.edu`

Abstract. We present preliminary results of an spectral-line survey at 1, 2, and 3 mm of the galactic center cloud Sgr B2(N). With the current data, several simple prebiotic molecules have been conclusively identified, while several more complex molecules have not. When complete, this survey will provide an accurate database of the gas-phase organic inventory in Sgr B2(N).

Keywords. ISM: molecules, astrochemistry, astrobiology, submillimeter, ISM: clouds

1. Introduction

Life on Earth is postulated to have started from simple chemical compounds. Where did this organic starting material come from? It is possible that the organic matter on the early Earth was delivered by comets, meteorites, and interplanetary dust particles. A diverse array of organic compounds have been found in meteorites, including sugars and amino acids (Cooper *et al.* 2001, Pizzarello *et al.* 2001). The most likely source of the meteoritic organic compounds is the interstellar medium (ISM). In the ISM, large quantities of organic matter have been found in giant molecular clouds. About half of the 140 molecules found in space have been detected in the Galactic center cloud Sgr B2(N), including the most complex organic species. Hence, the overall chemical composition of this source is of great interest. Therefore, we are conducting a spectral-line survey of this source to determine the identify and abundance of each molecule.

2. Observations

Observations towards Sgr B2(N) cover the millimeter atmospheric windows at 3 (65-116 GHz), 2 (130-180 GHz), and 1 (210-280 GHz) mm in wavelength. The observations were performed at the two radio telescopes of the Arizona Radio Observatory, the 12 m telescope on Kitt Peak and the Submillimeter Telescope (SMT) on Mt. Graham. The 12 m operates at 2 and 3 mm and the SMT performs measurements at 1 mm. The temperature scale is given as T_R^* for the data from 65-180 GHz, and T_A^* from 210-280 GHz. Spectral coverage is currently 78%, 66%, and 40% at 3, 2, and 1 mm, respectively, see Figure 1. This survey is conducted to the confusion limit in SSB mode, where the spectrum consists of a continuum of spectral lines, and maximum information can be obtained.

3. Results

One molecule of great interest is glycolaldehyde, CH_2OHCHO. The previous identification of this species has based on six transitions, five of which were contaminated by other molecules (Hollis *et al.* 2000). In the current data set, 40 transitions of glycolaldehyde were measured. Emission was present at all transitions, eight of which were uncontaminated. There were no "missing" lines. Therefore, glycolaldehyde is unambiguously present

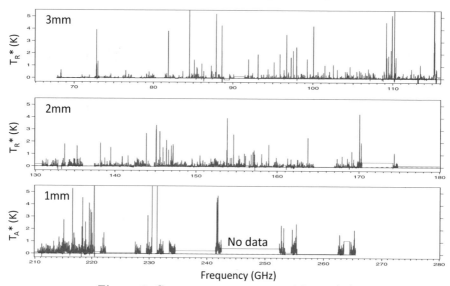

Figure 1. Current survey coverage of Sgr B2(N)

in Sgr B2(N), with an abundance of 2×10^{-11} (Halfen *et al.* 2006). Another compound of interest is formamide, NH_2CHO, the smallest species to contain a peptide bond, i. e., the important linkage between amino acids in proteins. Over 60 lines of formamide were observed, and these data give an abundance wrt H_2 of 1×10^{-10}. The next most complex simple protein analog is acetamide, CH_3CONH_2. For this molecule, 335 transitions are present in our current data, and emission is detected at all of these frequencies. Acetamide is present in Sgr B2(N) with an abundance of 1×10^{-10}. The simplest molecule with an amine group is methyl amine. Over 220 transitions of this compound have been detected in our survey, yielding an abundance in Sgr B2(N) of 1×10^{-9} wrt H_2. We also searched for ethyl amine. However, there are 20 transitions of this molecule where no emission is detected, and therefore, it is not present in Sgr B2(N) with an upper limit wrt H_2 of $< 3 \times 10^{-11}$ (Apponi *et al.* 2008). Several other complex molecules were not apparent in our survey. The 3-carbon sugar dihydroxyacetone (DHA), $HOCH_2COCH_2OH$, has been claimed to be present in Sgr B2(N) by Widicus *et al.* (2005). However, in our survey, we have 20 transitions where there is no emission (Apponi *et al.* 2006a). In addition, our data indicate that the sugar-like molecule hydroxyacetone, CH_3COCH_2OH, does not exist in Sgr B2(N) with an upper limit of 2×10^{-12} (Apponi *et al.* 2006b).

This research is supported by NASA through the Astrobiology Institute under Cooperative Agreement No. CAN-02 OSS02 and by an NSF grant AST-0602282.

References

Apponi, A. J., Halfen, D. T., Ziurys, L. M., Hollis, J. M., Remijan, A. J., & Lovas, F. J. 2006a, *ApJ* (Letters), 634, L29

Apponi, A. J., Hoy, J. J., Halfen, D. T., Ziurys, L. M., & Brewster, M. A. 2006b, *ApJ*, 652, 1787

Apponi, A. J., Sun, M., Halfen, D. T., Ziurys, L. M., & Müller, H. S. P. 2008, *ApJ*, 673, 1240

Cooper, G., Kimmich, N., Belisle, W., Sarinana, J., Brabham, K., & Garrel, L. 2001, *Nature*, 414, 879

Halfen, D. T., Apponi, A. J., Woolf, N. J., Polt, R., & Ziurys, L. M. 2006, *ApJ*, 639, 237

Hollis, J. M, Lovas, F. J., & Jewell, P. R. 2000, *ApJ* (Letters), 540, L107

Pizzarello, S., Huang, Y., Becker, L., Poreda, R. J., Nieman, R. A., Cooper, G., & Williams, M. 2001, *Science*, 293, 2236

Widicus Weaver, S. L. & Blake, G. A. 2005, *ApJ* (Letters), 624, L33

Organic Matter in Space
Proceedings IAU Symposium No. 251, 2008
S. Kwok & S. Sandford, eds.

© 2008 International Astronomical Union
doi:10.1017/S1743921308021108

Organic molecules in the spectral line survey of Orion KL with the Odin Satellite from 486–492 GHz and 541–577 GHz

N. Koning[1], S. Kwok[1,2], P. Bernath[3], Å. Hjalmarson[4], and H. Olofsson[4]

[1] Department of Physics and Astronomy, University of Calgary,
Calgary, AB, T2N 1N4, Canada
email: nkoning@iras.ucalgary.ca, sunkwok@hku.hk

[2] Department of Physics, University of Hong Kong, Pokfulam Road, Hong Kong, China

[3] Department of Physics, University of Waterloo, Waterloo, Ontario, N2L 3G1, Canada

[4] Onsala Space Observatory, Chalmers University of Technology, 439 92, Onsala, Sweden

Abstract. A spectral line survey of Orion KL has been performed over the frequency range of 486–492 GHz and 541–577 GHz using the Odin satellite. Over 1000 lines have been identified from 40 different molecular species, including several organic compounds such as methyl cyanide (CH_3CN), methanol (CH_3OH, $^{13}CH_3OH$), and dimethyl ether (CH_3OCH_3).

Keywords. Astrochemistry, submillimeter

1. Introduction

Although there have been over 20 ground-based spectral surveys of the Orion Kleinmann-Low Nebula, observations between 500 GHz and 600 GHz have not been possible from the ground due to atmospheric absorption by H_2O and O_2. The Odin Satellite (Hjalmarson *et al.* 2003) is capable of observing frequencies in the range of 486-504 GHz and 541–581 GHz via 4 tunable receivers. Odin therefore provides a great opportunity to observe this unexplored region of Orion KL's spectrum.

2. Observations and analysis

Odin performed a spectral line survey of the Orion KL region over a 1.5 year period from February 2004 to November 2005, spanning 1100 orbits. All four tunable receivers were used to observe between 486–492 GHz and 541–577 GHz. The observations were made with a beam size of 2.1′ (at 557 GHz) in position switching mode where the entire telescope physically moved 15′ between the target and an off position with a cycle period of 1 minute.

The data were processed and analyzed in parallel by the Space Astronomy Laboratory (SAL) in Calgary and the Onsala Space Observatory in Sweden. A part of the surveyed spectrum between 541–577 GHz is shown in Figure 3. The JPL Molecular Spectroscopy catalogue (Pickett *et al.* 1998) and the Cologne Database for Molecular Spectroscopy (CDMS, Müller *et al.* 2005) were used in the line identification. In the case of methanol, a separate list of transitions supplied by Eric Herbst was used. Gaussian profiles were fit to each observed feature using a least squares algorithm. In the case where more than one transition contributed to a single feature, special routines were used to fit the sum of several profiles. In some situations the number of blended lines was excessive and no fits were attempted due to substantial ambiguities.

Table 1. Percent differences between calculated and observed values

Computation\Conformer	I	II	III	IV
aug-cc-pVDZ	1.96%	3.86%	2.26%	2.51%
CCSD(T)	1.95%	2.56%	2.38%	2.07%
Stephanian et. al.	4.17%	4.45%	4.6%	N/A

3. Conclusions

We have provided the highest level of accuracy calculations for the four most stable conformers of glycine. We predict these calculations will help aid the debate about the presence of glycine in the interstellar medium by providing researchers with the best information available concerning the spectrum of glycine. We will also entertain the notion that computational methods for calculating vibrational spectra are becoming so reliable as to render experiments in this area obsolete.

References

Kuan, Y.-J., Charnley, S. B., Huang, H.-C., Tseng, W.-L., & Kisiel, Z. 2003, *ApJ*, 593, 848

Gordon, M. S. & Schmidt, M. W. 2005, in: C.E.Dykstra, G. Frenking, K. S. Kim, & G.E.Scuseria (eds.), *Theory and Applications of Computational Chemistry, the first forty years*, (Amsterdam: Elsevier)

Schmidt, M. W., *et al.* 1993, *J. Comput. Chem.*, 14, 1347

Snyder, L., Lovas, F., Hollis, J., Friedel, D., Jewell, P., Remijan, A., Ilyushin, V., Alekseev, E., & Dyubko, S. 2005, *ApJ*, 619, 914

Stephanian, S., Reva, I., Radchenko, E., Rosado, M., Duarte, M., Fausto, R., & Adamowicz, L. 1998, *J. Phys. Chem. A*, 102, 1041

Woon, D. E. & Dunning, T. H. 1993, *J. Chem. Phys.*, 98, 1358

Organic Matter in Space
Proceedings IAU Symposium No. 251, 2008
S. Kwok & S. Sandford, eds.

A search for interstellar CH_2D^+

Alwyn Wootten[1] and Barry E. Turner[1]

[1]National Radio Astronomy Observatory
520 Edgemont Road, Charlottesville, Virginia 22903, USA
email: awootten@nrao.edu

Abstract. We report on a search for interstellar CH_2D^+. Four transitions occur in easily accessible portions of the spectrum; we report on emission at the frequencies of these transitions toward high column density star-forming regions. While the observations can be interpreted as being consistent with a detection of the molecule, further observations will be needed to secure its identification. The CH_2D^+ rotational spectrum has not been measured to high accuracy. Its lines are weak, as the dipole moment induced by the inclusion of deuterium in the molecule is small. Astronomical detection is favored by observations toward strongly deuterium-fractionated sources. However, enhanced deuteration is expected to be most significant at low temperatures. The sparseness of the available spectrum and the low excitation in regions of high fractionation make secure identification of CH_2D^+ difficult. Nonetheless, owing to the importance of CH_3^+ to interstellar chemistry, and the lack of rotational transitions of that molecule owing to its planar symmetric structure, a measure of its abundance would provide key data to astrochemical models.

Keywords. Astrochemistry, line: identification, ISM: molecules, radio lines: ISM, submillimeter

1. Introduction

The symmetric species CH_3^+ is a reactant of extreme importance in interstellar organic chemistry, as it initiates the formation of more complex hydrocarbons. Unfortunately CH_3^+ cannot be observed through its rotational lines, as it is symmetric. CH_2D^+ is asymmetric, however, and emits a rotational spectrum. Because the deuterium is bound to CH_2D^+ more tightly than hydrogen to CH_3^+, and because the binding energies are similar to typical molecular cloud temperatures, CH_2D^+ becomes more abundant relative to its undeuterated counterpart in cold clouds. This binding energy is higher for the deuterium in CH_2D^+ than for that in H_2D^+, so that in warmer clouds, CH_2D^+ remains heavily fractionated to much higher temperatures than H_2D^+. This expectation is borne out by observations of its deuterated derivatives. Since DCO^+ derives from reaction of CO with H_2D^+, the temperature dependence of its fractionation mimics that of $[H_2D^+]/[H_3^+]$, as demonstrated in observations of a cross-section of clouds by Wootten, Loren & Snell (1982). The observed $[DCO^+]/[HCO^+]$ ratio reaches high values only in the coldest clouds; DCO^+ is practically unobservable in a warm cloud such as OMC1. Persistence of a high $[DCN]/[HCN]$ ratio to high temperatures, as observed by Greason (1986) and discussed by Wootten (1987), by the same effect mimics the behavior of $[CH_2D^+]/[CH_3^+]$. Both because it remains abundant at relatively high temperatures, and because rotational lines are accessible, CH_2D^+ is an ideal candidate for observation and confirmation of deuterium fractionation theory. The three accessible mm-wave transitions of CH_2D^+ are $J_{K_{-1}K_1} = 1_{01}$-0_{00} near 280 GHz, 2_{11}-2_{12} near 200 GHz, and 1_{10}-1_{11} near 67 GHz. All should be detectable in warm molecular clouds, given model abundances of CH_3^+, the expected degree of deuteration, and the temperatures of appropriate sources.

Good frequency estimates for rotational lines are available from Rosslein *et al.* (1991) and Jagod *et al.* (1992). The estimated accuracy of the frequencies is estimated by

Rosslein *et al.* (1991) to be $\pm 2 \times 10^{-5}$ (1 σ) times the frequency. Four lines are relatively easily observed by radiotelescopes; we have attempted to detect all four. The lowest frequency but highest excitation line, at 23.01595 GHz, was sought in 1992 April at the NRAO 43 m telescope; no emission was detected. The $J_{K_{-1}K_1} = 1_{01} \to 0_{00}$ line was observed at a center frequency of 278.69162 GHz in 1992 at the 10.4 m CSO telescope on Mauna Kea, Hawaii. The $J_{K_{-1}K_1} = 1_{01} \to 0_{00}, 1_{11} \to 1_{10}$ and $2_{11} \to 2_{12}$ lines were observed at center frequencies of 278.69162 GHz, 67.27371 GHz and 201.76264 GHz at various times between 1992 April and 1997 September at the 12 m NRAO telescope at Kitt Peak, Arizona. At the frequencies where emission from CH_2D^+ was expected, emission was detected in , NGC6334N, SgrB2OH N and W51M. The detections were most convincing for NGC 6334N.

The NGC 6334N region is complex, with several sources appearing within the $30''$ beam of our telescopes. The dominant sources within our beam include SMA1, SMA2, SMA3 and SMA4. In high excitation ammonia studies, SMA1 and SMA2 dominate the emission; emission was not detected toward SMA3 and SMA4. Column densities on the order of $N_{Tot} = 2 \times 10^{14}$ are indicated for $T_{rot} \sim 50$ K. The study of deuterium fractionation in warm clumps recently published by Roueff, Parise and Herbst (2007) predicts $[CH_2D^+]/[CH_3^+] \sim 0.03$ for 50 K, declining by a factor of a few as T rises. Our data would suggest then that $N_{Tot}(CH_3^+)$ exceeds 7×10^{15}. Additional laboratory data is urgently needed to secure the identity of the lines we report.

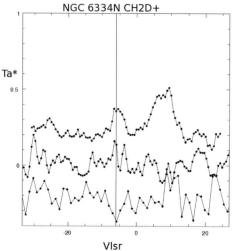

Figure 1. Spectrum of NGC 6334N in the vicinity of the $J_{K_{-1}K_1} = 2_{11} \to 2_{12}$ (201 GHz, top, 12m SSB data, center, offset by 0.2 K), $1_{01} \to 0_{00}$ (278 GHz, center, CSO DSB data, upper) and $J_{K_{-1}K_1} = 1_{11} \to 1_{10}$ (67GHz, lower, 12m SSB data, lower, offset by -0.2 K) lines of CH_2D^+. The center spectrum has been multiplied by a factor of four. The pointing center was $17^h\ 17^m\ 32.0^s\ -35°\ 44'\ 20''$. (B1950); $V_{lsr} = -6.0$.

References

Greason, M. R. 1986, *M.S. Thesis*, U. of Va.

Jagod, M., Roesslein, M., Gabrys, C. M., & Oka, T. 1992, *J. Molecular Spectroscopy*, 55, 153, 666

Roesslein, M., Jagod, M., Gabrys, C. M., & Oka, T. 1991, *ApJ (Letters)*, 382, L51

Roueff, E., Parise, B., & Herbst, E. 2007, *A&A*, 464, 245

Wootten, A. 1987, in: Vardya, M. S. and Tarafdar, S. P. (eds.), *IAU Symposium 120, Astrochemistry* (Dordrecht, D. Reidel Publishing Co.), p. 311

Wootten, A., Loren, R. B., & Snell, R. L. 1982, *ApJ*, 255, 160

Organic Matter in Space
Proceedings IAU Symposium No. 251, 2008
S. Kwok & S. Sandford, eds.

© 2008 International Astronomical Union
doi:10.1017/S1743921308021133

The origin and evolution of interstellar organics

Jean E. Chiar[1] and Yvonne J. Pendleton[2]

[1]SETI Institute, Carl Sagan Center, 515 N. Whisman Road, Mountain View, CA 94043
email: jchiar@seti.org

[2]NASA Headquarters
email: yvonne.pendleton@nasa.gov

Abstract. Over the last decade, we have made great strides in better understanding dust composition and evolution in dense clouds and the diffuse interstellar medium (ISM). Thanks to improvements in IR detector sensitivity on ground-based telescopes and the Spitzer Space Telescope mission, we are no longer limited to a handful of bright background stars in order to study dust composition in quiescent dense clouds and the diffuse ISM. More thorough sampling of lines of sight in these regions has highlighted the dichotomy of the nature and composition of dust in these environments. In addition, successes in recreating interstellar processes and dust-analogs in the laboratory have helped us to understand the differences in dust absorption features we observe in the ISM. In this article, we focus on the organic components of interstellar dust, reviewing past work and highlighting the most recent observations and laboratory experiments.

Keywords. Astrochemistry, molecular processes, ISM: dust, extinction, ISM: lines and bands, ISM: molecules

1. Introduction

Observationally, interstellar dust is known to consist mainly of amorphous silicates as evidenced by the 9.7 and 18.5 μm Si-O stretching and bending vibrations of this material, and graphitic carbon, responsible for the strong 2175 Å bump. Detailed models have been developed that link the measured optical properties of these materials with the grain size distribution to the observed interstellar extinction. Typically, these models conclude that both materials contribute approximately equal volumes (per H-atom) to the interstellar grain population. The well-known Draine & Lee (1984) model illustrates that graphitic and silicate grain components dominate the extinction in different wavelength regimes. Specifically, while silicates dominate the extinction beyond about 8 μm through their strong resonances, graphitic dust controls the extinction in the visual through near-IR.

In a general sense, interstellar dust models can be separated in two distinct classes. First, in some models dust is primarily injected by stellar sources such as Asymptotic Giant Branch (AGB) stars in the form of high temperature condensates such as oxides, carbides, silicates, and graphite. These grains are slowly eroded and destroyed in the diffuse interstellar medium (ISM) by strong shock waves driven by supernova shocks. Inside dense clouds, these grains may temporarily acquire a thin ice mantle of simple molecules such as H_2O and CO, but from an evolutionary viewpoint, these ice mantles have little relevance because they are rapidly photodesorbed in the diffuse ISM when they are eventually recycled (Draine & Lee 1984). In the second class of models, stardust is still an important source of dust cores. However, the ice mantles accreted by these cores are thought to be transformed by penetrating and/or cosmic-ray-created far-UV photons into an organic mantle while still residing inside the molecular cloud. In these models, the organic mantle is thought to be a hardy, refractory material that survives both dense and diffuse cloud environments during the lifecycle of an interstellar dust grain, as does the core material itself (Li & Greenberg 2003). Hence, these two classes of models differ in the assumed character of the carbonaceous grain material and the processes that form them: soot-like dust formed at high temperatures in stellar ejecta versus polymeric organic carbon formed at low temperatures by energetic processing of simple molecules.

Besides the strong UV resonance at 2175 Å – characteristic for aromatic carbon – carbonaceous dust is the carrier of a number of infrared absorption features: specifically at 3.4, 6.8 and 7.2 μm. In the Milky Way, hydrocarbon bands – due to the CH stretching and deformation modes of methyl (CH_3) and methylene (CH_2) groups in aliphatic hydrocarbon materials – are detected along lines of sight that probe the local diffuse ISM within 3 kpc of the Sun (Adamson *et al.* 1990, Sandford *et al.* 1991, Pendleton *et al.* 1994, Whittet *et al.* 1997) and along the line of sight toward the Galactic center (Butchart *et al.* 1986, McFadzean *et al.* 1989, Sandford *et al.* 1991, Pendleton *et al.* 1994, Chiar *et al.* 2000, 2002). These bands have also been detected in one young planetary nebula, AFGL 618 (Chiar *et al.* 1998). In recent years, ground-based and Spitzer studies have revealed that these IR absorption features are also common in the spectra of dense and dusty galactic nuclei such as associated with UltraLuminous InfraRed Galaxies (Imanishi 2000, Spoon *et al.* 2002, Mason *et al.* 2004, Imanishi *et al.* 2006, Armus *et al.*, 2007, Dartois & Muñoz-Caro 2007). A wide variety of materials have been considered as the carrier of these interstellar aliphatic bands mainly differing in their production methods (c.f., Pendleton & Allamandola 2002). These materials generally have in common that they contain aromatic and aliphatic moieties and the best-studied material which also provides the best fit to the IR observations is hydrogenated amorphous hydrocarbon (HAC, Duley 1994, Duley *et al.* 1998, Mennella *et al.* 2002). Laboratory studies have shown that HAC grains may also be responsible for the 2175 Å feature (Schnaiter *et al.* 1999).

2. Carbonaceous Solids in the Diffuse ISM

Aliphatic hydrocarbons are confirmed to be present in the interstellar media in our own and other galaxies, through observations of vibration absorption features centered at 3.4 μm (CH stretch), 6.85, and 7.25 μm (deformation modes) (Figure 1). Within the profile of the 3.4 μm feature, the relative depths of subfeatures at 3.338, 3.419, and 3.484 μm are indicative of short-chained hydrocarbons, with other perturbing chemical groups (Sandford *et al.* 1991). The individual chain lengths are not likely to be much longer than 4 or 5, and they are attached to electronegative chemical groups (Sandford *et al.* 1991, Pendleton & Allamandola 2002). Together, the profile of the 3.4 μm feature, the CH stretch-to-deformation mode optical depth ratio, and the absence of evidence for other types of chemical subgroups, provide indirect evidence that aromatic, sp^2 hybridized carbon domains (aromatics) are an important component of the interstellar material. Recent analysis of the dust features in the diffuse ISM toward the active galactic nucleus, IRAS 08572+3915, shows that the hydrocarbons are highly aliphatic: the aliphatic bonds outnumber the aromatic bonds by a ratio of 12.5:1 (Dartois *et al.* 2007). Based on the depth of the corresponding deformation features at 6.85 and 7.25 μm, the best candidate material is likely to be hydrogenated amorphous carbon (HAC, Chiar *et al.* 2000, Pendleton & Allamandola 2002, Mennella *et al.* 2002). Furthermore, the HAC material in the diffuse ISM possesses little nitrogen or oxygen (Pendleton & Allamandola 2002, Figure 2).

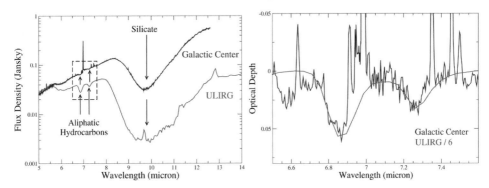

Figure 1. [left] Mid-IR spectrum of the line of sight toward the Galactic Center taken with the Infrared Space Observatory's SWS (upper spectrum, Chiar *et al.* 2000). Mid-IR spectrum of the ULIRG F00183-7111 taken with the Spitzer Space Telescope's IRS (lower spectrum, Spoon *et al.* 2004). Both the galactic and extragalactic spectra show absorption features due to HACs (6.85 and 7.25 μm) and silicates (9.7 μm). [right] The 6.5 to 7.5 μm region, showing the HAC absorption features in more detail. The ULIRG spectrum (smooth line) has been divided by a factor of 6 for presentation purposes.

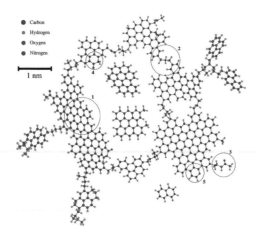

Figure 2. The basic structural and molecular character of carbonaceous, interstellar dust in the diffuse interstellar medium. The specific geometries of the aromatic plates and aliphatic components simply represent what is likely. The structure is somewhat splayed out to reveal the molecular structural details; we envision the actual structure somewhat more closed in. The approximate volume of this fragment is on the order of 10^{-19} cm^3. Thus, a typical 0.1 μm DISM carbonaceous dust grain would contain approximately 10^4 of these fragments. The encircled sections signify the different structural unite: 1. aromatic network, 2. aliphatic bridge, 3. aliphatic carbonyl, 4. aromatic carbonyl, 5. aromatic nitrogen. From Pendleton & Allamandola (2002).

The 3.4 μm band resulting from experiments that produce HAC material in the laboratory show excellent agreement with the astronomical feature profile. The differences are in the production route of the HACs. Experiments show that the IR spectral behavior of hydrocarbons produced by laser pyrolysis of acetylene is similar to that observed in the diffuse ISM (Schnaiter *et al.* 1999). These results point toward production of hydrocarbon grains in the AGB/PN phase of stellar evolution. The detection of the aliphatic hydrocarbon absorption feature in the outflow of an evolved star (Lequeux & Jourdain de Muizon 1990, Chiar *et al.* 1998) is indirect evidence that the diffuse ISM hydrocarbons originate, at least partially, in these environments (i.e. as stardust). On the other hand, the 3.4 μm

Scott Sandford showing a piece of aerogel like that flown on the Stardust spacecraft during the press conference (photo by Sze-Leung Cheung).

Organic Matter in Space
Proceedings IAU Symposium No. 251, 2008
S. Kwok & S. Sandford, eds.

© 2008 International Astronomical Union
doi:10.1017/S1743921308021145

Dicarbon molecule in the interstellar clouds†

Maja Kaźmierczak[1], Mirosław Schmidt[2], and Jacek Krełowski[1]

[1]Center for Astronomy of Nicolaus Copernicus University,
ul. Gagarina 11, 87-100 Toruń, Poland
email: maja.kazmierczak@gmail.com

[2]Nicolaus Copernicus Astronomical Center, ul. Rabiańska 8, 87-100 Toruń, Poland

Abstract. We present high-resolution and high signal-to-noise spectroscopic observations of interstellar molecular lines of C_2 towards early-type stars. C_2 is particularly interesting because it is the simplest multicarbon molecule and its abundances give information on the chemistry of interstellar clouds, especially on the pathway of formation of (hydro)carbon chains and PAHs which may be considered as possible carriers of diffuse interstellar bands (DIBs). Homonuclear diatomic molecules have negligible dipol moments and hence radiative cooling of excited rotational levels may go only trough the slow quadrupole transitions (van Dishoeck & Black 1982). In C_2, pumped by galactic average interstellar field rotational levels are excited effectively much above the gas kinetic temperature and a rotational ladder of electronic transitions is usually observed from high rotational levels. Relations between abundances of the dicarbon and other simple interstellar molecules are considered as well.

Keywords. ISM: molecules, ISM: abundances

1. Introduction

The observational material consists of a set of high-quality echelle spectra (with resolution of 110,000 and signal-to-noise ratio above 200) from the high-resolution spectrograph UVES, acquired at the ESO Paranal Observatory (Bagnulo *et al.* 2003). Dicarbon is easily available observationally through its electron-vibration-rotation transitions from ground electronic level $X^1\Sigma_g^+$ to excited electronic levels $A^1\Pi_u$ (Phillips band) in optical and near-infrared range of the spectrum (6900–12000Å).

2. Results

We measured equivalent widths of (1-0) 10133-10262Å, (2-0) 8750-8849Å, (3-0) 7714-7793Å and in one case (HD 147889) also (4-0) 6909-6974Å bands of the C_2 Phillips system. The list of observed objects with some basic parameters is presented in Table 1. There are also determined total column densities of C_2 toward each star and excitation temperatures, which were extracted from the fit to the first four rotational levels ($J'' = 0, 2, 4, 6$).

In cases of optically thin clouds, i. e. for weak lines the column densities are derived from the relationship: $N_{col}(J'') = 1.13*10^{17}EW/(f_{ij}*\lambda)$, where λ is the wavelength in [Å], f_{ij} is the absorption oscillator strength and EW equivalent widths in [mÅ]. Total column density is a sum of column densities determined from each low rotational level $N_{col}(J'')$. Error of column densities depends on EW, which follows quality of spectrum: resolution and signal-to-noise ratio. Column density depends also on the oscillator strengths. We compare their theoretical values with our observations (to check their correctness), building relations between equivalent widths for transitions originated in each low rotational level, according to equation: $EW1/EW2 = f1/f2$. Observationally determined ratios of oscillator strengths are very similar to theoretical ones (from Bakker *et al.* 1997). Some

† Based on observations collected at the European Southern Observatory, Paranal, Chile

Table 1. Basic parameters of program stars and results for C_2 toward these objects

object	$T_{exc}[K]$	$N_{col}[10^{13}/cm^2]$	Sp/L	E(B-V)
HD 76341	66 ± 16	1.20 ± 0.2	O 9 Ib	0.49
HD 96917	66 ± 16	1.36 ± 0.4	O 8.5 Ib(f)	0.32
HD 97253	73 ± 68	1.27 ± 0.3	O 5.5 IIIf,cl,el	0.45
HD 115363	60 ± 1	4.01 ± 0.5	B 1 Ia	0.69
HD 136239	56 ± 11	3.66 ± 0.9	B 1.5 Ia v	0.94
HD 147933	63 ± 35	5.86 ± 0.3	B 2.5 V	0.45
HD 147889	73 ± 11	12.6 ± 0.8	B 2 V	1.07
HD 148184	67 ± 18	2.8 ± 0.3	B 2 Vn,el	0.52
HD 148379	81 ± 22	1.47 ± 0.2	B 1.5 Ia el	0.61
HD 151932	87 ± 12	1.57 ± 0.2	W N7A wr	0.35
HD 152003	76 ± 2	2.62 ± 0.6	O 9.7 Iab v	0.53
HD 152270	67 ± 27	2.56 ± 0.8	W C7 wr	0.53
HD 154368	53 ± 6	5.6 ± 0.7	O 9.5 Iab	0.78
HD 161056	80 ± 18	3.18 ± 0.5	B 1.5 V	0.60
HD 163800	59 ± 22	2.60 ± 0.6	O 7 IIIf	0.60
HD 168607	89 ± 10	1.35 ± 0.2	B 9 Iape	1.61
HD 168625	70 ± 24	1.32 ± 0.5	B 6 Iap	1.49
HD 169454	39 ± 3	5.80 ± 0.6	B 1 Iael	0.93
HD 179406	48 ± 9	4.60 ± 0.6	B 3 V	0.33
BD-14 5037a	53 ± 5	6.38 ± 0.6	B 1.5 Ia	1.59
BD-14 5037b	64 ± 4	4.26 ± 0.6		

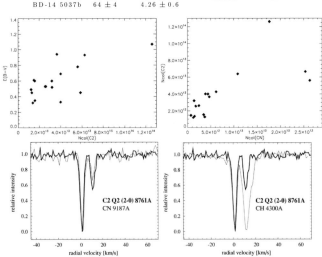

Figure 1. (a) [Top left] $N_{col}(C_2)$-E(B-V) (b) [Top right] $N_{col}(CN)$-$N_{col}(C_2)$ (c) [Bottom left] C_2-CN (d) [Bottom right] C_2-CH

departures likely follow errors in measuring C_2 lines. For the strongest lines, especially from (2-0) vibrational band, the relations evidently confirm theoretical expectations.

We measured dicarbon column densities toward above 20 stars. After that we tried to correlate C_2 abundance with those of other interstellar molecules (CN and CH) and with extinction (Figure 1a). Such relations may constrain chemical pathways which lead to the formation of these species.

Figure 1b shows relation between column densities of C_2 and CN. A majority of the considered objects suggest a tight correlation, but two points are considerably discrepant. These represent HD 169454 and HD 154368. Physico-chemical conditions in the intervening interstellar clouds (e.g., very low T_{exc}) likely vary considerably in these objects while compared to average conditions. Figures 1c and 1d compare Doppler-splitted profiles of molecular features in the spectrum of BD-14 5037. We compare lines profiles of C_2 (Q(2) 8761Å of (2-0) band) and CN (9187Å) which suggests that C_2 is spatially correlated with CN. Figure 1d compares C_2 and CH (4300Å) in the spectrum of BD-14 5037. CH does not share the profile shape with C_2.

References

Bagnulo, S., Jehin, E., Ledoux, C., Cabanac, R., Melo, C., Gilmozzi, R., & The ESO Paranal Science Operations Team 2003, The Messenger, 114, 10

Bakker, E. J., van Dishoeck, E. F., Waters, L. B. F. M., & Schoenmaker, T. 1997, *A&A*, 323, 469

van Dishoeck, E. F., & Black, J. 1982, *ApJ*, 258, 533

Organic Matter in Space
Proceedings IAU Symposium No. 251, 2008
S. Kwok & S. Sandford, eds.

Comparing ice composition in dark molecular clouds

C. Knez[1,2], M. Moore[2], S. Travis[3], R. Ferrante[3], J. Chiar[4], A. Boogert[5], L. Mundy[1,2], Y. Pendleton[6], A. Tielens[7], E. van Dishoeck[8], and N. Evans[9]

[1]University of Maryland, email: claudia@astro.umd.edu, [2]Goddard Center for Astrobiology, [3]U.S. Naval Academy, [4]SETI Institute, [5]Herschel Science Center, [6] NASA Headquarters, [7]NASA Ames, [8]Leiden University, [9]University of Texas at Austin

Abstract. We present 5–20 μm Spitzer/IRS spectroscopy toward stars behind dark molecular clouds. We present preliminary results from the Serpens dark cloud to show the variation between environments within a cloud. We are surveying 3 clouds with varying levels of star formation activity. Serpens has the highest level of activity from our 3 clouds. We show that location as well extinction can cause variations in ice composition. We also find that some lines of sight contain organic molecules such as methane and methanol, and the first detection of acetylene ice in the interstellar medium. We believe the high extinction lines of sight have been enriched by star formation activity near those lines of sight.

Keywords. ISM: molecules, astrochemistry, techniques: spectroscopic

1. Introduction

Infrared spectroscopy of protostars and extincted background stars has revealed the presence of numerous features due to simple molecules frozen on refractory dust grains: H_2O, CO_2, CO, CH_3OH, CH_4, HCOOH, OCN^-, and possibly NH_3 and NH_4^+ (e.g., Whittet *et al.* 1996, Schutte *et al.* 1996). Several key species (H_2O, CH_3OH) are formed by chemical reactions on the grains, while others (CO) accrete inertly from the gas phase. Evidently, the composition of these icy grain mantles is directly related to the composition of the gas (e.g., the C/CO ratio), which in turn is related to cloud history. Laboratory irradiation studies of simple ices have shown that quite complex species can be formed this way (PAHs, amino acids, hydrocarbons; Bernstein *et al.* 2002, Greenberg *et al.* 2000). Eventually, the ices may be released back in the gas phase, driving a chemistry capable of forming complex species in the 'hot cor(ino)es' surrounding protostars (Cazaux *et al.* 2003). Alternatively, they may be incorporated in icy bodies and deliver volatiles to planetesimals in circumstellar disks. Knowledge of the ice composition and the relevance of the various molecule formation and destruction processes is thus key to understanding interstellar and protoplanetary chemistry.

The composition of interstellar ices is expected to reflect the local conditions in the dense regions of the clouds. Specifically, gas density and dust temperature are thought to be important drivers of grain surface chemistry (Tielens & Hagen 1982, Hasegawa & Herbst 1993, Bergin *et al.* 1997). These physical conditions drive the star formation activity in the cloud core and in turn newly formed stars affect the density and dust temperature. Hence, the composition of the accreted ice mantles is expected to vary with the star formation activity of the cloud.

As part of a Spitzer GO-4 program, we are obtaining 5–20 μm spectra toward stars behind the Serpens, Perseus, and Lupus dark clouds. Spitzer has greatly enhaced the

Table 1. Abundances towards two lines of sight with high A_V

Species	Unit	CK 2	Ser BG1
CO_2	% H_2O	33	35
CH_3OH	% H_2O	<2.1	17
CH_4	% H_2O	<3	~3
C_2H_2	% H_2O	...	~3
HCOOH	% H_2O	1.9	2.6

study of ice composition toward molecular clouds by being able to probe lines of sight toward faint young stellar objects as well as toward faint stars behind the clouds. We conducted follow up observations of the brightest extincted stars behind three of the clouds mapped by the c2d legacy team: Serpens (high star formation activity), Perseus (medium activity), and Lupus (low activity).

2. Observations and results

In this paper, we analyze a sample of 10 stars in Serpens to study the ice composition to determine what factors play a role in the icy grain mantle composition. The preliminary sample shows that the 6 μm feature increases in depth with extinction but it is not necessarily larger at the highest extinction. The depth of the 6 μm feature generally increases with A_V but there are variations especially at low and at high extinctions. Some variations in the 6 μm feature may due to the contribution of other species such as HCOOH, especially at high extinction.

We compared two highly extincted lines of sight in Serpens to see if both sight-lines showed similar composition. CK 2 has been analyzed by Knez *et al.* (2005). It has an A_V greater than 30 mag and shows many ice features with only upper limits on organics such as CH_4 and CH_3OH (see Table 1). The other sight-line is toward Ser BG1 with an $A_V \sim 25$ mag. This sources has a lower extinction than CK 2 yet shows higher column densities of the following organics: CH_4 and CH_3OH. In addition, this is the first source in which solid C_2H_2 has been detected in space.

3. Conclusions

We present the first detection of C_2H_2 ice in the interstellar medium. It has been detected with comparable abundance to CH_4. Extinction is not the only variable in determining the composition of ices in molecular clouds. Even toward lines of sight where there are no protostars, the effects of star formation can affect the grain composition.

References

Bergin, E. A., *et al.* 1997, *ApJ*, 482, 285

Bernstein, M. P., *et al.* 2002, *ApJ*, 576, 1115

Cazaux, S., *et al.* 2003, *ApJL*, 593, 51

Greenberg, J. M., *et al.* 2000, *ApJL*, 531, 71

Hasegawa, T. I. & Herbst, E. 1993, *MNRAS*, 263, 589

Knez, C., *et al.*, 2005, *ApJL*, 635, 145

Schutte, W., *et al.* 1996, *A&A*, 309, 633

Tielens, A. G. G. M. & Hagen, W. 1982, *A&A*, 114, 245

Whittet, D. C. B., *et al.* 1996, *ApJ*, 458, 363

Organic Matter in Space
Proceedings IAU Symposium No. 251, 2008
S. Kwok & S. Sandford, eds.

Organic compounds as carriers of the diffuse interstellar bands

Peter J. Sarre

School of Chemistry, The University of Nottingham,
University Park, Nottingham, NG7 2RD, United Kingdom
email: Peter.Sarre@Nottingham.ac.uk

Abstract. The diffuse interstellar bands appear as absorption features in spectra of reddened stars and lie mostly in the visible region of the spectrum. The first examples were recorded photographically nearly one hundred years ago and despite a huge amount of observational, theoretical and laboratory effort the spectra remain unassigned. Most researchers believe that organic material is responsible for the absorptions, the most popular form being polycyclic aromatic (hydro)carbon (PAH) structures. This article reviews briefly the main characteristics of the spectrum, describes some current research and outlines some lines of inquiry.

Keywords. ISM: clouds, line: identification, ISM: individual (HD 44179, V854 Cen, IRC+10216)

1. Introduction

The diffuse interstellar band problem is one that has fascinated astronomical observers for many decades and has stimulated much debate and original thinking among researchers from a wide range of scientific backgrounds. The diffuse bands are a large set of absorption features seen towards stars that are partly obscured by interstellar material in the Galaxy, in external galaxies, and towards quasars lying behind red-shifted galaxies with z up to $c.$ 0.5. It is surprising that it has been so difficult to identify the carriers given that spectra of other components of the diffuse interstellar medium (ISM) such as H, Na, K and Ca^+, diatomic molecules including H_2, CO, CH^+, CH and CN, and the triatomic molecules H_3^+ and C_3, are easily obtained in the laboratory. However, after much experimental and theoretical effort not one of the $c.$ 300 bands has been assigned. Using CCDs and large telescopes, spectra can be recorded towards bright moderately-reddened Galactic stars which are essentially noise-free for modest exposure times. A number of reviews have been written on the subject including those by Herbig (1995), Fulara & Krełowski (2000) and Sarre (2006).

2. Is the diffuse band problem important?

Although intriguing, one might ask if the problem is significant. A precise answer to this question is difficult to give and there will probably be some surprises; these may well include the simplicity of the eventual solution. There remains the interesting possibility that the spectrum arises from new forms of matter or dust in the interstellar medium and it is notable that new forms of carbon including fullerenes, nanotubes and graphene have only relatively recently become experimentally accessible. The diffuse bands are a good spectroscopic tracer of extinction by dust grains and this characteristic is being exploited in connection with the GAIA mission (Munari 2000, Vidrih & Zwitter 2005, Wallerstein *et al.* 2007). Particularly important in the context of this Symposium is that the diffuse band carriers may be large *organic* molecules or small grains with possible links to the chemistry of life. Grains are also important more generally in astronomy in the formation of H_2 and larger molecules, in a range of physical processes in the interstellar medium, and in star and planet formation.

3. Diffuse band characteristics

Over three hundred bands have been recorded in the near-UV, visible and near-IR spectral regions with widths ranging between *c.* 2 and 100 cm^{-1}. There is no doubt that most of the reported diffuse band absorptions are interstellar rather than stellar or circumstellar in origin, and this has been confirmed through observations towards binary stars (Merrill 1936, Hobbs *et al.* 2008). There are reasonably good correlations of diffuse band strengths with E_{B-V} but with statistically significant deviations, with H I abundance, and, for some bands, with the column density of the C_2 molecule (Thorburn *et al.* 2003). Another notable property is the relative weakening of diffuse bands with respect to E_{B-V} in denser clouds, which suggests that the column density of the carriers is not determined simply by the total abundance of material but is influenced by, e.g. the UV flux. A very striking characteristic is their common occurrence; the bands are seen towards numerous Galactic and, increasingly, extragalactic optical sources. Their ubiquity is shared by the carriers of the 'unidentified' infrared (UIR) emission bands that are a very strong indicator of widespread polycyclic aromatic hydrocarbon material. This in itself provides circumstantial support for the idea of PAHs being closely linked with the diffuse band problem.

In spectroscopic terms, there can be little doubt that the bands arise from electronic transitions, which for gas-phase molecular carriers would generally be expected to be accompanied by vibrational and rotational fine structure. Attributes commonly used in the analysis of spectra are line widths, which should be the same for one or more transitions accessing a common upper level, fine structure which can act as a label, spectroscopic patterns, combination differences, and profile variation according to rotational and vibrational temperature. To date use of these well-established approaches has not led to assignment of the diffuse band spectra. It should be noted that the widths, which for most bands greatly exceed the velocity dispersion in the cloud(s), are now generally attributed to rapid intramolecular vibrational relaxation (IVR) in the absorber following photon absorption.

4. Carrier proposals

The history of proposals as to the nature of the carriers is as long as the subject itself. It would appear that virtually every structure, charge state and size of gaseous, solid, embedded or adsorbed species has at some time been considered including H^-, H_2, porphyrins and colour centres in crystals to name just a few. In terms of organic (i.e., carbon-based) carriers that have been suggested, these may be discussed with reference to three limiting symmetry classes: 1-dimensional carbon chains, 2-dimensional planar polycyclic aromatic (hydro)carbons or rings (i.e., necklaces), and 3-dimensional graphitic particles, nanodiamonds, fullerenes or nanotubes. Numerous variations can be considered including partially hydrogenated, dehydrogenated, ionised, organometallic and locally (partially) aromatic structures. Key issues that need be addressed are satisfying elemental cosmic abundance constraints and the fact that the observed spectrum is dominated in appearance by the order of a dozen strong 'special' bands, the identification of which

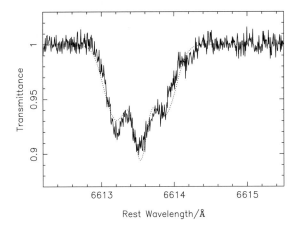

Figure 1. Profile of the $\lambda 6614$ diffuse interstellar band recorded towards μ Sgr with the Ultra-high Resolution Facility on the Anglo-Australian Telescope, with the result of molecular rotational contour fitting (Kerr *et al.* 1996).

would surely unlock much of the rest of the spectrum. There is a growing list of candidates that have appeared to be attractive possibilities but which now can be ruled out at least for the strongest diffuse bands. These include C_5, the C_{18} necklace molecule, the PAH coronene ($C_{24}H_{12}$), and probably most neutral molecules with $n_C \leqslant 50$; a recent list is given by Snow & McCall (2006) in their Table 3. However, there remain a wide range of organic structures for which rather little is known of their electronic spectra including large, radical (hydrogenated and dehydrogenated) and protonated PAHs, polyenes and nanotubes. Of those molecules for which laboratory data are suggestive of a 'match', in the opinion of the writer only two are possibly viable: C_{60}^+, for which low-temperature matrix data lie close to diffuse bands near 9577 and 9632 Å (Foing & Ehrenfreund 1997), and CH_2CN^- which has a gas-phase origin band transition in reasonable agreement with a single weak diffuse band at 8037 Å (Cordiner & Sarre 2007).

5. Fine structure in some of the bands

The discovery that some of the narrower bands have a high level of fine structure and asymmetric profiles (Sarre *et al.* 1995, Ehrenfreund & Foing 1996) is probably one of the best indicators that the carriers are molecular in nature. One example is shown in Figure 1 where the $\lambda 6614$ diffuse band shows three (possibly four) components. The dashed line is a χ^2 minimisation fit assuming an oblate symmetric top molecule about the size of coronene at a derived rotational temperature of 20 K and this is discussed further by Kerr *et al.* (1996). The band shape is reminiscent of P, Q and R rotational branch structure which is a common attribute of molecular electronic spectra. Such a fit is by no means unique to an oblate top as illustrated by others (Schulz *et al.* 2000), and an alternative description in terms of isotope structure has been put forward (Webster 1996). An even greater challenge to interpretation is presented by the $\lambda 5797$ band which exhibits considerable ultra-fine structure (Kerr *et al.* 1998) as illustrated in Figure 2.

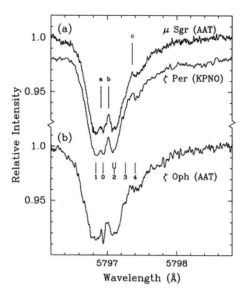

Figure 2. Observations of the λ5797 diffuse band obtained with the Ultra-high Resolution Facility on the Anglo-Australian Telescope (AAT) and at Kitt Peak National Observatory (KPNO). Data recorded towards three stars are shown and illustrate the consistency of the fine structure of the features (Kerr *et al.* 1998).

6. Circumstellar Environments

Although the diffuse band spectrum is commonly observed in absorption along lines-of-sight towards reddened stars, the data obtained often represent the superposition of absorptions due to more than one cloud, each with its own density, ionisation and velocity structure. Consequently some attention has been paid to more localised carbon-rich circumstellar and nebular environments in the hope that not only would the formation regions for diffuse band carriers be identified, but that spectra might be observed under different and warmer conditions than hold for the diffuse ISM. Three cases are of particular interest: carbon-rich circumstellar shells such as IRC+10216, the Red Rectangle, and the R Coronae Borealis (RCB) star V854 Cen.

6.1. *The circumstellar shell of IRC+10216*

Given that most researchers consider that the diffuse band carriers are carbon-based molecules, it has been widely thought that these could be formed initially in the outflows of mass-losing carbon stars. Based on observations this does not appear to be the case, at least in the sense of diffuse band carriers being present in exactly the same chemical form as seen in the diffuse ISM. A search for absorptions arising in the circumstellar shell of the mass-losing carbon star IRC+10216 using a background star revealed no evidence for circumstellar diffuse band carriers (Kendall *et al.* 2002), and a similar result has been found for observations directly towards post-AGB stars (see contribution by Luna *et al.*, this volume). Additionally no diffuse band spectra were seen towards stars in the background of circumstellar material of the Helix nebula (Kendall & Mauron 2004).

6.2. *The Red Rectangle*

The Red Rectangle displays a strong set of UIR (PAH) bands which provides very strong evidence for organic aromatic material in the nebula. There is also emission from silicates which is thought to be confined largely to a circumbinary disk. Prominent unidentified optical bands appear near 5800, 6380 and 6615 Å and have a pronounced X-shaped spatial distribution in the nebula, falling along the bicone interfaces (Schmidt & Witt 1991). The peak wavelengths of the optical bands (and also ERE) shift to shorter wavelength with increase in offset as would be expected in moving towards cooler conditions.

Apart from the intrinsic importance of assigning these bright emission features, the peak wavelengths of the most prominent bands fall close to the absorption wavelengths of some of the diffuse interstellar absorption bands (Sarre 1991, Fossey 1991). The wavelengths and widths of the bands both decrease with offset *towards* the individual diffuse band characteristics as discussed by Scarrott *et al.* (1992) and Sarre, Miles & Scarrott (1995). However, even at the highest offsets of *c.* 15″ (Glinksi & Anderson 2002) and 22″ (Van Winckel *et al.* 2002) there remains a small difference between the peak wavelength of, e.g. the 5800 Å emission feature and the λ5797 diffuse interstellar band. In their paper Glinski & Anderson (2002) concluded that the hypothesis that the same molecule may be the carrier of the 5800 Å Red Rectangle band and the λ5797 DIB was contradicted by their observations. In Figure 3, high-quality data from Van Winckel *et al.* (2002) has been taken and, under the assumption of linear behaviour, least-squares fitted for the region beyond 10″. By extrapolation it is found that the peak wavelength of the Red Rectangle emission band would reach the diffuse absorption band wavelength at an offset of 58 ± 19″ from HD 44179. Hence the best available published data are not inconsistent with the same carrier being responsible for the 5800 Å emission feature and the λ5797 diffuse interstellar band, with the ultimate proof or otherwise likely only to be obtained when the spectra are recorded in the laboratory and assigned.

The major part of the shift in peak wavelength occurs within 10″ of the star and may be due to overlapping *vibrational* sequence structure as mentioned by Sarre *et al.* (1995) and more recently taken up by Sharp *et al.* (2006). A possible explanation for the band behaviour beyond 10″ could be that it reflects the influence of a slowly decreasing *rotational* 'temperature' with offset. The likely link between the Red Rectangle optical emission bands and a sub-set of diffuse interstellar bands provides the third example of spectral evolution of emission bands of the Red Rectangle towards an absorption counterpart in the ISM, the other two being the 3.3 and 11.2 μm PAH infrared bands.

It may be of interest to note that a long-standing close association between the λ6614 diffuse band as recorded in absorption (Herbig 1975) and the Red Rectangle (Cohen *et al.* 1975) can be found at the back of the hard copy of the Astrophysical Journal volume 196 (1975). Plate 2 from Herbig (1975) shows the λ6614 diffuse band recorded towards bright stars and on the facing page is Plate 3 which contains a Red Rectangle image from Cohen *et al.* (1975). On closing the journal volume the information in the two plates lies in touching distance and continues to do so in many libraries worldwide!

6.3. *V854 Cen*

To date the only object other than the Red Rectangle to exhibit the unidentified optical emission bands is the RCB star V854 Cen observed at minimum light (Kameswara-Rao & Lambert 1993). V854 Cen also presents a display of UIR (PAH) bands though these are unusual in their relative strengths with features at 3.3, 6.3, 11.3 and 13.5 μm being the most prominent; the commonly observed UIR bands at 7.7 and 8.6 μm are very weak in V854 Cen (Lambert *et al.* 2001). The optical bands in the, e.g. 5800 Å region are broad compared with their equivalents at high offset in the Red Rectangle and also have different

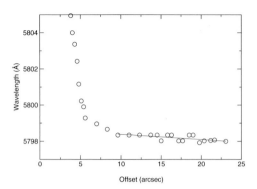

Figure 3. Peak wavelength of the '5800' Å (F1) emission feature of the Red Rectangle as a function of offset from HD 44179. The data are taken from Van Winckel *et al.* (2002).

relative strengths, the strongest in the 5800 Å set being the 5829 Å feature. The reason for this is unclear. Lambert *et al.* (2001) noted that the UIR band spectrum of V854 Cen is broadly similar to that of the carbon-rich PPN IRAS 22272+ 5435. However, there is no evidence for optical diffuse bands arising in absorption from circumstellar material of this star (Luna *et al.* 2008).

7. Some lines of inquiry

In this section three possible contributors to the diffuse band spectrum are considered. It has been suggested that PAH-based anions could play a role as many of these molecules are predicted to have electronic absorption spectra that lie in the visible spectral region. The suggested transitions, not yet observed experimentally for PAH anions, occur between a stable ground electronic state and a shallow dipole-bound higher electronic state which lies just below the electron detachment threshold (Sarre 2000). This proximity in energy means that the electron affinity of the neutral molecule (i.e., the ionisation energy of the anion) is a very good approximation to the transition energy for the ground to dipole-bound state excitation. As an illustration, DFT B3LYP calculations with X. Liu on the phenoxy radical, C_6H_5O, yield an electron affinity of 2.15 eV in reasonable agreement with the experimental value of 2.253(6) eV; this falls in the diffuse band range at approximately 550 nm. A high electric dipole moment for the neutral is a prerequisite for the existence of the dipole-bound state and this is met for C_6H_5O with a calculated value of 4.0 Debye. These properties of C_6H_5O extend to numerous larger polar polyaromatic molecules and laboratory spectra are needed to explore this proposal. One argument against molecular anions as candidate carriers is their susceptibility to electron photodetachment.

Secondly, although discussed at various times over the years in the literature, the possible rôle of heteroatoms incorporated in or on the periphery of PAH molecules has probably been underestimated but there is little experimental data with which to make comparison with astronomical data.

Finally, the discovery of carbon nanotubes warrants some consideration. A recent paper by Zhou *et al.* (2006) presents calculated transition energies for nanotubes with $n_C \sim$ 80–200. These transitions are polarised along the tube axis and are found to have a high f value and fall in the 1-2 eV range. An alternative type of transition in carbon nanotubes has been observed in the condensed phase through excitation spectra (Weisman & Bachilo 2003) resulting in over sixty (n,m) assigned spectra for single-walled nanotubes, many of which fall in the visible spectral region. A complementary set of DFT calculations on both semiconducting and metallic nanotubes reveals spectra in the diffuse band spectral region (Barone *et al.* 2005b, Barone *et al.* 2005a). However, an examination of the available experimental data suggests there is no consistent correspondence of these transitions in carbon nanotubes with the diffuse band spectrum.

Acknowledgements

We thank the Royal Society and the Symposium organisers for financial support and Radmila Topalovic for assistance in generating Figure 3.

References

Barone, V., Perlta, J. E., & Scuseria, G. E. 2005a, *NanoLett*, 5, 1830

Barone, V., *et al.* 2005b, *NanoLett*, 5, 1621

Cohen, M., *et al.* 1975, *ApJ*, 196, 179

Cordiner, M. A. & Sarre, P. J. 2007, *A&A*, 472, 537

Ehrenfreund, P. & Foing, B. H. 1996, *A&A*, 307, L35

Foing, B. H. & Ehrenfreund, P. 1997, *A&A* (Letters), 317, L59

Fossey, S. J. 1991, *Nat*, 353, 393

Fulara, J. & Krełowski. J. 2000, *NewAR*, 44, 581

Glinski, R. J. & Anderson, C. M. 2002, *MNRAS*, 332, L17

Herbig, G. 1975, *ApJ*, 196, 129

Herbig, G. 1995, *ARA&A*, 33, 19

Hobbs, L. M., *et al.* 2008, *ApJ*, 680, 1256

Kameswara-Rao, N. & Lambert, D. L. 1993, *MNRAS* (Letters), 263, L27

Kendall, T. R., *et al.* 2002, *A&A*, 387, 624

Kendall, T. R. & Mauron, N. 2004, *A&A*, 428, 535

Kerr, T. H., *et al.* 1996, *MNRAS* (Letters), 283, L105

Kerr, T. H., *et al.* 1998, *ApJ*, 495, 941

Lambert, D. L., Rao, N. K., Pandey, G., & Ivans, I. I. 2001, *ApJ* 555, 925

Luna, R., *et al.* 2008, *A&A* 480, 133

Merrill, P. W. 1936, *ApJ*, 83, 126

Munari, U. 2000, *Molecules in Space and in the Laboratory*, Conference Proceedings, held 2-5 June 1999 in Carloforte, Cagliari, Italy. Edited by I. Porceddu, S. Aiello. v67, 179

Sarre, P. J. 1991, *Nat*, 351, 356

Sarre, P. J. 2000, *MNRAS* (Letters), 313, L14

Sarre, P. J. 2006, *JMolSpec*, 238, 1

Sarre, P. J., Miles, J. R., & Scarrott, S. M. 1995, *Sci*, 269, 674

Sarre, P. J., *et al.* 1995, *MNRAS* (Letters), 277, L41

Scarrott, S. M., Watkin, S., Miles, J. R., & Sarre, P. J. 1992, *MNRAS*, 255, 11P

Schmidt, G. D. & Witt, A. N. 1991, *ApJ*, 383, 698

Schulz, S. A., King, J. E., & Glinski, R. J. 2000, *MNRAS*, 312, 769.

Sharp, R. G., Reilly, N. J., Kable, S. H., & Schmidt, T. W. 2006, *ApJ*, 639, 194

Snow, T. P. & McCall, B. J. 2006, *ARA&A*, 44, 367

Thorburn, J. A., *et al.* 2003, *ApJ*, 584, 339

Van Winckel H., Cohen M., & Gull T.R. 2002, *A&A*, 390, 147

Vidrih, S., & Zwitter, T. 2005, in: C. Turon, K. S. O'Flaherty, M. A. C. Perryman (eds.),
    *Proceedings of the Gaia Symposium 'The Three-Dimensional Universe with Gaia' (ESA
    SP-576)*, held at the Observatoire de Paris-Meudon, 4-7 October 2004. 576, p. 201
Wallerstein, G., Sandstrom, K., & Gredel, R. 2007, *PASP*, 119, 1268
Webster, A. 1996, *MNRAS*, 282, 1372
Weisman, R. B. & Bachilo, S. M. 2003, *NanoLett*, 3, 1235
Zhou, Z., *et al.* 2006, *ApJ* (Letters), 638, L105

Discussion

HENNING: If your nanotube explanation would be correct, why should nature select the special diameter for your nanotube? Because if you change the diameter, the band position would immediately change. So nature would have to fine tune the nanotube size distribution.

SARRE: I think that's really a key issue. I put it in for completeness. There is the issue of variability of the diameter, the question of length and there's the question of chirality. As you probably know when you try to make nanotubes, you make a soup of them, so nature would have to select. But on the other hand, there would be equivalent selection problems for PAHs when you get to very large systems. So I don't think we wish to rule it out at this point.

SALAMA: I have a question about your search for diffuse bands in carbon stars. It is based on a search for one band or a few bands?

SARRE: No, we looked for about six in all. It was done as a broad band survey, not just for a particular section. So we looked for 5797, 5780, 6284, etc., all those standard candidates.

SPECK: Have you looked at any of the carbon stars that actually show UIR bands? Hardly any carbon stars have them, but at least three of them show UIR bands.

SARRE: I have to say no, we haven't. There has been some work on detection of diffuse bands *towards* carbon stars, but that's not actually the question. So it sounds like that would be worth doing if they optically bright enough.

GARCIA-LARIO: Regarding DIBs in the circumstellar envelopes of evolved stars, I would like to draw your attention to our poster paper (p. 217) where we surveyed a number of post-AGB stars with various chemistry conditions. We show the lack of this feature as formed in the circumstellar shells; all the features we have found can be explained as from the interstellar medium between the stars and ourselves. So there is a real depletion of these features in the envelope of evolved stars, independent of whether they are carbon rich or oxygen rich.

Organic Matter in Space
Proceedings IAU Symposium No. 251, 2008
S. Kwok & S. Sandford, eds.

Fullerenes as carriers of extinction, diffuse interstellar bands and anomalous microwave emission

Susana Iglesias-Groth

Instituto de Astrofísica de Canarias, C/Via Láctea sn, 38200, La Laguna, Spain
email: sigroth@iac.es

Abstract. According to semiempirical models, photoabsorption by fullerenes (single and multishell) could explain the shape, width and peak energy of the most prominent feature of the interstellar absorption, the UV bump at 2175 Å. Other weaker transitions are predicted in the optical and near-infrared providing a potential explanation for diffuse interstellar bands. In particular, we find that several fullerenes could contribute to the well known strong DIB at 4430 Å. Comparing cross sections and available data for this DIB and the UV bump we estimate a density of fullerenes in the diffuse interstellar medium of 0.1–0.2 ppm. These molecules could then be a major reservoir for interstellar carbon. We also study the rotation rates and electric dipole emission of hydrogenated icosahedral fullerenes. We investigate these molecules as potential carriers of the anomalous (dust-correlated) microwave emission recently detected by several cosmic microwave background experiments.

Keywords. Ultraviolet: ISM, (ISM:) dust, extinction, ISM: lines and bands

1. Introduction

In 1985 Kroto and Smalley proposed the existence of a new allotropic form of carbon: the fullerenes (Kroto *et al.* 1985). Their research on samples of vaporized graphite using laser beams, initially aimed to reproduce the chemistry of the atmospheres of carbon enriched giant stars, gave as a result the unexpected discovery of the C_{60} (the 60 carbon atoms fullerene). Subsequent experiments by Kroto and Smalley, and others, showed the existence of carbon aggregates with a larger number of atoms (C_{70}, C_{84}, C_{240}) and established that for an even number of atoms larger than 32, these aggregates were stable. While other molecules have serious difficulties to survive in the interstellar medium, the robustness of C_{60} and of fullerenes in general, strongly support a long survival in the harsh conditions of interstellar space.

Several laboratory experiments (Chhowalla *et al.* 2003) and theoretical studies (Henrard *et al.* 1997, Iglesias-Groth 2004) of the photoabsorption by fullerenes and buckyonions (multishell fullerenes) in the UV suggest that these molecules could be responsible of the most intense feature of interstellar absorption, the so-called ultraviolet bump located at 2175 Å. A significant fraction of interstellar carbon (10–30%) could then reside in fullerene related molecules.

It has been suggested (Webster 1991, 1992, 1993) that fullerenes could also be a carrier of diffuse interstellar bands (DIBs). According to theoretical spectra obtained using semiempirical models (Iglesias-Groth *et al.* 2002, 2003), icosahedric fullerenes and buckyonions (from C_{60} to C_{6000}) present numerous low-intensity bands in the optical and near-infrared, several with wavelengths very similar to well known DIBs. Fullerenes deserve further study as potential carriers of DIBs. For the moment, only two DIBs may have been identified as caused by the cation of C_{60} (Foing & Ehrenfreund 1994).

is close to the solar atmosphere value, individual fullerenes may lock up 20-25% of the total carbon in the diffuse interstellar space.

Our computations also show that fullerenes and buckyonions present weaker transitions in the optical and near infrared with their number decreasing towards longer wavelengths. These transitions may be responsible for some of the known but unexplained diffuse interstellar bands. It would be very important to obtain high sensitivity, high resolution laboratory spectra of these molecules in the optical and near infrared for a more precise comparison with the very detailed observations of DIBs. Finally, hydrogenated fullerenes and buckyonions are expected to produce rotationally based electric dipole microwave radiation under the conditions of the diffuse interstellar medium. These molecules are potential carriers for the anomalous Galactic microwave emission recently detected by several cosmic microwave experiments.

References

Becker, L. & Bunch, T. E. 1997, *Meteorit. Planet. Sci.*, 32, 479

Casassus, S., Cabrera, G., Forster, F., Pearson, T. J., Readhead, A. C. S., & Dickinson, C. 2006, *ApJ*, 639, 951

Chhowalla, M., Wang, H., Sano, N., Teo, K. B. K., Lee, S. B., & Amaratunga, G. A. J. 2003, *Phys. Rev. Lett.*, 90, 155504

Draine, B. T. & Lazarian, A. 1998a, *ApJ* (Letters), 494, L19

de Oliveira-Costa, A. *et al.* 1999, *ApJ* (Letters), 527, L9

de Oliveira-Costa, A. *et al.* 2004, *ApJ* (Letters), 606, L89

Désert, F. X., Jenniskens, P., & Dennefeld, M. 1995, *A & A*, 303, 223

Finkbeiner, D. P., Langston, G. I., & Minter, A. H. 2004,*ApJ*, 617, 350

Foing, B. F. & Ehrenfreund, P. 1994, *Nature*, 369, 296

Herbig, G. H. 1995, *ARAA*, 33, 19

Henrard, L., Lambin, Ph., & Lucas, A. A. 1997, *ApJ*, 487, 719

Hildebrandt, S. R., Rebolo, R., Rubiño-Martín, J. A.,Watson, R. A., Gutiérrez, C. M., Hoyland, R. J., & Battistelli, E. S. 2007, *MNRAS*, 382, 594

Iglesias-Groth, S. 2004, *ApJ* (Letters), 608, L37

Iglesias-Groth, S. 2005, *ApJ* (Letters), 632, L25

Iglesias-Groth, S. 2006, *MNRAS*, 368, 1925

Iglesias-Groth, S. 2007, *ApJ* (Letters) 661, L167

Iglesias-Groth, S. & Bretón, J. 2000, *A&A*, 357, 782

Iglesias-Groth, S., Ruiz, A., Bretón, J., & Gómez Llorente, J. M. 2002, *J. Chem. Phys.*, 116, 1648

Iglesias-Groth, S., Ruiz, A., Bretón, J., & Gómez Llorente, J. M. 2003, *J. Chem. Phys.*, 118, 7103

Kogut, A. *et al.* 1996, *ApJ* , 460, 1

Kroto, H. W., Heath, J. R., Obrien, S. C., Curl, R. F., & Smalley, R. E. 1985, *Nature*, 318, 162

Leitch, E. M., Readhead, A. C. S., Pearson, T. J., & Myers, S.T. 1997, *ApJ* (Letters), 486, L23

Watson, R. A. *et al.* 2005, *ApJ*, 624, L89

Webster, A. S. 1991, *Nature*, 352, 412

Webster, A. S. 1992, *A&A*, 257, 750

Webster, A. S. 1993, *MNRAS*, 263, 385

Webster, A. S. 1995, *MNRAS*, 277, 1555

Discussion

Cox: The question is about the ionized fullerenes. Could you tell us anything about the possibilities that these are carriers of the diffuse bands?

IGLESIAS-GROTH: Yes, these molecules could be carriers of DIBs. In fact two transitions of cation of C_{60} were possibly identified in spectra of reddened stars.

COX: But it's not possible to do any calculations?

IGLESIAS-GROTH: Yes, it is possible to predict their transitions.

COX: Yes, I mean your calculations on fullerenes, right? So can you do anything on your ionized fullerenes?

VAN DISHOECK: Have you calculated any ionized fullerenes?

IGLESIAS-GROTH: No, it is not possible using these models. It's an ab-initio calculation.

VAN DISHOECK: Actually a related question to that is, if you compare your calculated positions of the fullerenes bands, for example, diffuse interstellar bands of 4430 Å, you must have some sort of uncertainty bar on your calculations because there is a relatively simple model of the calculation. Can you estimate what are the uncertainties in your calculations in terms of positions of the bands? For example, for comparison with the data of smaller ones?

IGLESIAS-GROTH: Indeed, these models give approximate wavelengths for the transitions. I think the accuracy is of order 10 Å.

VAN DISHOECK: What is the uncertainty in the positions of your calculations - you say you have a band of 5 eV or 1 eV or 2 eV, plus or minus. I'm just wondering what's the uncertainties if you're calculating energies.

IGLESIAS-GROTH: In terms of energy they are accurate to the order of 0.01 eV.

VAN DISHOECK: I suppose you compared and benchmarked them against known observed transitions, but what are the uncertainties for other transitions?

IGLESIAS-GROTH: The available experimental data for C_{60} allows a detailed comparison for this molecule. The models fit the stronger transitions within 0.01 eV.

UNKNOWN: Do you expect the strong correlation between 4430 Å and the UV bump? We looked. It is not there. There does not exist any correlation.

IGLESIAS-GROTH: No correlation in your opinion? But several published papers propose that there is some correlation. I just point out that fullerenes could be responsible of both the UV bump and the 4430 Å DIB.

Getting ready to start on Monday morning.

Organic Matter in Space
Proceedings IAU Symposium No. 251, 2008
S. Kwok & S. Sandford, eds.

© 2008 International Astronomical Union
doi:10.1017/S1743921308021182

Observing ultraviolet signatures of interstellar organics with the *Hubble Space Telescope*

Theodore P. Snow

Center for Astrophysics and Space Astronomy, University of Colorado
389 UCB
Boulder, CO 80309-0389
U.S.A.
e-mail: `tsnow@casa.colordao.edu`

Abstract. The *Cosmic Origins Spectrograph* (*COS*) will be more sensitive for ultraviolet spectroscopy than either the *GHRS* or the *STIS*, especially in the far UV where many absorption lines and bands formed by atoms and molecules have electronic transitions from the ground state. Here we outline our plans for using the *COS* to observe interstellar gas and dust in the cold ISM, along with a report on the results of preliminary archival *HST* search for UV diffuse interstellar bands.

Keywords. ISM: clouds, ISM: molecules, ultraviolet: ISM, astrobiology, space vehicles: instruments

1. Preface

At this writing, the *Hubble Space Telescope* (*HST*) was to be serviced by Space Shuttle astronauts in late October, 2008. By the date of this volume's publication, that servicing mission should have taken place. In this paper, we assume that the mission has been successful.

2. Introduction to UV Observations of the ISM

The value of making ultraviolet spectroscopic observations of the interstellar gas and dust was first conceptualized by Spitzer (1946), who envisioned a large telescope in space, to observe (among other things) interstellar UV absorption lines in the spectra of hot stars lying behind interstellar clouds. Spitzer's dream has been realized in the form of the *Hubble Space Telescope*.

In the very low densities of interstellar space, even in the densest clouds, collisions between particles are so rare that essentially all atoms and molecules remain in their ground electronic and vibrational state. In the energy-level structures of most universally abundant atoms and a number of common organic molecules, a wide gap exists between the ground and the first electronically excited states. Hence, most electronic transitions from the ground state lie at UV wavelengths. Despite the many challenges for would-be UV interstellar absorption-line observers, a lot has been learned about the diffuse ISM from UV observations, including the general composition of the gas (and hints about the composition of the dust), the physical and chemical conditions; and the kinetics of the gas and dust. But what is missing is detailed knowledge about denser interstellar clouds, because of the difficulties in making UV spectroscopic observations there, due to the high dust extinction.

Our *COS* Science Team observing plan for the cold ISM is centered around UV spectra of stars seen behind translucent clouds. These clouds are defined by their total extinction (ranging between A_V 1.5 and 10 mag). Figure 1 shows, as a function of line-of-sight column density, the expected relative abundances of atomic and molecular hydrogen and carbon, of translucent clouds (see Snow & McCall 2006 for references). The shaded area on the figure shows what parameter space *COS* can cover in observing absorption lines. Note that this figure, and the general definition of translucent clouds, refers to the local conditions in a single cloud, not to be confused with the average conditions integrated over the entire line of sight.

3. The Capabilities of the *Cosmic Origins Spectrograph*

The *COS* instrument and its capabilities are described in the paper by Green (2000), and on the Space Telescope Science Institute web site (www.stsci.edu/hst/cos). The emphasis in designing the instrument was placed on the far-UV, so a one-reflection, Rowland-circle design was chosen for that spectral region (between ∼1,150 and ∼1,800 Å), to achieve 10 to 20 times greater sensitivity than *STIS* at comparable spectral resolution. For the near-UV (∼2,000 to ∼3,200 Å) additional reflections are used, but the sensitivity to point sources is still greater than that of *STIS*. The *COS* has two resolution modes: the medium-resolution (R ∼ 20,000) gratings, comparable to the M-gratings in *STIS*; and low-resolution gratings (R ∼ 5,000).

4. An Archival Search for Unidentified UV Absorption Features

In starting to plan for our Guaranteed Time Observations, we (J. D. Destree and T. P. Snow) initiated an archival project using existing spectra obtained by *STIS*. This study led to some interesting results, which are outlined below.

4.1. *The detection of CS absorption in the UV*

In a blind search for unidentified absorption features, we found one near 1,400 Å, which we then able were able to identify as an absorption band of the diatomic molecule CS.

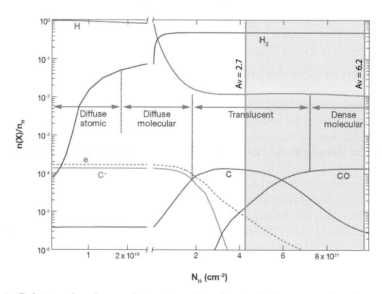

Figure 1. Relative abundance of atomic and molecular hydrogen and carbon of translucent clouds. The shaded area shows the parameter space the *COS* can explore and observe in absorption lines against hot background stars.

This was the first verified detection of this molecule in the UV (but a detection had been tentatively made before; Smith *et al.* 2001). Of course, CS is commonly detected by radio observations in dark and translucent clouds (Scappini *et al.* 2007 and references cited therein), and theoretical models of dense and translucent clouds show that CS is formed easily by ion-neutral reactions (Drdla *et al.* 1989), so the UV detection was not a surprise when stars with high-enough reddening could be observed. We found CS in 16 out 21 lines of sight we checked; these results, including a correlation analysis, can be found in Destree *et al.* (2008).

4.2. *The search for diffuse interstellar bands*

The primary goal of our archival *STIS* project was to search for ultraviolet diffuse interstellar bands (i.e., UV DIBs), because that is a high-priority item in our *COS* GTO program, and we wanted to test the waters. The visible-wavelength DIBs have been known since the early 1920s (Heger 1922), but their carriers remain unidentified, almost 90 years later. After several false starts, now the consensus is that the DIBs are formed by a family of complex organic molecules. The paper by Peter Sarre in this volume elaborates on this point, as do previous reviews (e.g., Snow 2001).

Searches for UV DIBs have been made previously (Snow *et al.* 1977, Seab & Snow 1985, Clayton *et al.* 2003, Jenkins *et al.* 2008), but no unequivocal DIBs were found. One UV DIB was detected in another study (Tripp *et al.* 1993), but that feature was later identified as CH (Sheffer & Federman 2007 and references cited therein).

Our search, however, yielded several absorption features that conform to what we expect of UV DIBs: they are found only in at least moderately-reddened stars; they are truly interstellar; and they are unidentified. Further criteria were that any thought-to-be DIB must be observed in more than one line of sight; it must be broader than a typical interstellar atomic line while having a significantly different width than stellar lines; and where possible, we also tried find a feature in separate exposures on the same star (preferably with different grating settings, to rule out detector glitches). In our first attempt, we found ten or more features that survived all of our tests, and there are probably several more. The results of our archival search will be published by Destree & Snow (2008).

5. The *COS* Science Team's Observing Program on the Cold ISM

As a member of the *COS* Science Team, the author is responsible for planning the Team's GTO observations of the cold ISM. Our adopted program is designed to explore the parameter space shown in the shaded region in Figure 1. We hope to measure column densities of atomic, ionic, and molecular gas-phase species in more heavily-reddened and denser lines of sight than previously possible to probe in the UV. The specific observing target list can be found on the internet, at www.stsci.edu/hst/proposing/docs/COS-GTO/APTProposal_Snow.pdf.

As seen in Figure 1, much of the previously unexplored region represents translucent clouds. In these clouds, hydrogen becomes almost wholly molecular and carbon makes the transitions from C $_{\rm II}$ to C $_{\rm I}$ to CO (from C^+ to C to CO, in terms used by chemists). While we can't observe H_2, we can easily observe carbon in all its abundant forms, so we can trace carbon as it goes through the transitions we expect to see in these clouds.

Our scientific goals include: estimating elemental depletions (of gas-phase species onto dust) in translucent clouds, to see how the gas-dust interaction changes with increasing volume density and decreasing radiation field intensity; to test and improve chemical models of translucent clouds; to see whether and how the dust extinction curve is affected by the conditions in these clouds; and to search for new molecular species in translucent clouds, including species whose UV spectra are known, as well as searching for new unidentified species (i.e., UV DIBs). We will specifically search for PAHs, because that family is among the leading candidates for the carriers of the visible-wavelengths DIBs, and in their neutral form, many PAHs have electronic spectra in the UV. Some work has been done to measure the UV spectrum of these species in the lab (e.g., Salama *et al.* 1995, Salama 1999), so identifications should eventually be possible.

6. *HST* Observations of Organics in Comets

We (K. O'Malia and T.P. Snow) plan to carry out using *HST* absorption-line observations of cometary comae by stellar occultation measurements. When a comet passes in front of a hot star, we expect to see absorption lines caused by gas in the coma. This method has been tried in the past using visible light, without success (see Herbig & McNally 1999) — but UV observations are a lot more likely to produce positive results, for all the same reasons that UV spectra are more productive than visible spectra for observing interstellar absorption lines. Despite the lack of success in previous absorption line searches, in a recent study (O'Malia *et al.* 2008), even visible-wavelength absorption lines were tentatively detected by comparing very high S/N on-comet spectra with off-comet spectra of the same stars. Thus we are very hopeful that we will detect a host of atomic, ionic, and molecular species, including organics, when we apply UV spectroscopy to future comets.

References

Clayton, G. C., Gordon, K. D., Salama, F., Allamandola, L. J., Martin, P. G., Snow, T. P., Whittet, D. C. B., Witt, A. N., & Wolff, M. J. 2003, *ApJ*, 592, 947
Destree, J. D. & Snow, T. P. 2008, in preparation
Destree, J. D, Snow, T. P., & Black, J. H. 2008, *ApJ*, submitted
Drdla, K., Knapp, G. R., & Van Dishoeck, E. F. 1989, *ApJ*, 345, 815
Green, J. 2000, *Proc. Soc. Photo-Opt. Eng.*, 4013, 352
Heger, M. L. 1922, *Bull. Lick Obs.*, No. 337, 141
Herbig, G. H. & McNally, D. 1999, *MNRAS*, 340, 951
Jenkins, E. B., Snow, T. P., & Rachford, B. L. 2008, in preparation
O'Malia, K., Snow, T. P., York, D. G., Thorburn, J. D., *et al.*, 2008, in preparation
Salama, F. 1999, in: L. d'Hendecourt, C. Joblin, & A. Jones (eds.), *Solid Interstellar Matter: The ISO Revolution*, (New York: Springer), p. 65
Salama, F., Joblin, C., & Allamandola, L. J. 1995, *Planet. Space Sci.*, 43, 1165
Scappini, F., Cecchi-Pestellini, C., Casi, S., & Olberg, M. 2007, *A&A*, 466, 243
Seab, C. G. & Snow, T. P. 1985 *ApJ*, 295, 485
Sheffer, Y. & Federman, S. R. 2007, *ApJ*, 659, 1352
Smith, A. M., Lyu, C.-H., & Bruhweiler, F. C. 2001, *Bull. AAS*, 33, 1452
Snow, T. P. 2001, *Spectrochimica Acta*, Part A, 57, 615
Snow, T. P. & McCall, B. J. 2006, *ARAA*, 44, 367
Snow, T. P., York, D. G., & Resnick, M. 1977, *PASP*, 89, 758
Spitzer, L. 1946, Report To Project RAND: Astronomical Advantages of an Extra-Terrestrial Observatory (reprinted in *Astr. Quarterly*, 7, 131, 1990)
Tripp, T. M., Cardelli, J. A., & Savage, B. D. 1994, *AJ*, 107, 645

Discussion

ZIURYS: This detection of CS is pretty interesting. You care to speculate what other diatomics may triatomics maybe detectable in the UV?

IRVINE: May be you are in a better position to tell me the answer to this question.

Organic Matter in Space
Proceedings IAU Symposium No. 251, 2008
S. Kwok & S. Sandford, eds.

Testing the attribution of selected DIBs to dehydrogenated coronene cations

G. Mulas[1], G. Malloci[1], I. Porceddu[1], and C. Joblin[2]

[1] Istituto Nazionale di Astrofisica–Osservatorio Astronomico di Cagliari,
Strada n.54, Loc. Poggio dei Pini, I–09012 Capoterra (CA) (Italy)
email: [gmulas,gmalloci,iporcedd]@ca.astro.it

[2] Centre d'Etude Spatiale des Rayonnements, CNRS et Université Paul Sabatier Toulouse 3,
Observatoire Midi-Pyrénées, 9 Avenue du Colonel Roche, 31028 Toulouse cedex 04 (France)
email: chistine.joblin@cesr.fr

Abstract. Dehydrogenated coronene molecules have been proposed as the source of the UV-bump in the interstellar extinction curve as well as of some of the diffuse interstellar bands (DIBs). To test this hypothesis we have recently undertaken a combined (a) modelling, and (b) observational work on the subject. (a) In the framework of a global approach to the photophysics of a PAH–like species in space, we used combined theoretical calculated properties, obtained with (time–dependent) density functional theory, and a Monte–Carlo model simulating the time evolution of the population of levels of a given molecule, to obtain the detailed ro–vibrational spectral structure of selected electronic transitions. (b) From the observational point of view, we compare our predictions with observations of the well–known $\lambda6284$ and $\lambda5780$ DIBs.

Keywords. Astrochemistry, ISM: lines and bands, line: profiles, molecular processes

1. Introduction

The carrier of the well–known bump at about 2175 Å in the interstellar extinction curve seems to require a free–flying form of carbon, either in the form of size–restricted graphite pieces (Draine 1985) or of single or stacked PAHs (Duley & Seahra 1998). Based on scattering calculations (Duley & Seahra 1998) in the discrete–dipole–approximation, Duley (2006a) suggested the possible assignment of this feature to a $\pi \rightarrow \pi^\star$ plasmon resonance in dehydrogenated coronene molecules ($C_{24}H_n$, $n \leqslant 3$), and their corresponding cations. The same class of molecules has been proposed to explain some of the diffuse interstellar bands (DIBs, Duley 2006b).

2. Methods

To test the attribution of the $\lambda6284$ and $\lambda5780$ DIBs to the completely dehydrogenated coronene cation we have undertaken a combined modelling and observational work.

From the modelling side, we first computed the absorption cross–section, the complete vibrational spectrum (frequencies and intensities of the IR–active modes), and the rotational constants in the states involved of planar C_{24}^+. We then used a Monte–Carlo model of the photophysics of isolated PAHs (Mulas 1998, Malloci *et al.* 2003) to obtain the expected rotational band profiles under diffuse ISM conditions.

From the observational side we restricted ourselves to lines of sights characterized by very low color excess ($E_{B-V} < 0.15$ mag). These data, unpublished, have been obtained with the ESO 1.4 m CAT Telescope.

3. Results

We show in Figure 1 the comparison between the computed rotational profile of the "DIBs" of planar C_{24}^+ at 5780 and 6284 Å. According to the proposal by Duley (2006b) they should be the $0 \to 0$ and $0 \to 1$ bands in a vibronic progression of the $\pi_{-4} \to \pi_0^\star$ electronic transition. Figure 2 shows the correlation plot for the equivalent widths of the 5780 and 6284 Å DIBs measured for selected lines of sight.

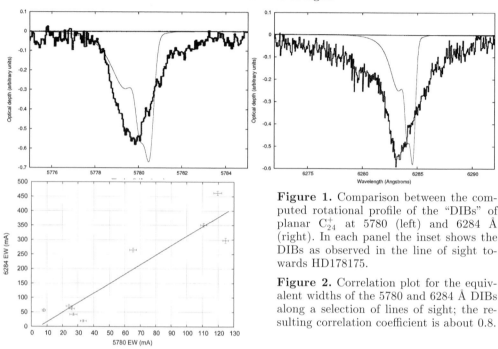

Figure 1. Comparison between the computed rotational profile of the "DIBs" of planar C_{24}^+ at 5780 (left) and 6284 Å (right). In each panel the inset shows the DIBs as observed in the line of sight towards HD178175.

Figure 2. Correlation plot for the equivalent widths of the 5780 and 6284 Å DIBs along a selection of lines of sight; the resulting correlation coefficient is about 0.8.

4. Discussion and future work

We find a relatively low correlation between the λ6284 and λ5780 DIBs. This is inconsistent with the hypothesis that they are due to different vibronic transitions from the same state of the same molecule, which would require a perfect correlation, with scatter only due to measurement errors. The higher correlation previously reported between these two bands on more reddened lines of sight (Moutou *et al.* 1999) may be due to statistical effects and/or from the two DIBs arising from different carriers that respond in a similar way to environmental conditions. Moreover, the specific transitions proposed in this hypothesis as tentative identifications of the two bands, are predicted to be blue–shaded, in stark contrast with the observed red–shaded profiles of the λ6284 and λ5780 DIBs. This excludes their common origin in C_{24}^+, unless upon dehydrogenation this species relaxes to an isomer different from the planar, aromatic one we considered.

References

Draine, B. T. 1985, *ApJS*, 57, 587

Duley, W. W. 2006a, *ApJ* (Letters), 639, L59

Duley, W. W. 2006b, *ApJ* (Letters), 643, L21

Duley, W. W. & Seahra, S. 1998, *ApJ*, 507, 874

Malloci, G., Mulas, G., & Benvenuti, P. 2003, *A&A*, 410, 623

Mulas, G. 1998, *A&A*, 338, 243

Moutou, C., Krelowski, J., d'Hendecourt, L., & Jamroszczak, J. 1999, *A&A*, 351, 680

Organic Matter in Space
Proceedings IAU Symposium No. 251, 2008
S. Kwok & S. Sandford, eds.

© 2008 International Astronomical Union
doi:10.1017/S1743921308021200

Depleted diffuse bands in circumstellar envelopes of post-AGB stars

N. L. J. Cox[1], R. Luna[2], M.A. Satorre[2], D.A. García Hernández[3], O. Suárez[4], and P. García Lario[1]

[1]Herschel Science Centre, European Space Astronomy Centre, European Space Agency, Villanueva de Cañada, Madrid, Spain; email: nick.cox@sciops.esa.int

[2]Laboratorio de Astrofísica Experimental, Escuela Politécnica Superior de Alcoy, Universidad Politécnica de Valencia, Alcoy, Alicante, Spain

[3]W. J. McDonald Observatory, The University of Texas at Austin, Austin, Texas, USA

[4]UAN, Université de Nice Sophia Antipolis, Nice, France

Abstract. We report on new results in the search for diffuse bands, signatures of still unknown origin, in the circumstellar envelopes of evolved (post-AGB) stars.

Keywords. Circumstellar matter, stars: AGB and post-AGB, ISM: lines and bands

1. Diffuse interstellar bands

The diffuse interstellar bands (DIBs) constitute a group of ~300 interstellar (IS) optical absorption features (Galazutdinov *et al.* 2000) observed ubiquitously throughout the Universe (see Cox & Cordiner 2008). DIB strengths correlate roughly with the observed IS reddening by dust (e.g. Herbig 1995). Although not one of the DIBs has been identified there is strong evidence for a molecular organic nature of the carriers (Sarre 2008).

It is well known that evolved stars replenish the interstellar medium (ISM) during their mass loss phases. Depending on the chemistry, carbon-rich compounds can form in these outflows, which thus act as cosmic factories of complex organic matter. Despite unavoidable processing of such organic matter when exposed to the harsh UV conditions of the ISM one could foresee a direct connection with the carriers of the DIBs. Diffuse band (DB) carriers, or their pre-cursors, may be formed already in the circumstellar (CS) envelopes of evolved stars. In order to investigate this scenario we started a program to search for DBs in CS envelopes of post-AGB stars.

2. Search for diffuse *circumstellar* bands

Post-AGB stars are stars in a short-lived transition phase between AGB stars and protoplanetary nebulae. In this phase the remnant AGB shell is seen superimposed on the stellar spectrum (from early to late spectral type) of the central star. The lines-of-sight toward these targets usually probe both interstellar and CS matter. Therefore, it is important to derive the relative contribution of each component to the total observed reddening. Once the line-of-sight IS reddening is derived (e.g. with the extinction map by Drimmel *et al.* 2003) one can use the Galactic relationship between DIB strength and IS reddening to correct the observed DB strength and reddening to obtain the corrected DCB strength and CS reddening.

We divided our sample of 33 stars/sightlines in two groups; those dominated by CS and IS reddening, respectively. Figure 1 shows the 5780 Å DB strength versus reddening.

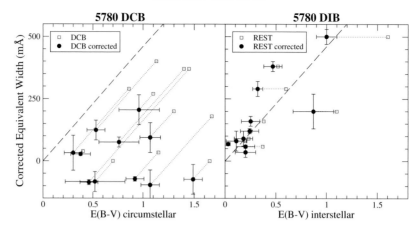

Figure 1. The corrected equivalent widths of the 5780 DCBs and DIBs versus the inferred CS and IS reddening, respectively. The thick dashed line represents the Galactic relationship: $EW(5780)/E(B-V) = 460$ mÅ. For the CS correction (right panel) we assume a total absence of DCBs: $EW(5780)/E(B-V) = 0$. Error bars are identical for uncorrected and corrected values. Uncertainties in the Galactic relationship are not considered.

Uncorrected equivalent widths and reddening are indicated by squares, while corrected equivalent widths (for IS and CS contribution in the left and right panel, resp.) and reddening are plotted as filled circles (with error bars).

3. Results and future work

For the IS group the observed DB strengths are in line with the Galactic DIB relationship with IS reddening. On the other hand, the inferred DCB strengths are systematically weaker for the inferred CS reddening, than expected from the IS trend. Note that positive values represent upper limits to the possible presence of DCBs since stellar line contamination has not been taken into account. A full description of the sample, data analysis and results, also for additonal DBs, is given in Luna *et al.* (2008).

Our work shows that (statistically) the carriers of IS DBs are not present – at least not in a detectable state – in CS envelopes of post-AGB stars. Processing of the organic CS matter by UV photons may be required to give rise to DIBs. Thus, although DIB carriers are not present in CS envelopes their pre-cursors could be. It would therefore be interesting to search for their absorption features in the UV/optical spectra of post-AGB stars. To detect these weak CS features we will need to accurately remove or separate the interstellar and atmospheric "contamination" from the CS spectrum. One possibility is to select targets with high radial velocities (such that IS and CS components have separate velocities) and whose line-of-sight extinction is dominated by the CS envelope.

References

Cox, N. L. J. & Cordiner, M. A. 2008, *this volume*

Drimmel, R., Cabrera-Lavers, A., & López-Corredoira, M. 2003, *A&A*, 409, 205

Galazutdinov, G. A., Musaev, F. A., Krelowski, J., & Walker, G. A. H. 2000, *PASP*, 112, 648

Herbig, G. H. 1995, *ARA&A*, 33, 19

Luna, R., Cox, N. L. J., Satorre, M., García-Hernández, A., Suárez. O., & García-Lario, P. 2008, *A&A*, 480, 133

Sarre, P. J. 2008, *this volume*

Organic Matter in Space
Proceedings IAU Symposium No. 251, 2008
S. Kwok & S. Sandford, eds.

On buckyonions as a carrier of the 2175 Å interstellar extinction feature

Juhua Chen[1,2], Moping Li[2], Aigen Li[2], and Yongjiu Wang[1]

[1]Department of Physics, Hunan Normal University, Changsha, Hunan 410081, China
email: jhchen@hunnu.edu.cn, wyj@hunnu.edu.cn

[2]Department of Physics & Astronomy, University of Missouri, Columbia, MO 65211, USA
email: limo@missouri.edu, lia@missouri.edu

Abstract. In recent years buckyonions have been suggested as a carrier of the 2175 Å interstellar extinction feature, based on the close similarity between the electronic transition spectra of buckyonions and the 2175 Å interstellar extinction feature. We examine this hypothesis by calculating the interstellar extinction with buckyonions as a dust component. It is found that dust models containing buckyonions (in addition to amorphous silicates, PAHs, graphite or amorphous carbon) can closely reproduce the observed interstellar extinction curve. However, a more severe challenge to the buckyonion hypothesis is provided by the non-detection of the $\sim$7–8 μm C–H stretching bands expected from buckyonions in the diffuse interstellar medium. This will allow us to place an upper limit on the abundance of buckyonions.

Keywords. Dust, extinction, ISM: lines and bands, ISM: molecules

1. Introduction

In the interstellar medium (ISM), the strongest spectroscopic extinction feature is the 2175 Å bump. Since Stecher & Donn (1965) first detected this ultraviolet (UV) extinction feature through rocket observations, the origin of this feature and the nature of its carrier(s) are still an enigma. Many candidate materials, including graphite, amorphous carbon, graphitized (dehydrogenated) hydrogenated amorphous carbon, nano-sized hydrogenated amorphous carbon, quenched carbonaceous composite, coals, PAHs, and OH^- ion in low-coordination sites on or within silicate grains have been proposed, while no single one is generally accepted (see Li & Greenberg 2003 for a review).

Recently, Ruiz *et al.* (2005) theoretically simulated the UV/optical photo-absorption spectra of buckyonions. They found that the calculated absorption spectra of buckyonions *alone* almost perfectly fit the 2175 Å interstellar extinction feature. Therefore, they proposed buckyonions as its carrier.

However, it is well recognized that in the ISM, in addition to the 2175 Å extinction carrier, there must exist other dust components as well – there must be a population of amorphous silicate dust, as indicated by the strong, ubiquitous 9.7 and 18 μm interstellar absorption features; there must be a population of aromatic hydrocarbon dust (presumably polycyclic aromatic hydrocarbon [PAH] molecules), as indicated by the distinctive set of "unidentified" infrared (UIR) emission bands at 3.3, 6.2, 7.7, 8.6, and 11.3 μm ubiquitously seen in the ISM; there must also exist a population of aliphatic hydrocarbon dust, as indicated by the 3.4 μm C–H absorption feature which is also ubiquitously seen in the diffuse ISM of the Milky Way and external galaxies.

Although buckyonions are able to closely reproduce the 2175 Å extinction feature, it is not clear if dust models consisting of buckyonions, amorphous silicates, PAHs, and other carbon dust species (amorphous carbon, hydrogenated amorphous carbon, or graphite) are capable of fitting the 2175 Å extinction feature since one may intuitively expect the almost perfect fit by buckyonions would easily be distorted by the addition of other dust components. It is the purpose of this short report to examine this issue.

2. Extinction of Multi-component Dust Models

We consider dust models consisting of amorphous silicate dust, graphite, PAHs, and buckyonions. We take the size distributions of silicate and graphite dust to be that of Weingartner & Draine (2001; hereafter WD). The quantity of silicate dust is taken to be consistent with the interstellar depletion: we assume that all cosmically available Si, Mg, and Fe are locked up in amorphous silicate dust. We assume the interstellar Si and C abundances (relative to H) to be solar (Si/H = 35 ppm, C/H = 355 ppm). With C/H = 140 ppm in the gas phase, we have C/H = 215 ppm left for carbon dust: buckyonions, PAHs, and graphite. The "UIR" bands require C/H ≈ 60 ppm to be in PAHs (Li & Draine 2001). So we only have C/H = 155 ppm for buckyonions and graphite. As shown in Figure 1, with C/H = 20 ppm in graphite and C/H = 130 ppm in bukyonions (for which the absorption spectrum is taken from Ruiz *et al.* 2005), the silicate-graphite-PAH-buckyonion model provides an excellent match to the interstellar extinction curve at $\lambda^{-1} \sim 3.5$–$7 \, \mu m^{-1}$ (at present the photo-absorption spectra of buckyonions are available only in this wavelength range).

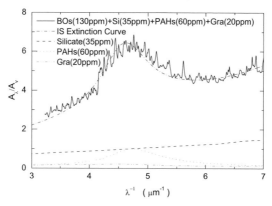

Figure 1. Comparison of the interstellar extinction (dot-dashed line) with the model extinction (solid line) obtained by summing up the contributions of buckyonions (C/H = 130 ppm), PAHs (C/H = 60 ppm), graphite (C/H = 20 ppm), and amorphous silicate dust (Si/H = 35 ppm).

3. Future Work

A powerful test of the buckyonions hypothesis would be in the IR. In the ISM, buckyonions (with their surface hydrogenated) will be stochastically heated by single UV photons and emit at the ~ 7–$8 \, \mu m$ C–H stretching bands. The non-detection of these bands will allow us to place an upper limit on the abundance of buckyonions. Moreover, it would be interesting to see if the silicate-graphite-PAH-buckyonion model is able to reproduce the ~ 2–$3000 \, \mu m$ overall IR emission of the Galactic ISM.

Acknowledgements

This project is supported by Hunan Provincial Natural Science Foundation of China.

References

Li, A. & Draine, B. T. 2001, *ApJ*, 554, 778
Li, A. & Greenberg, J. M. 2003, *Solid State Astrochemistry*, p. 37
Ruiz, A., Bretón, J., & Gomez Llorente, J. M. 2005, *Phys. Rev. Lett.*, 94, 105501
Stecher, T. P. & Donn, B. 1965, *ApJ*, 142, 1681
Weingartner, J. C. & Draine, B. T. 2001, *ApJ*, 548, 296 (WD)

Organic Matter in Space
Proceedings IAU Symposium No. 251, 2008
S. Kwok & S. Sandford, eds.

© 2008 International Astronomical Union
doi:10.1017/S1743921308021224

A study of 2175 Å absorption feature with TAUVEX: An Indo-Israeli UV mission

C. Muthumariappan[1], G. Maheswar[2,3], C. Eswaraiah[3], and A. K. Pandey[3]

[1] Vainu Bappu Observatory, Indian Institute of Astrophysics, Kavalur India
email: muthu@iiap.res.in

[2] Korea Astronomy & Space Science Institute, Daejeon, R. O. Korea
email: maheswar@kasi.re.kr

[3] ARIES, Nainital, India
email: eswarbramha@aries.ernet.in, pandey@aries.ernet.in

Abstract. TAUVEX is an Indo-Israeli collaborative mission to make photometric observations in the UV region. Using the narrowband filters positioned near 2175 Å feature, we plan to construct the UV bump in the extinction curve towards B type stars brighter than $m_v = 14$. Archival data of TD1, Galex missions are used for stars brighter than $m_v = 10$. We sample the distance to obtain the extinction at different location of the local ISM upto 2 kpc. Using a dust model having silicate, graphite and PAH as components, the extinction curve at different location of the Galaxy is fitted to constrain the dust characteristics.

Keywords. (ISM:) dust, extinction

1. Introduction

The absorption feature at 2175 Å is widely observed in the ISM which implies that its carrier is made of commonly available material in the ISM. The width and the extinction peak of the feature show large variation with line-of-sight ($\pm 12\%$) indicating a large chemical dispersion in the Galaxy, however, the peak value always accurs at 2175 Å (Fitzpatrick & Massa 2007). From the observations of thermal emission from ISM dust in the 3–60 μm range, and the microwave emission from ISM, it is now recognized that the interstellar grain population can include a substantial amount of ultra-small grains with PAH composition. PAH has strong absorption in the 2000 to 2500 Å region and the expected carrier of the 2175 Å feature is a mixture of PAHs. Moreover, ultra-small grain population is needed to reproduce the Far-UV rise in the extinction curve. The observed variation in the width of the feature would result from differences in the PAH mix.

2. TAUVEX mission and data resource

TAUVEX has three equivalent 20 cm aperture UV telescopes with f/8 beam and an effective focal length of 1600 mm which will be put in its orbit by India in mid 2008. It will image large parts of the sky in the wavelength region between 1400 and 3200 Å with a spatial resolution of about 6″ to 10″, depending on the wavelength. TAUVEX will be used to measure the fluxes at 1770 Å, 2180 Å, 2680 Å and 2220 Å using narrowband filters SF1, SF2, SF3 and NBF1 respectively. These observations will provide us with three points for the interstellar extinction in the UV domain, including at 2175 Å bump. TAUVEX can detect un-reddened early B stars brighter than $m_v = 14$ (S/N ~ 10) in one scan. Details on the mission can be found in http://tauvex.iiap.res.in.

The previous all sky survey in UV was performed by the TD1 satellite with a controlled scan of the whole sky. It measured the absolute UV flux distribution between 2740 Å and 1350 Å of unreddened early B stars with brightness upto $m_v = 10$, and are compiled

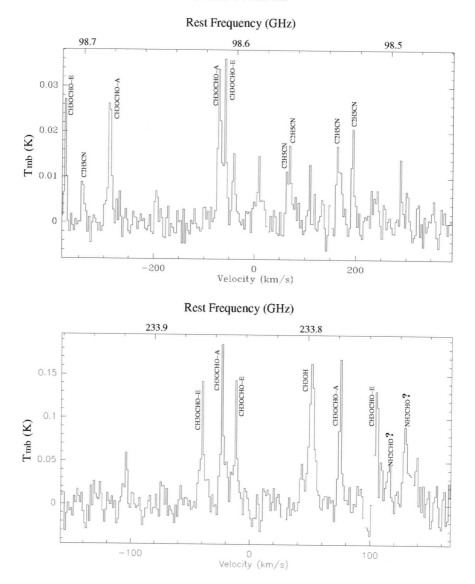

Figure 2. Sample spectra of the Hot Corino IRAS16293-2422 (courtesy by E. Caux).

the 3, 2, 1 and 0.8 millimeter bands with the largest existing telescopes: IRAM and JCMT respectively. Figure 2 shows a sample of this survey and the richness of the spectrum. For the first time, several COMs have been detected in a Hot Corino with dozens of lines. This not only permits to make sure identifications of the molecule, but also to derive reliable abundances in the outer envelope and Hot Corino of the source respectively. Here we report the example of propyne (CH_3CCH), an hydrocarbon of the alkynes family, extremely reactive on Earth and a building block of more complex organic molecules. Figure 3 shows the rotational diagram of this molecules, with 44 detected lines. Cazaux *et al.* (2003) previously reported the detection of this molecule based on four lines and estimated an abundance in the Hot Corino of about 3×10^{-7} with respect to H_2. The one order larger number of detected lines in the new survey by Caux and colleagues allows to have a much more sophisticated analysis, where the contribution

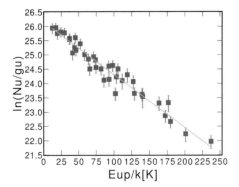

Figure 3. Rotational diagram of propyne in IRAS16293-2422 (from Caux *et al.*, in preparation). The diagram reports the log column density of each rotational level as function of the energy of the level. Note that no beam dilution factor has been applied here to take into account for the different emitting sizes.

from the outer envelope and the Hot Corino can be disentangled. The comparison of the measured line fluxes with those predicted by a model that takes into account the density and temperature profiles derived by several previous observations (Ceccarelli *et al.* 2000) gives a propyne abundance of $\sim 3 \times 10^{-10}$ and $\sim 1 \times 10^{-7}$ in the outer and Hot Corino regions respectively (Ceccarelli *et al.*, in preparation). Therefore, propyne is present in both regions, although its abundance is 300 times larger in the Hot Corino region. The relative abundance between the regions give us an hint on how this molecule is formed.

Finally, the systematic study of the abundance of COMs in the outer envelope and Hot Corino will likely allow to shed light on the mechanisms of formation and destruction of COMs in the Inter-Stellar Medium and, consequently, on the potential heritage passed throughout the various phases of the formation of a planetary system as the Solar System.

4. Conclusions

Despite the harsh conditions, low temperatures and densities, presence of FUV photons, shocks etc, Star Forming Regions are nurseries not only of stars but also of COMs. Several COMs have already been detected, but we know that many other await to be discovered. Very likely we have just seen the tip of the iceberg so far. Digging into the deep to search for the rest of the iceberg is one of the challenges of the next decade. We will be aided in this adventure by two new powerful telescopes: Herschel and ALMA. Herschel, a 3.5 mt space telescope to be launched in 2009, will allow observations in regions of the spectrum mostly unexplored, between 500 and 2,000 GHz, because blocked by the Earth atmosphere (http://herschel.esac.esa.int/home.shtml). ALMA (http://www.eso.org/sci/facilities/alma/), the 64 telescopes millimeter interferometer which will start operation in 2010, will provide unprecedent sensitivity observations, with a spatial resolution which will allow to image the COMs-rich regions. No doubts that next decade will give us the opportunity to answer some fundamental questions on COMs: what COMs are formed, where and why. In turn, this will also likely provide the answer whether Earth received the heritage of the ancient eons.

References

Agundez, M., Cernicharo, J., & Goicoechea, J. R. 2008, *A&A*, 483, 831
Bacmann, A., Lefloch, B., Ceccarelli, C., *et al.* 2003, *ApJ* (Letters), 585, L55

Catherine Cesarsky, Philippe Brechignac, and Cecilia Ceccarelli enjoying the buffet during the harbour cruise.

Organic Matter in Space
Proceedings IAU Symposium No. 251, 2008
S. Kwok & S. Sandford, eds.

The birth and death of organic molecules in protoplanetary disks

Thomas Henning and Dmitry Semenov

Max Planck Institute for Astronomy
Koenigstuhl 17, D-69117 Heidelberg, Germany
email: `henning,semenov@mpia.de`

Abstract. The most intriguing question related to the chemical evolution of protoplanetary disks is the genesis of pre-biotic organic molecules in the planet-forming zone. In this contribution we briefly review current observational knowledge of physical structure and chemical composition of disks and discuss whether organic molecules can be present in large amounts at the verge of planet formation. We predict that some molecules, including CO-bearing species such as H_2CO, can be underabundant in inner regions of accreting protoplanetary disks around low-mass stars due to the high-energy stellar radiation and chemical processing on dust grain surfaces. These theoretical predictions are further compared with high-resolution observational data and the limitations of current models are discussed.

Keywords. Astrochemistry, line: formation, molecular data, molecular processes, radiative transfer, circumstellar matter, planetary systems: protoplanetary disks, stars: pre–main-sequence, submillimeter, X-rays: stars

1. Introduction

The origin and evolution of life as we know it are tightly related to the chemistry of complex carbon-bearing molecules. While the transition from macromolecules to the simplest living organisms has likely proceeded on Earth, we do not know yet organic molecules of what complexity have been available during build-up phase of the primordial/secondary Earth atmosphere and oceans. During the last few decades a multitude of species, including alcohols (e.g. CH_3OH), ethers (e.g. CH_3OCH_3), acids (e.g. $HCOOH$) have been discovered in interstellar space, with atomic masses up to a few hundred† (for a recent review see Snyder 2006). A precursor of amino acids, amino acetonitrile, and the simplest sugar, glycolaldehyde, have been found toward the star-forming region Sagittarius B2(N) (Hollis *et al.* 2004, Belloche *et al.* 2008). Thus, many simple "blocks" of prebiotic molecules do exist in space. It is natural to ask what happens to these species during the prestellar/hot core phase, passage of an accretion shock, and inside a protoplanetary disk. Are organic molecules present in large amounts in circumstellar disks at the verge of planet formation? Could they form and survive in such a harsh environment as an accretion disk?

Despite the variety of "interstellar" molecules, only formaldehyde (H_2CO) and a few other non-organic species have been detected and spatially resolved with interferometers in several nearby protoplanetary disks (e.g., Dutrey *et al.* 1997, Kastner *et al.* 1997, Aikawa *et al.* 2003, Qi *et al.* 2003, Dutrey *et al.* 2007b). These multi-molecule, multi-transition studies allowed to constrain basic disk parameters like radii, masses, kinematics, temperature and density profiles, ionization degree and depletion factors

† *http://astrochemistry.net/*

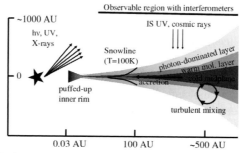

Figure 1. Physical and chemical structure of a protoplanetary disk. In the dark, dense and cold midplane most of molecules reside on dust grains, and chemical evolution is dominated by ion-molecule and surface reactions. This region has the lowest ionization degree, with dust grains being the most abundant charged species. A warmer intermediate layer is located above midplane. It is heated by mild UV radiation. Many reactions with barriers can occur and a rich variety of molecules exist in the gas phase. This is the zone where most of the molecular lines are formed. The ionization fraction in the intermediate layer is determined by a multitude of molecular ions, in particular HCO^+. Further above, a hot, highly ionized disk atmosphere is located, where only the simplest radicals and ions (apart from H_2) survive. This is the region where ionized carbon is abundant and C^+ emission lines are excited.

There is strong evidence that in many evolved disks, with ages of a few Myrs, dust grains grow until at least pebble-like sizes. The results from the *Infrared Space Observatory* and *Spitzer* reveal the presence of a significant amount of frozen material and a rich variety of amorphous and crystalline silicates and PAHs in disks (e.g., van den Ancker *et al.* 2000, van Dishoeck 2004, Bouwman *et al.* 2008). The PAH emission features at near- and mid-infrared wavelengths are excited by the incident stellar radiation field and as such depend on disk vertical structure and turbulent state (Dullemond *et al.* 2007). These lines are more easily observed in disks around hot, intermediate-mass Herbig Ae/Be stars as compared to cool, Sun-like T Tauri stars (e.g., Acke & van den Ancker 2004, Geers *et al.* 2007, Sicilia-Aguilar *et al.* 2007).

Various solid-state bands observed at 10–30 μm in emission belong to amorphous and crystalline silicates at $T \gtrsim$ 100–300 K with varying Fe/Mg ratios and grain topology/sizes (e.g., van Boekel *et al.* 2004, Natta *et al.* 2007, Bouwman *et al.* 2008, Voshchinnikov & Henning 2008). The composition of the hot gas in the inner disk as traced by ro-vibrational emission lines from CO, CO_2, C_2H_2, HCN and recently H_2O and OH, suggests that complex chemistry driven by endothermic reactions is at work there (Brittain *et al.* 2003, Lahuis *et al.* 2006, Eisner 2007, Salyk *et al.* 2008). At larger distances from the star the disk becomes colder and most of these molecules stick to dust grain, forming icy mantles. The main mantle component is water ice with trace amount of other more volatile materials like CO, CO_2, NH_3, CH_4, H_2CO, HCOOH and CH_3OH (Zasowski *et al.* 2007). Typical relative abundances of these minor constituents are about 0.5–10% of that of water.

3. Chemical structure of a disk

The chemical evolution of protoplanetary disks has been investigated in detail by using robust chemical models (Willacy *et al.* 1998, Aikawa & Herbst 1999, Willacy & Langer 2000, Aikawa *et al.* 2002, Bergin *et al.* 2003, van Zadelhoff *et al.* 2003, Ilgner *et al.* 2004, Semenov *et al.* 2004, Willacy *et al.* 2006, Ilgner & Nelson 2008). The current theoretical picture based on a steady-state prescription of the disk structure divides the disk into three zones, see Figure 1. Before planets have formed and disk gas is dispersed, the dense midplane is well shielded from stellar and interstellar high-energy radiation. While its inner part can be heated up by accretion, the outer zone is cold, $T \sim 10$–20 K. The only ionization sources are cosmic ray particles and decay of short-living radionuclides, and thus matter remains almost neutral, with a low degree of turbulence. The molecular complexity in the midplane is determined by ion-molecule and surface reactions, with most molecules sitting on the grains. Adjacent to the midplane a warmer zone is located, which is partly shielded from stellar and interstellar UV/X-ray radiation. The complex cycling between efficiently formed gas-phase molecules, accretion onto dust surfaces, rapid surface reactions, and non-negligible desorption result in a rich chemistry (see for review Bergin *et al.* 2007). The inner part ($\lesssim 10$–20 AU) of this region in the disks around T Tauri stars can be substantially ionized by stellar X-ray radiation. The intermediate molecular layer is sufficiently dense ($\sim 10^5$–10^6 cm^{-3}) to excite most of the observed emission lines. Atop a hot and heavily irradiated surface layer is located, where C^+, light hydrocarbons, their ions, and other radicals like C_2H and CN are able to survive. This is the region where PAH and silicate emission features are produced.

4. Theoretical constrains

Here we discuss a puzzling observation of a putative H_2CO inner cavity in the disk of DM Tau. The Plateau de Bure interferometric image of the DM Tau disk at the 1.5″ resolution in the H_2CO (3_{12}-2_{12}) line is shown in Figure 2 (left panel). Despite high noise level, the H_2CO emission appears as asymmetric ring-like structure, with a dip in southern direction. To make a proper analysis of these data, a consistent combination of disk physical and chemical models along with radiative transfer in molecular lines is used. To simulate the disk physical structure we utilize a 1+1D flared disk model which is similar to the model of D'Alessio *et al.* (1999) with a vertical temperature gradient. The dust grains are modeled as compact amorphous silicate spheres of uniform 0.1 μm radius, with the opacity data taken from Semenov *et al.* (2003) and a dust-to-gas mass ratio of 100. The accretion rate is assumed to be 2×10^{-9} M$_\odot$ yr^{-1}, $\alpha = 0.01$, and the disk outer radius is 8,000 AU. We focus on the observable disk structure beyond the radius of ~ 10 AU. The total disk mass is 0.07 M$_\odot$ and the disk age is 5 Myr (Piétu *et al.* 2007).

We assumed that the disk is illuminated by UV radiation from the central star with an intensity $\chi = 410\,\chi_0$ at 100 AU and by interstellar UV radiation with intensity χ_0 in plane-parallel geometry (Draine 1978, van Dishoeck *et al.* 2006, Dutrey *et al.* 2007b). We model the attenuation of cosmic rays (CRP) by Eq. 3 from Semenov *et al.* (2004) with an initial value of the ionization rate $\zeta_{CRP} = 1.3 \cdot 10^{-17}$ s^{-1}. In the disk interior ionization due to the decay of short-living radionuclides is taken into account, assuming an ionization rate of $6.5 \cdot 10^{-19}$ s^{-1} (Finocchi & Gail 1997). The X-ray ionization rate in a given disk region is computed according to the results of Glassgold *et al.* (1997a), Glassgold *et al.* (1997b) with parameters for their high-metal depletion case and a total X-ray luminosity of $\approx 10^{30}$ erg cm^{-2} s^{-1} (Glassgold *et al.* 2005). The gas-phase reaction rates are taken from the RATE 06 database (Woodall *et al.* 2007), while surface reactions together with desorption energies were adopted from the model of Garrod & Herbst (2006). A standard rate approach to the surface chemistry modeling, but without H and H_2 tunneling was utilized (Katz *et al.* 1999).

Using the time-dependent chemical code "ALCHEMIC"†, we simulated 5 Myr years of evolution in the DM Tau disk, followed by 2D non-LTE line radiative transfer modeling

† *www.mpia.de/homes/semenov*

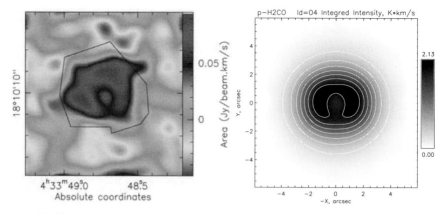

Figure 2. (Left) The observed integrated intensity map of $H_2CO\,(3-2)$ in the disk of DM Tau shows an asymmetric shell-like distribution with a chemical inner "hole" of ~ 100 AU in radius. (Right) The same features are present in the synthetic integrated intensity map of para-$H_2CO\,(4-3)$ that is produced with a realistic disk physical and chemical model and line radiative transfer.

with "URANIA" (Pavlyuchenkov *et al.* 2007), see Figures 2 and 3. The observed ring of H_2CO emission with a depression is fully reproduced by our model. We found two possible explanations why such a large-scale chemical hole can exist in the disk around DM Tau. The ~ 100 AU hole in the H_2CO emission is fully reproduced by both a disk model with X-ray driven chemical processes and somewhat less markedly in the model without surface chemistry. These two models predict different spatial distributions of molecular species, which can be tested by future interferometric observations.

Destruction of formaldehyde has important consequences for organic chemistry. The X-ray chemical model leads to the clearing of an inner hole of ~ 100 AU radius in all chemically related CO-bearing species, including HCO^+, by converting gas-phase CO into heavier CO_2-containing and chain-like hydrocarbon molecules. In contrast to CO, these heavier species are locked on dust surfaces in the inner disk region, where temperatures are lower than about 35–50 K (Figure 3, solid line). This implies substantially different initial conditions with respect to the presence of complex organic molecules inside the planet-forming zone of protoplanetary disks, if this X-ray driven chemistry is important.

The less realistic model without surface chemistry shows the inner depression in column densities of highly saturated molecules only, like H_2O, NH_3, and to some extent H_2CO (see Figure 3, dashed line). These species are formed on dust surfaces in a sequence of hydrogen addition reactions. Though at current stage we cannot fully distinguish between these two scenarios, for our understanding of the evolution of organic species in protoplanetary disks it will be of great importance to verify which of these explanations are valid.

5. Summary

We briefly overview recent progress in our understanding of chemical evolution in protoplanetary disks, from both the theoretical and observational perspective. A puzzling observation of the chemical inner hole visible in the spatial distribution of the H_2CO emission in the disk of DM Tau is addressed theoretically. We found that such a hole can be explained either by the absence of efficient hydrogenation reactions on dust surfaces or efficient processing of disk matter by stellar X-ray radiation in the inner disk region, which was overlooked in previous studies. In future, when the Atacama Large Millimeter

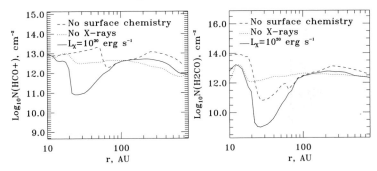

Figure 3. (Left) The radial distribution of the HCO$^+$ column density in the DM Tau disk as computed with 3 different chemical models: 1) the stellar X-ray luminosity is assumed to be close to the observed value of 10^{30} erg s^{-1} (solid line), 2) no X-ray radiation penetrates into the inner disk (dotted line), and 3) the model without surface reactions but with $L_X = 10^{30}$ erg s^{-1} (dashed line). (Right) The same calculations but for the chemically related H$_2$CO molecule.

Array will become operational, the planet-forming zone of disks will be observable and this hypothesis can be verified. In general, chemo-dynamical models of disks together with interferometric observations well lead to a comprehensive understanding of the molecular inventory of protoplanetary disks.

References

Acke, B. & van den Ancker, M. E. 2004, *A&A*, 426, 151

Aikawa, Y. & Herbst, E. 1999, *A&A*, 351, 233

Aikawa, Y., Momose, M., Thi, W.-F., *et al.* 2003, *PASJ*, 55, 11

Aikawa, Y., van Zadelhoff, G. J., van Dishoeck, E. F., & Herbst, E. 2002, *A&A*, 386, 622

Belloche, A., Menten, K. M., Comito, C., *et al.* 2008, ArXiv e-prints, 801

Bergin, E. A., Aikawa, Y., Blake, G. A., & van Dishoeck, E. F. 2007, in: B. Reipurth, D. Jewitt, & K. Keil (eds.), *Protostars and Planets V*, p. 751

Bergin, E., Calvet, N., D'Alessio, P., & Herczeg, G. J. 2003, *ApJ (Letters)*, 591, L159

Bouwman, J., Henning, T., Hillenbrand, L. A., *et al.* 2008, ArXiv e-prints, 802

Brittain, S. D., Rettig, T. W., Simon, T., *et al.* 2003, *ApJ*, 588, 535

D'Alessio, P., Calvet, N., Hartmann, L., Lizano, S., & Cantó, J. 1999, *ApJ*, 527, 893

Dartois, E., Dutrey, A., & Guilloteau, S. 2003, *A&A*, 399, 773

d'Hendecourt, L. B., Allamandola, L. J., Baas, F., & Greenberg, J. M. 1982, *A&A (Letters)*, 109, L12

Draine, B. T. 1978, *ApJS*, 36, 595

Dullemond, C. P., Henning, T., Visser, R., *et al.* 2007, *A&A*, 473, 457

Dutrey, A., Guilloteau, S., & Guelin, M. 1997, *A&A (Letters)*, 317, L55

Dutrey, A., Guilloteau, S., & Ho, P. 2007a, in: B. Reipurth, D. Jewitt, & K. Keil (eds.), *Protostars and Planets V*, p. 495

Dutrey, A., Henning, T., Guilloteau, S., *et al.* 2007b, *A&A*, 464, 615

Eisner, J. A. 2007, *Nature*, 447, 562

Finocchi, F. & Gail, H.-P. 1997, *A&A*, 327, 825

Garrod, R. T. & Herbst, E. 2006, *A&A*, 457, 927

Garrod, R. T., Wakelam, V., & Herbst, E. 2007, *A&A*, 467, 1103

Geers, V. C., van Dishoeck, E. F., Visser, R., *et al.* 2007, *A&A*, 476, 279

Glassgold, A. E., Feigelson, E. D., Montmerle, T., & Wolk, S. 2005, in: A. N. Krot, E. R. D. Scott, & B. Reipurth (eds.), *Astronomical Society of the Pacific Conference Series, Vol. 341, Chondrites and the Protoplanetary Disk*, p. 165

Glassgold, A. E., Najita, J., & Igea, J. 1997a, *ApJ*, 480, 344

Glassgold, A. E., Najita, J., & Igea, J. 1997b, *ApJ*, 485, 920

Glassgold, A. E., Najita, J. R., & Igea, J. 2007, *ApJ*, 656, 515

Hassel, Jr., G. E. 2004, PhD thesis, AA(RENSSELAER POLYTECHNIC INSTITUTE)

Hollis, J. M., Jewell, P. R., Lovas, F. J., & Remijan, A. 2004, *ApJ* (Letters), 613, L45

Igea, J. & Glassgold, A. E. 1999, *ApJ*, 518, 848

Ilgner, M., Henning, T., Markwick, A. J., & Millar, T. J. 2004, *A&A*, 415, 643

Ilgner, M. & Nelson, R. P. 2008, ArXiv e-prints, 802

Isella, A., Testi, L., Natta, A., *et al.* 2007, *A&A*, 469, 213

Kastner, J. H., Zuckerman, B., Weintraub, D. A., & Forveille, T. 1997, *Science*, 277, 67

Katz, N., Furman, I., Biham, O., Pirronello, V., & Vidali, G. 1999, *ApJ*, 522, 305

Lada, C. J. 1985, *ARAA*, 23, 267

Lahuis, F., van Dishoeck, E. F., Boogert, A. C. A., *et al.* 2006, *ApJ* (Letters), 636, L145

Leger, A., Jura, M., & Omont, A. 1985, *A&A*, 144, 147

Najita, J., Bergin, E. A., & Ullom, J. N. 2001, *ApJ*, 561, 880

Natta, A., Testi, L., Calvet, N., *et al.* 2007, in: B. Reipurth, D. Jewitt, & K. Keil (eds.), *Protostars and Planets V*, p. 767

Öberg, K. I., Fuchs, G. W., Awad, Z., *et al.* 2007, *ApJ* (Letters), 662, L23

Pascucci, I., Hollenbach, D., Najita, J., *et al.* 2007, *ApJ*, 663, 383

Pavlyuchenkov, Y., Semenov, D., Henning, T., *et al.* 2007, *ApJ*, 669, 1262

Piétu, V., Dutrey, A., & Guilloteau, S. 2007, *A&A*, 467, 163

Qi, C., Kessler, J. E., Koerner, D. W., Sargent, A. I., & Blake, G. A. 2003, *ApJ*, 597, 986

Qi, C., Wilner, D. J., Aikawa, Y., Blake, G. A., & Hogerheijde, M. R. 2008, ArXiv e-prints, 803

Qi, C., Wilner, D. J., Calvet, N., *et al.* 2006, *ApJ* (Letters), 636, L157

Rodmann, J., Henning, T., Chandler, C. J., Mundy, L. G., & Wilner, D. J. 2006, *A&A*, 446, 211

Salyk, C., Pontoppidan, K. M., Blake, G. A., *et al.* 2008, *ApJ* (Letters), 676, L49

Semenov, D., Henning, T., Helling, C., Ilgner, M., & Sedlmayr, E. 2003, *A&A*, 410, 611

Semenov, D., Pavlyuchenkov, Y., Schreyer, K., *et al.* 2005, *ApJ*, 621, 853

Semenov, D., Wiebe, D., & Henning, T. 2004, *A&A*, 417, 93

Semenov, D., Wiebe, D., & Henning, T. 2006, *ApJ* (Letters), 647, L57

Shalabiea, O. M. & Greenberg, J. M. 1994, *A&A*, 290, 266

Sicilia-Aguilar, A., Hartmann, L. W., Watson, D., *et al.* 2007, *ApJ*, 659, 1637

Snyder, L. E. 2006, Proceedings of the National Academy of Science, 103, 12243

van Boekel, R., Min, M., Leinert, C., *et al.* 2004, *Nature*, 432, 479

van den Ancker, M. E., Bouwman, J., Wesselius, P. R., *et al.* 2000, *A&A*, 357, 325

van Dishoeck, E. F. 2004, *ARAA*, 42, 119

van Dishoeck, E. F., Jonkheid, B., & van Hemert, M. C. 2006, in: I. R. Sims & D. A. Williams (eds.), *Chemical evolution of the Universe*, Faraday discussion, Vol. 133, p. 231

van Zadelhoff, G.-J., Aikawa, Y., Hogerheijde, M. R., & van Dishoeck, E. F. 2003, *A&A*, 397, 789

Voshchinnikov, N. V. & Henning, T. 2008, ArXiv e-prints, 803

Willacy, K., Klahr, H. H., Millar, T. J., & Henning, T. 1998, *A&A*, 338, 995

Willacy, K., Langer, W., Allen, M., & Bryden, G. 2006, *ApJ*, 644, 1202

Willacy, K. & Langer, W. D. 2000, *ApJ*, 544, 903

Woodall, J., Agúndez, M., Markwick-Kemper, A. J., & Millar, T. J. 2007, *A&A*, 466, 1197

Zasowski, G., Markwick-Kemper, F., Watson, D. M., *et al.* 2007, ArXiv e-prints, 712

Discussion

ZIURYS: You talked about formaledehyde forming a hole and you attributed that to formaldehyde being destroyed. Could it also be attributed to formaldehyde being converted to something else? What if it's converted to some kind of polymer.

HENNING: We have developed a model which is quite complete in the treatment of disk chemistry, including X-Ray processes. However, potential polymerization processes are not considered. In addition, I want to stress that we consider a steady-state model for the disk structure, but do not treat transport of species for the chemical model. This may allow the hole to be "filled" by material. Furthermore, the reality of the presence of this hole has to be considered with caution until observations with higher signal-to-noise ratio become available.

VAN DISHOECK: Do you see evidence for this hole in other molecules as well?

HENNING: There is no strong indication for a hole in the dust continuum and CO data.

UNKNOWN: Why do you think the X-rays can penetrate into the inner disk while the surface density is quite high?

HENNING: In a flared disk model the X-rays are able to penetrate into the surface and intermediate layer, but not into the midplane. However, the vertical peak density of formaldehyde is reached in the intermediate layer and the X-rays are able to effect the abundance in this layer. An additional possibility may be irradiation by X-ray jets which practically would serve as a lamp irradiating the disk from above.

Thomas Henning and Ewine van Dishoeck in deep discussion (photo by Dale Cruikshank).

Lucy and Bruce Hrivnak at Repulse Bay.

Organic Matter in Space
Proceedings IAU Symposium No. 251, 2008
S. Kwok & S. Sandford, eds.

Models and observations of deuterated molecules in protostellar cores

H. Roberts

School of Mathematics and Physics, The Queen's University of Belfast
Belfast, BT9 1NN, UK
email: h.roberts@qub.ac.uk

Abstract. Measuring the deuterium fractionation in different molecules can allow one to determine the physical conditions in the gas and to differentiate between gas-phase and grain surface chemical processing. Observations of molecular D/H ratios in different species towards the dense gas surrounding low-mass protostars are presented and are compared with model simulations. These consider gas-phase chemistry, accretion and desorption, and reactions on grain surfaces during the initial stages of core collapse.

Keywords. Astrochemistry, molecular processes, stars: formation, ISM: molecules

1. Introduction

Many different organic species have been detected in regions of both low and high mass star formation. This has led researchers to ask whether the building blocks for life on Earth were formed in the protosolar nebula. Could organic matter on the early Earth even have come from the interstellar medium? Observations of isotopic species can help to determine under what conditions molecules formed and deuterium (^{2}H) is a particularly useful isotope for this purpose. Although the underlying D/H ratio is $\sim 10^{-5}$, D is preferentially incorporated into molecules at low temperatures. This fractionation is mainly temperature dependent, but can also be affected by density, by electron abundance, and by gas-grain interactions.

The primary fractionation reaction in dark clouds and prestellar cores is H_3^+ reacting with HD (the deuterium reservoir) to give H_2D^+ and H_2. At temperatures $<$25 K, H_2D^+ is not destroyed by H_2 so the H_2D^+/H_3^+ ratio becomes enhanced (Millar *et al.* 1989). In dark clouds, H_3^+ and H_2D^+ are primarily destroyed by proton transfer to neutral molecules (e.g., CO, O, N$_2$) and by dissociative recombination recombination with electrons. Another destruction channel for H_2D^+ is further reaction with HD to form D_2H^+ and D_3^+, but this is minor compared with destruction by CO *et al.* If a core within the dark cloud begins to contract under gravity, however, the density increases, the temperature and ionisation fraction decrease, and heavier species begin to freeze onto dust grains. In this case the conversion of H_3^+ into its deuterated isotopologues becomes much more efficient and the relative abundances of H_2D^+, D_2H^+ and D_3^+ can become similar to or even exceed the H_3^+ abundance. This leads to a very efficient transfer of deuterium to other species (see, e.g., Roberts *et al.* 2003, Walmsley *et al.* 2004). This theory is supported by various observational results (Caselli *et al.* 1999, Roueff *et al.* 2000, Vastel *et al.* 2003, Vastel *et al.* 2004, Crapsi *et al.* 2005, etc.).

Different molecules are fractionated via different processes. Simple molecular ions like N_2D^+ and DCO^+ form directly from D_3^+ (and isotopologues) via proton transfer. Neutral hydrogen-bearing species (e.g., H_2O, HCN, NH_3) are deuterated by successive

99

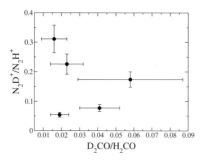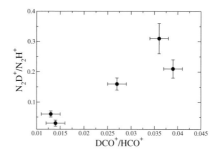

Figure 1. A comparison of the deuterium fractionation in selected species towards 5 low mass protostellar sources: HH211, L1448mms L1448NW, L1527, and L151 IRS5. LEFT: D_2CO/H_2CO vs N_2D^+/N_2H^+; RIGHT: DCO^+/HCO^+ vs N_2D^+/N_2H^+.

protonation followed by dissociative recombination reactions, but for H_2CO recent theoretical work has shown that this route is inefficient (Osamura *et al.* 2005). Extremely high D_2CO/H_2CO ratios have been observed towards both prestellar and protostellar cores and are increasingly taken to be evidence of active grain surface chemistry and desorption (e.g., Ceccarelli *et al.* 2001, Bacmann *et al.* 2003).

The formation of methanol in the gas-phase is extremely inefficient under interstellar conditions (Geppert *et al.* 2005). The likely route for CH_3OH production is on the grain surfaces via successive additions of H and D atoms to CO molecules, which also forms formaldehyde as an intermediate product. The surface molecular D/H ratios should reflect the relative abundances of H and D atoms accreting from the gas phase (Brown & Millar 1989, Stantcheva & Herbst 2003). In the prestellar core phase, this atomic D/H ratio is also enhanced due to H_3^+ fractionation (Roberts *et al.* 2003). Once H_2CO and CH_3OH form they may be further fractionated by exchange reactions with atomic D (Watanabe *et al.* 2004, Hidaka *et al.* 2005). Also, exothermic hydrogen abstraction reactions (e.g., $H + H_2CO \rightarrow HCO + H_2$) may enhance deuterium fractionation in the later stages of freeze-out by destroying the non-deuterated formaldehyde (Rodgers & Charnley 2002).

2. Observations and model results

As part of an ongoing project to compare deuterium fractionation in low and high mass star forming regions, DCN, HDCO, D_2CO, N_2D^+ and DCO^+ have been observed using the Arizona Radio Observatory (ARO) 12 m telescope, the 15 m JCMT, and the IRAM 30 m telescope (Roberts *et al.* 2002, Roberts & Millar 2007). The comparison of N_2D^+ with D_2CO is particularly interesting: N_2D^+ traces gas-phase deuteration and depletion, while D_2CO fractionation appears to trace grain surface chemistry followed by evaporation, so one might expect an anti-correlation between the two. N_2D^+/N_2H^+ will increase as the gas evolves from a dark cloud to a prestellar core, then decrease once the protostar forms and begins to heat its environment. Conversely, D_2CO/H_2CO will increase as molecules which formed via low-temperature surface chemistry evaporate.

Figure 1 compares ratios for 5 low-mass protostellar cores (L1527 and L1551 IRS5 in Taurus; L1448mms, L1448NW, and HH211 in Perseus). DCO^+ and N_2D^+ ratios appear to be fairly well correlated (as we expect since they both reflect the H_3^+ fractionation) but the N_2D^+/N_2H^+ ratios in all sources are significantly higher than the DCO^+/HCO^+ ratios. The N_2D^+/N_2H^+ ratios are all >0.05, which suggests the presence of cold, dense, CO-depleted gas; the DCO^+/HCO^+ ratios are also indicative of cold gas, but not necessarily heavy molecular depletion.

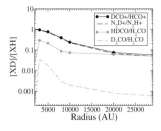 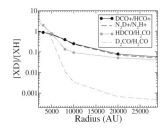 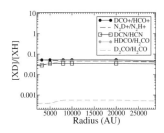

Figure 2. Molecular D/H ratios predicted as a function of radius in a cold core which has collapsed from a constant density to a centrally peaked distribution over 5×10^5 yrs. LEFT: a model containing gas-phase chemistry plus accretion onto grains; CENTRE: as at left, but with non-thermal desorption due to exothermic reactions occurring on the grain surfaces; RIGHT: as at left, but with photo-desorption of CO by cosmic-ray induced photons.

The $80''$ beam of the ARO 12 m telescope equates to 10000 AU at the distance of the Taurus Molecular Cloud (15000 AU at Perseus), so it makes sense that the observations are probing the cold envelopes which have yet to be disrupted by the central protostar. Unfortunately, there are not enough sources to make a convincing case for anti-correlation between N_2D^+ and D_2CO. It is interesting, though, that the D_2CO/H_2CO ratios appear to be 0.01–0.06 even in this cold, gas. Of course, the observations are averaging over the whole protostar-core-envelope system, so is the enhanced D_2CO fractionation tracing warmer gas close to the protostar?

Maret *et al.* (2004) carried out a multiline survey of formaldehyde and methanol towards a sample of low-mass protostellar cores which overlaps with ours. Using radiative transfer modelling, they found evidence for significant increases in species abundances close to the protostars where the gas temperature exceeds 50 K. The radii of these warmer regions, though, were only $\sim$100 AU. This is clearly too small for our observations to be sensitive to, even if the density of a warm core is several orders of magnitude larger than the extended envelope. As it does not appear that ion-molecule chemistry and accretion can significantly enhance D_2CO/H_2CO ratios, the high fractionation seen in the envelopes is evidence that grain-surface chemistry can affect the gas-phase deuteration, even at low temperatures.

Figure 2 shows predicted molecular D/H ratios in a cold, collapsing envelope. The model includes both gas-phase and grain surface reactions. The model is based on that presented in Roberts *et al.* (2004), but with updated reaction rates based on the latest release of the UMIST database for astrochemistry, RATE06 (www.udfa.net). The rate of accretion of molecules onto grains is governed by the density, which increases from 10^4 cm^{-3} at the outer radius to 10^6 cm^{-3} at the inner radius. The temperature is 10 K throughout, so thermal desorption is inefficient except for very light species (H, He, H_2). The figure compares the effect of different non-thermal desorption mechanisms on the deuterium chemistry.

The left-hand plot of Figure 2 shows results when only thermal desorption is used. The fractionation of the singly deuterated species increases from $\sim$a few $\times 10^{-2}$ in the outer envelope to >0.1 at 5000 AU. There is no difference between the N_2D^+ and DCO$^+$ fractionation and D_2CO fractionation at 10000 AU is significantly lower than observed. The central plot shows the effect of desorption due to reactions on the grain surfaces. The bulk of the energy released by an exothermic reaction will be dissipated in the grain, but there is a possibility that some of the reaction products will desorb. This has been modelled in detail by Garrod *et al.* (2007) using classical unimolecular rate theory.

Here the simple assumption is that 1% of the product species are returned to the gas phase. It is clear that this has a large effect on HDCO and D_2CO fractionation, so that the D_2CO/H_2CO ratios are even higher than the observations at radii $\leqslant 6000$ AU. This method, therefore, returns enough deuterated formaldehyde to the gas-phase to enhance beam averaged column densities. It does not, however, affect fractionation of the molecular ions. The right-hand plot shows the effect of including CO photo-desorption. The visual extinction in the models is high enough that direct photo-desorption is inefficient, but there is a cosmic ray induced photon field even in the high density regions. Using rates derived from experiments by Öberg *et al.* (2007), this mechanism significantly increases the amount of CO in the gas-phase. As Figure 2 shows, this drastically reduces the deuterium fractionation at smaller radii for all species, because CO destroys D_3^+ and isotopologues. Direct cosmic-ray desorption is not yet included, but if it is efficient enough, it could further reduce the gas-phase fractionation.

Roberts *et al.* (2007) presented a more comprehensive comparison of the effects of non-thermal desorption mechanisms in dark clouds, constraining the efficiency of the methods by a comparison with the observed CO depletion, but one can only determine a lower limit on this quantity. Molecular D/H ratios are very sensitive to the overall level of molecular depletion and, as shown above, different species are affected by different desorption mechanisms. A comparison of the effects of different desorption mechanisms on deuterium fractionation in prestellar cores is now underway, particularly looking for mechanisms which can explain both the enhanced D_2CO/H_2CO ratios and the difference between N_2D^+ and DCO^+.

References

Bacmann, A., Lefloch, B., Ceccarelli, C., *et al.* 2003, *ApJ* (Letter), 585, L55
Brown, P. B. & Millar, T. J. 1989, *MNRAS*, 237, 661
Caselli, P., Walmsley, C. M., Tafalla, M., Dore, L., & Myers, P. C. 1999, *ApJ* (Letter), 523, L165
Ceccarelli, C., Loinard, L., Castets, A., *et al.* 2001, *A&A*, 375, 40
Crapsi, A., Caselli, P., Walmsley, C. M. *et al.* 2005, *ApJ*, 619, 379
Garrod, R. T., Wakelam, V., & Herbst, E. 2007, *A&A*, 467, 1103
Geppert, W. D., Hellberg, F., Österdahl, F., *et al.* 2005, in: D. C. Lis, G. A. Blake & E. Herbst (eds.), *Proc. IAU 231*
Hidaka, H., Watanabe, N., & Kouchi, A. 2005, in: D. C. Lis, G. A. Blake & E. Herbst (eds.), *Proc. IAU 231*
Maret, S., Ceccarelli, C., Caux, E., *et al.* 2004, *A&A*, 416, 577
Millar, T. J., Bennett, A., & Herbst, E. 1989, *ApJ*, 340, 906
Öberg, K. I., Fuchs, G. W., Awad, Z., *et al.* 2007, *ApJ*, 662, 23
Osamura, Y., Roberts, H., & Herbst, E. 2005, *ApJ* 621, 348,
Roberts, H., Fuller, G., Millar, T. J., Hatchell, J., Buckle, J. V. 2002, *A&A*, 381, 283
Roberts, H., Herbst, E., & Millar, T. J. 2003, *ApJ* (Letter), 591, L41
Roberts, H., Herbst, E., & Millar, T. J. 2004, *A&A*, 424, 905
Roberts, H. & Millar, T. J. 2007, *A&A*, 471, 849
Roberts, J., Rawlings, J. M. C., Viti, S., & Williams, D. A. 2007, *MNRAS*, 382, 733
Rodgers, S. D. & Charnley, S. B. 2002, *P&SS*, 50, 1215
Roueff, E., Tiné, S., Coudert, L. H., *et al.* 2000, *A&A*, 354, L63
Stantcheva, T. & Herbst, E. 2003, *MNRAS*, 340, 983
Vastel, C., Phillips, T. G., Ceccarelli, C., & Pearson, J. 2003, *ApJ* (Letter), 593, L97
Vastel, C., Phillips, T. G., & Yoshida, H. 2004, *ApJ* (Letter), 606, L127
Walmsley, C. M., Flower, D. R., & Pineau des Forêts, G. 2004, *A&A*, 418, 1035
Watanabe, N., Nagaoka, A., Shiraki, T., & Kouchi, A. 2004, *ApJ*, 616, 638

Discussion

ZIURYS: When determining the DCO^+/HCO^+ ratio, how do you account for high opacity in HCO^+?

ROBERTS: We actually observed $H^{13}CO^+$ and estimated the HCO^+ column densities based on a constant ^{13}C/^{12}C ratio.

VAN DISHOECK: But $H^{13}CO^+$ lines can sometimes also have significant optical depth.

ZIURYS: Yes, with the HCN you could use the hyperfine structure from the nitrogen quadropole. I just wonder how much uncertainty will be introduced if the $H^{13}CO^+$ lines had significant optical depth.

ROBERTS: That is an interesting point: if the $H^{13}CO^+$ lines are not optically thin then we will be underestimating the HCO^+ abundance, making the DCO^+/HCO^+ ratios in Figure 1 upper limits. This would mean that the difference between N_2D^+ and DCO^+ fractionation is even more pronounced. Actually, I am currently using the radiative transfer code, RATRAN, to model the $H^{13}CO^+$ line profiles and better determine the HCO^+ abundance.

VAN DISHOECK: I have a question about your ammonia, ND_3. It wasn't quite clear to me from your conclusion whether you now favor grain surface or gas phase formation, because it has been sort of going back and forth in the literature.

ROBERTS: We believe that deuterated ammonia forms relatively efficiently via both gas-phase and grain surface reactions at low temperatures and high depletions. The relative abundances of the different isotopomers, however, may be different depending on which route dominates (Rodgers & Charnley 2002). I haven't yet compared results for the different desorption mechanisms for every molecule, but they are likely to have different efficiencies with respect to evaporating ammonia and so could alter the gas-phase deuterium fractionation to different degrees. Observations show that ammonia resists depletion in prestellar cores to higher densities than CO, but is this due to a lower binding energy/sticking coefficient for N followed by formation of NH_3 in the gas-phase? Or has NH_3 formed from N frozen onto the grain surfaces and then evaporated via non-thermal desorption? Looking at the deuterium fractionation could distinguish between the two.

MUMMA: Very interesting. I want to ask you to comment on why you would think that if a molecule forms on the grain surface which is hydrogen bonded weakly (the bond energy that has to be dissipated being far larger than this binding strength); why would you not expect this to work all the time so all the products desorb?

ROBERTS: The formation of methanol on the grain surfaces is a multi-step process, so all the products cannot desorb at each step otherwise it would never form. From an observational point of view; the abundances of 'surface formed' species (e.g. water, methanol, hydrogen sulphide) are significantly higher towards warm regions (where shocks and/or thermal desorption have occurred) than towards cold regions (where non-thermal desorption mechanisms operate). Also, water and methanol ice are detected along many lines of sight. These observations all suggest that the bulk of the species which form on the grains remain there until the whole mantle is evaporated. The original calculations using this method (Garrod *et al.* 2007) took into account the rate at which the energy is lost to the surface, compared to the binding energy of the species onto the grain. My simple assumption, that 1% of the products evaporate, is based on their work and does give a reasonable agreement with observations towards cold, dark clouds.

GARROD: The energy dissipation to the surface is fast, so only a small proportion of the energy released by bond formation is available to desorb the product molecule, so you end up with this approximately one percent fraction.

VAN DISHOECK: That must be different for different molecules.

GARROD: It is, but that is a second order effect. To first order, the results for most reactions are very similar, even taking larger bonding into account.

VAN DISHOECK: I think it's good to remember there probably some other desorption mechanisms which will also operate at that level of efficiency.

ROBERTS: Yes, that's definitely true. I initially looked at desorption due to surface reactions because the species we want to return into the gas phase to improve agreement with observations (CH_3OH, D_2CO, N_2, and NH_3) are exactly the ones which are formed on the grains. Figure 2 compares the effect on the D/H ratios of this method vs. CO photo-desorption, which has a dramatically different effect on the deuterium fractionation. Future work will look at the effects of combining both methods, along with other mechanisms, like cosmic-ray desorption, which may act to desorb more volatile species.

Helen Roberts presents her talk on deuterated molecules.

Organic Matter in Space
Proceedings IAU Symposium No. 251, 2008
S. Kwok & S. Sandford, eds.

© 2008 International Astronomical Union
doi:10.1017/S1743921308021285

Precursors of complex organic molecules: NH_3 and CH_3OH in the ices surrounding low-mass protostars

Sandrine Bottinelli[1], Adwin C. A. Boogert[2], Ewine F. van Dishoeck[1,3] Martha Beckwith[1], Jordy Bouwman[1,4], Harold Linnartz[1,4], and Karin I. Öberg[1,4]

[1] Leiden Observatory, Leiden University
P.O. Box 9513, NL-2300 RA Leiden, The Netherlands
email: `sandrine@strw.leidenuniv.nl`

[2] IPAC, NASA Herschel Science Center
Mail Code 100-22, California Institute of Technology, Pasadena, CA 91125, USA
email: `aboogert@ipac.caltech.edu`

[3] MPE Garching
Postfach 1312, 85741 Garching, Germany
email: `ewine@strw.leidenuniv.nl`

[4] Sackler Laboratory for Astrophysics, Leiden University
P.O. Box 9513, NL-2300 RA Leiden, The Netherlands
email: `beckwith,bouwman,linnartz,oberg@strw.leidenuniv.nl`

Abstract. NH_3 and CH_3OH are key molecules in the chemical networks leading to the formation of complex N- and O-bearing organic molecules. However, despite a number of recent studies, there is still a lot to learn about their abundances in the solid state and how they relate to those of other N/O-bearing organic molecules or to NH_3 and CH_3OH abundances in the gas phase. This is particularly true in the case of low-mass young stellar objects (YSOs), for which only the recent advent of the *Spitzer* Space Telescope has allowed high sensitivity observations of the ices in their enveloppes. We present a combined study of *Spitzer* data (obtained within the Legacy program "From Molecular Cores to Planet-Forming Disks", *c2d*) and laboratory spectra, leading to the detections of NH_3 and CH_3OH in the ices of low-mass protostars. We investigate correlations with other ice features and conclude with prospects on further studies linking these two precursors of complex organic molecules with their gas-phase products.

Keywords. Line: identification, ISM: molecules, ISM: abundances, infrared: stars, stars: formation, astrochemistry

1. Introduction

NH_3 and CH_3OH are among the most ubiquitous and abundant molecules. They are found in the gas-phase in a variety of environments such as infrared dark clouds, ultra-compact H II regions, massive hot cores, hot corinos, and comets. Solid NH_3 and CH_3OH have also been observed in the ices of massive YSOs (e.g. Schutte *et al.* 1991, Lacy *et al.* 1998, Dartois *et al.* 2002) and more recently of low-mass protostars (for CH_3OH, Pontoppidan *et al.* 2003). These two molecules are also key species in gas-grain chemical networks since they are the reactants leading to the formation of complex N- and O-bearing organic molecules, such as CH_3CN and CH_3OCH_3 (e.g. Rodgers & Charnley 2001). Moreover, UV processing of solid NH_3- and CH_3OH-containing ices could produce amino-acids, (e.g. Muñoz Caro & Schutte 2003). Finally, the presence and amount of NH_3 in ices also has an impact on constraining the content of ions such as NH_4^+ and OCN^-,

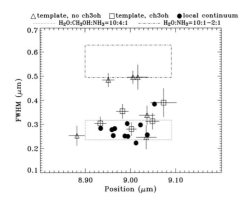

Figure 1. FWHM vs position for the NH_3 feature. Filled circles represent the values obtained with the local continuum method for subtracting the silicate absorption. Triangles and squares show the values obtained with the template method for CH_3OH-poor and CH_3OH-rich sources, respectively. The dash–dot and dotted lines indicate the range of values measured in the laboratory spectra of $H_2O:NH_3$ and $H_2O:CH_3OH:NH_3 = 10:4:1$ mixtures, respectively.

which are important species in solid-state chemical networks since they react more easily. Thus, a better knowledge of the NH_3 and CH_3OH content of interstellar ices, especially in enveloppes surrounding low-mass protostars, would help to constrain chemical models, and gain a better understanding of the formation of (pre-)biotic molecules in young solar analogs.

Both these species have been noted to have gas-phase abundances in hot cores/ corinos much larger than in cold dense clouds. This indicates that ices are an important reservoir of NH_3 and CH_3OH and that prominent features should be seen in the absorption spectra towards high- and low-mass protostars. Unfortunately, NH_3 and CH_3OH features are often blended with deep water and/or silicate absorptions, making the identifications and column density measurements difficult. Despite the blend with the so-called 10-μm silicate feature, the umbrella mode of NH_3 at $\sim$9 μm and the C–O stretch of CH_3OH at $\sim$9.7 μm appear as the most promising features to determine or better constrain the abundances of these two molecules. This kind of study has only become possible for low-mass (hence low-luminosity) protostars with the recent advent of *Spitzer* whose sensitivity enabled the observations of these objects. We present here a combined space and laboratory study of solid NH_3 and CH_3OH in the $\sim$ 8–10 μm region.

2. Observations and laboratory work

Astronomical data. The source sample consists of 41 low-mass protostars, out of which 35 fall in the embedded Class 0/I category, the remaining 6 objects being flat-type objects. We refer the reader to Boogert *et al.* (2008) for a complete description of the sample, data reduction process, and SED continuum determination.

Laboratory data. A laboratory study was carried out to spectroscopically characterize the NH_3 umbrella mode at $\sim$9.0 μm (1110 cm^{-1}) and the CH_3OH CO-stretch mode at $\sim$9.7 μm (1027 cm^{-1}) in circumstellar ice analogs. Using a Fourier Transform InfraRed (FTIR) spectrometer, we obtained transmission spectra for 12 binary and tertiary mixtures of H_2O, CH_3OH and/or NH_3 at different concentrations. The general procedure was to deposit the ice at 15 K, and to warm it up to 140 K, in 10 K steps. The range of concentrations and temperatures yielded a comprehensive dataset that allowed us, on the one hand, to investigate the behavior of the full-width at half-maximum (FWHM) and position of the features as a function of concentration and temperature, and on the other hand, to perform a direct comparison with the above astronomical observations

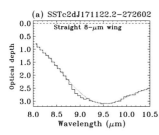 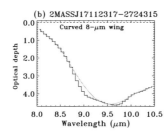

Figure 2. Examples of the two types of profiles for the 8-μm wing of the silicate feature: (a) straight, (b) curved. The gray lines show the local continuum.

in order to determine information on the relative compositions. The main trends in the behaviors are (see dotted and dash-dot lines in Figure 1):

- in $H_2O{:}NH_3$ mixtures, the NH_3 umbrella mode is located around 8.9-9.0 μm (1110–1128 cm^{-1}) and has FWHM between 0.5 and 0.6 μm.
- in $H_2O{:}CH_3OH{:}NH_3$ mixtures, the same values are found for the position and FWHM of the NH_3 umbrella mode, except that the FWHM in the strong CH_3OH mixture ($H_2O{:}CH_3OH{:}NH_3 = 10{:}4{:}1$) is reduced to 0.24–0.32 μm.
- in both $H_2O{:}CH_3OH$ and $H_2O{:}CH_3OH{:}NH_3$ mixtures, the CH_3OH CO-stretch mode is located at 9.7–9.8 μm (1020–1028 cm^{-1}) and has FWHM 0.2–0.27 μm.

3. Analysis: estimating the silicate contribution

Local continuum. The first approach was to derive a local continuum by fitting a fourth-order polynomial to the wavelength regions 8.25–8.75, 9.23–9.37, and 9.98–10.4 μm. The position of the NH_3 and the position and FWHM of CH_3OH features derived with this method agree with the values obtained from the laboratory spectra. However, the FWHM of the NH_3 feature is around 0.3 μm (regardless of the relative amounts of NH_3 and CH_3OH – see filled circles in Figure 1); for CH_3OH-rich sources, this narrow width is consistent with that seen in the laboratory data of the $H_2O{:}CH_3OH{:}NH_3 = 10{:}4{:}1$ mixture (dotted line in Figure 1), but there is a disagreement in the case of CH_3OH-poor sources since the FWHM of NH_3 in $H_2O{:}NH_3$ mixtures (dash-dot line in Figure 1) is about twice as large.

Template method. The local continuum method is a reliable indicator of the presence of NH_3 and CH_3OH, but another way is needed to determine the possible influence of the 10-μm silicate absorption. This was investigated using an empirical method. Upon examination of the 10-μm feature of the entire sample, the sources could be separated into two broad categories: sources with a "straight" profile between 8 and 8.5 μm (hereafter the 8-μm wing), and sources with a "curved" 8-μm wing (see Figure 2). In each category, a source with little NH_3 and CH_3OH (hence with an almost pure silicate absorption at 10 μm) was selected as template for the silicate feature, scaled to the optical depth at 9.7 μm and subtracted from the other spectra. We also used as an additional template the GCS3 spectrum observed by Kemper *et al.* (2004) towards the Galactic Center.

Figure 3 shows three examples of a comparison between the spectra obtained after subtracting the local continuum (black line) or the template spectrum (light gray line): a straight and curved 8-μm wing (left and middle panel respectively). We also include SVS 4-5, the only source for which the GCS3 spectrum was the best approximation of the silicate feature.

4. Results

Using the template method, we detected NH_3 at 9 μm and CH_3OH at 9.7 μm in 11 and 17 sources respectively, including 6 sources with both features being present. Using band

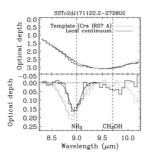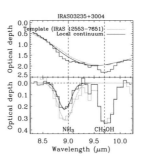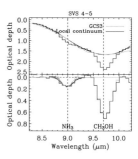

Figure 3. Determination of the contribution of the 10-μm silicate absorption: comparison between the local continuum (black lines) and template method (gray lines). The left and middle panels show sources with a straight and curved 8-μm wing respectively, and the right panel displays SVS 4-5, the only sources for which the best template is the GCS3 spectrum.

strengths of 1.3×10^{-17} and 1.8×10^{-17} cm molecule^{-1} for NH$_3$ and CH$_3$OH, respectively (Kerkhof *et al.* 1999, d'Hendecourt & Allamandola 1986), we derive column densities $N_{\text{NH}_3} = (1.3\text{–}19.6) \times 10^{17}$ cm^{-2} and $N_{\text{CH}_3\text{OH}} = (0.4\text{–}19.8) \times 10^{17}$ cm^{-2}, corresponding to abundances, with respect to water ice, in the ranges 3–12% and 1–25% for NH$_3$ and CH$_3$OH, respectively.

As can be seen from Figure 3, the relative amounts of NH$_3$ and CH$_3$OH vary greatly from source to source. No correlation was found between the NH$_3$ and CH$_3$OH abundances, but upon investigation of possible correlations with other ice components, we found one between NH$_3$ and CH$_4$, as measured by Öberg *et al.* (2008). Since all three species are expected to form via hydrogenation of C, N and CO (for CH$_4$, NH$_3$ and CH$_3$OH, respectively) on grain surfaces, possible explanations for the (non-)correlations could reside in (*i*) a preferential hydrogenation of atoms over molecules, combined with a time effect, or (*ii*) different C/CO, N/N$_2$ and/or H/H$_2$ ratios towards different line-of-sights. However, further investigation is needed, and in particular, a millimeter search of the possible gas-phase products of evaporated NH$_3$, CH$_3$OH and CH$_4$ (HNC, HNCO, C$_{n=3-5}$H$_2$, CH$_2$CN) is under way, aiming at paving the bridge between grain-mantle components and their offsprings.

References

Boogert, A., Pontoppidan, K., Knez, C., Lahuis, F., Kessler-Silacci, J., van Dishoeck, E., Blake, G., Augereau, J., Bisschop, S., Bottinelli, S., Brooke, T., *et al.* 2008, *ApJ*, in press

Dartois, E., d'Hendecourt, L., Thi, W., Pontoppidan, K. M., & van Dishoeck, E. F. 2002, *A&A*, 394, 1057

d'Hendecourt, L. B. & Allamandola, L. J. 1986, *A&AS*, 64, 453

Kemper, F., Vriend, W. J., & Tielens, A. G. G. M. 2004, *ApJ*, 609, 825

Kerkhof, O., Schutte, W. A., & Ehrenfreund, P. 1999, *A&A*, 346, 990

Lacy, J. H., Faraji, H. Sandford, S. A., & Allamandola, L. J. 1998, *ApJ*, 501, L105

Muñoz Caro, G. M. & Schutte, W. A. 2003, *A&A*, 412, 121

Öberg, K. I., Boogert, A. C. A., Pontoppidan, K. M., & van Dishoeck, E. F. 2008, *A&A*, in press

Pontoppidan, K. M., Dartois, E., van Dishoeck, E. F., Thi, W.-F., & d'Hendecourt, L. 2003, *A&A*, 404, 17

Rodgers, S. D. & Charnley, S. B. 2001, *ApJ*, 546, 324

Schutte, W. A., Tielens, A. G. G., & Sandford, S. A. 1991, *ApJ*, 382, 523

Discussion

KNEZ: Could you comment on the sources that you have both methanol and ammonia, do they have similar abundances?

BOTTINELLI: It actually varies a lot. Sometimes, they seem to have as much ...[searching for relevant slide]... You see, there is already one example [in the spectra I showed] where there is as much ammonia as methanol whereas [in this other source], there is a lot more methanol. It does not seem to be any specific trend.

VAN DISHOECK: So what is the total range in methanol abundances that you find? You go all the way from less than 1% to up to 25%?

BOTTINELLI: It's about the same for ammonia although there are slightly fewer sources with ammonia detection. I need to check those 20% sources; they could be an exception. Usually they are at the 10% range [compared to water].

CECCARELLI: The sources where you detected these ices, do you see any difference between Class 1 and Class 0 sources?

BOTTINELLI: I have tried to plot the ice abundances with respect to water as a function of the alpha which is telling us more or less the Class, but I couldn't see any trend.

VAN DISHOECK: Spitzer can actually observe Class 0 sources, not many but there are a few in the sample that are really in the very deeply embedded phase.

SIVARMAN: We did some experiments with ammonia and methanol in the ice phase and radiate it with 1 keV electrons. The only product that we see is just the OCN$^-$. We didn't see any other product apart from the OCN$^-$.

VAN DISHOECK: So this was an ice consisting of?

SIVARMAN: Of ammonia and methanol, 1 to 1.

The Indian delegation (from left to right: D. B. Vaidya, Amit Pathak, C. Muthumariappan, B. Sivaraman).

Organic Matter in Space
Proceedings IAU Symposium No. 251, 2008
S. Kwok & S. Sandford, eds.

© 2008 International Astronomical Union
doi:10.1017/S1743921308021297

Chemical changes during transport from cloud to disk

Ruud Visser[1], Ewine F. van Dishoeck[1,2], and Steven D. Doty[3]

[1]Leiden Observatory, Leiden University,
P.O. Box 9513, 2300 RA Leiden, the Netherlands
email: `ruvisser@strw.leidenuniv.nl`

[2]Max-Planck-Institut für Extraterrestrische Physik,
Giessenbachstrasse 1, 85748 Garching, Germany

[3]Department of Physics and Astronomy, Denison University,
Olin Hall, Granville, OH 43023, USA

Abstract. We present the first semi-analytical model that follows the chemical evolution during the collapse of a molecular cloud and the formation of a low-mass star and the surrounding disk. It computes infall trajectories from any starting point in the cloud and it includes a full time-dependent treatment of the temperature structure. We focus here on the freeze-out and desorption of CO and H_2O. Both species deplete towards the centre before the collapse begins. CO evaporates during the infall phase and re-adsorbs when it enters the disk. H_2O remains in the solid phase everywhere, except within a few AU of the star. Material that ends up in the planet- and comet-forming zones is predicted to spend enough time in a warm zone during the collapse to form complex organic species.

Keywords. Planetary systems: protoplanetary disks, circumstellar matter, stars: formation, astrochemistry

1. Introduction

The chemical composition of the building blocks of planets and comets depends on the chemical evolution throughout the formation process of a low-mass star and its circumstellar disk, where the planetary and cometary building blocks are to be found. The chemistry in pre-stellar cores–essentially molecular clouds that will collapse to form a star at some point–is relatively easy to model, because the dynamics and the temperature structure are simpler before the protostar is formed than afterwards. A key result from the pre-stellar core models [see Lee *et al.* (2004) for a short review] is the depletion of many carbon-bearing species towards the centre of the core.

Ceccarelli *et al.* (1996) were the first to model the chemistry in the collapse phase, and others have done so more recently (e.g. Rodgers & Charnley 2003, Lee *et al.* 2004, Aikawa *et al.* 2008). All of these models are one-dimensional, which necessarily leaves out the circumstellar disk. As the protostar turns on and heats up the surrounding material, all models agree that frozen-out species return to the gas phase if the dust temperature surpasses their evaporation temperature. The higher temperatures can further drive a hot-core-like chemistry, and complex molecules may be formed if the infall timescales are long enough.

If the disk is included, the system gains a large reservoir where infalling material from the cloud can be stored for a long time before accreting onto the star. The inner parts of the disk are shielded from direct irradiation by the star, so they are colder than the

111

disk's surface and the remnant cloud. Hence, molecules that evaporated as they fell in towards the star may freeze out again when they enter the disk.

We present here the first semi-analytical model that follows the chemical evolution from the pre-stellar core to the disk phase in two dimensions. We trace individual parcels of material as they fall in from the cloud into the disk, and we analyze the gas/ice ratios of CO and H_2O in these parcels. This gives us a first indication of what happens to complex organics during the infall process. Finally, we briefly discuss the model results in light of the chemical diversity found in comets.

2. Description of the model

Our model solves the chemistry in a two-dimensional axisymmetric Lagrangian frame. It takes a parcel of material at some starting point in the cloud and uses time-dependent velocity fields to follow it into the star or disk. The density and temperature are computed at each point. Photoprocesses are ignored, except that stellar photons are used as a heating source. A full description of the model will appear in an upcoming paper (Visser *et al.*, in prep.); only the main features are summarized here.

The densities and velocities in the cloud are taken from Terebey *et al.* (1984), who added rotation to the well-known inside-out Shu (1977) collapse. The model switches to the Cassen & Moosman (1981) solution inside the centrifugal radius, R_c. The infall trajectories are radial at large radii and deflect towards the midplane closer in; they finally intersect the midplane inside of R_c. This causes a build-up of material in a thin layer around the star: the circumstellar disk. Conservation of angular momentum causes part of the disk to spread out beyond R_c. As soon as this happens, accretion from the cloud onto the extended part of the disk is also allowed. The radial dynamics of the disk are treated as in Dullemond *et al.* (2006). The vertical density profile is taken to be a Gaussian at all times and the vertical velocities are chosen such as to accomplish this. The dust temperature in the entire system is calculated with full radiative transfer (Dullemond & Dominik 2004) and the gas temperature is assumed to be equal to the dust temperature everywhere. Shocks were found not to be important beyond 0.1 AU for material entering the disk more than 3×10^4 yr after the onset of collapse (Neufeld & Hollenbach 1994). A bipolar outflow, perpendicular to the disk, arises naturally from the Terebey *et al.* (1984) solution and is included as an evacuated cavity.

At first, the only chemistry included in the model is the adsorption and desorption (freeze-out and evaporation) of carbon monoxide (CO) and water (H_2O). Their overall abundance remains constant throughout the collapse. The reaction rates are from Rodgers & Charnley (2003), except for the H_2O desorption, which is from Fraser *et al.* (2001). A network including chemistry beyond adsorption and desorption will be included in the future. CO and H_2O begin fully in the gas phase and a 10^5-yr static pre-stellar core phase is included before the onset of collapse.

In our standard model, all CO evaporates at a single temperature. In reality, however, CO is mixed with the H_2O ice, and some of it will be trapped until the H_2O desorbs (Collings *et al.* 2004). This effect can be approximated by letting part of the CO evaporate at higher temperatures. Specifically, a scheme with four "flavours" of CO ice was used (Viti *et al.* 2004) as an alternative to the standard model.

Some organic species, e. g. methanol (CH_3OH), are likely to be formed in the pre-stellar core (Garrod & Herbst 2006). Other organics, primarily the more complex ones like methyl formate ($HCOOCH_3$), are only formed if an infalling parcel spends at least several 10^4 yr in a warm, 20–60 K, region. The binding energy of both classes of organics is similar to or higher than that of H_2O, so they desorb at the same or at higher temperatures.

Freeze-out occurs at about the same rate as for H_2O. Hence, the H_2O gas/ice ratio is an upper limit to the organics gas/ice ratio.

3. Results

Results are presented here for our standard model with an initial cloud mass of $1.0\,M_\odot$, a cloud radius of 6700 AU, an effective sound speed of 0.26 km s^{-1}, a solid-body rotation rate of 10^{-13} s^{-1}, and an initial uniform temperature of 10 K. After the onset of collapse, the outer parts of the cloud reach the disk in 2.5×10^5 yr. The centrifugal radius at that time is 500 AU and the disk has spread to about 700 AU.

During the 10^5-yr pre-stellar core phase, CO and H_2O freeze out inside of $\sim$4000 AU and remain in the gas phase further out. This is consistent with other pre-stellar core models. The disk becomes important and the system loses its spherical symmetry at $t \approx 2 \times 10^4$ yr. Up to that point, our results are the same as those from the 1D models. The collapsing region quickly heats up to a few tens of K, driving CO (evaporating around 20 K in the one-flavour model) into the gas phase, but keeping H_2O (evaporating around 100 K) on the grains.

After $\sim$2 $\times 10^5$ yr, the disk has become massive enough that the temperature drops below 20 K near the midplane. CO arriving in this region re-adsorbs onto the grains and another 5×10^4 yr later, there is a significant amount of solid CO in the disk (Figure 1). Some solid CO also exists in the remnant cloud just beyond the disk, which is shielded from direct irradiation. Outside of 1000 AU, the cloud is thin enough that scattering from higher altitudes causes a slightly higher temperature ($\sim$25 K), keeping most CO in the gas phase. The border between the solid and gaseous CO regions lies close to the 20-K surface, showing that the re-adsorption is a fast process. H_2O is still predominantly frozen out at this time, but there can be up to 20% in the gas phase in the warmer regions above the disk. Within a few AU from the star, all H_2O has evaporated.

In the model with four flavours of CO ice, 13% of the CO desorbs at 70 K and 6.5% desorbs at 100 K (Viti *et al.* 2004). This has the expected effect of keeping about a fifth of the total CO in the solid phase throughout the fully gaseous region of the one-flavour model. The region with at least 80% of solid CO extends to the same radius in the

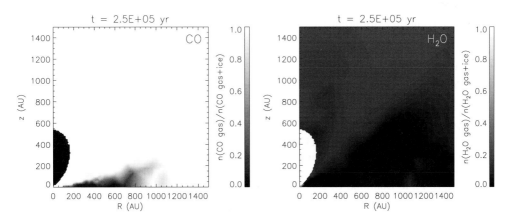

Figure 1. Fraction of gaseous CO (left) and H_2O (right) in the disk and remnant cloud 2.5×10^5 yr after the onset of collapse. The outflow region, extending out to 540 AU in the vertical direction, is indicated in black on the CO plot and in white on the H_2O plot. The one-flavour model was used for the CO desorption.

Kasandra O'Malia (photo by Dale Cruikshank).

Organic Matter in Space
Proceedings IAU Symposium No. 251, 2008
S. Kwok & S. Sandford, eds.

High-resolution observations of CH$_3$CN in the hot corino of NGC1333-IRAS4A

Sandrine Bottinelli[1,2,3,4], Cecilia Ceccarelli[1,3], Roberto Neri[5], and Jonathan P. Williams[2,3]

[1]Laboratoire d'Astrophysique de Grenoble
B.P. 53, 38041 Grenoble Cedex 9, France
email: cecilia.ceccarelli@obs.ujf-grenoble.fr

[2]Institute for Astronomy, University of Hawai'i
2680 Woodlawn Drive, Honolulu, HI 96822, USA
email: jpw@ifa.hawaii.edu

[3]NASA Astrobiology Institute

[4]Current address: Leiden Observatory, Leiden University
P.O. Box 9513, NL-2300 RA Leiden, The Netherlands
email: sandrine@strw.leidenuniv.nl

[5]Institut de RadioAstronomie Millimétrique
Saint-Martin d'Hères, France
email: neri@iram.fr

Abstract. The formation and evolution of complex organic molecules in the early stages of solar-type protostars (Class 0 objects) is crucial as it sets the stage for the content in pre-biotic molecules of the subsequent proto-planetary nebula. In order to understand the chemistry of these Class 0 objects, it is necessary to perform interferometric observations which allow us to resolve the hot corino, that is the warm, dense inner region of the envelope of a Class 0 object, where the complex organic molecules are located. Such observations exist for only two objects so far, IRAS16293-2422 and NGC1333-IRAS2A and we present here Plateau de Bure interferometric maps of a third hot corino, NGC1333-IRAS4A, which show emission of the complex organic molecule CH$_3$CN arising from a region of size $\sim 0.8''/175$ AU, that is, of the order of the size of the Solar System. Combining these high-angular resolution maps with prior single-dish observations of the same transitions of CH$_3$CN indicates that extended emission is also present, and we investigate the implications for organic chemistry in hot corinos.

Keywords. Stars: formation, stars: individual (NGC1333-IRAS4A), ISM: abundances, ISM: molecules

1. Introduction

One of the most fascinating questions which remains almost entirely open in Astrophysics is the chemical complexity reached during the birth of stars like our own Sun, and its impact on the later stages, when a planetary system forms. Complex organic molecules have been found in the innermost regions of the envelopes surrounding solar-type protostars, in the so-called hot corinos (Cazaux *et al.* 2003, Bottinelli *et al.* 2004a, b, Bottinelli *et al.* 2007). Both gas-phase and grain-surface reactions have been proposed as formation mechanisms of the complex organics and determining the location of the emission of these molecules can help constraining their formation path. In this context, we present here new interferometric, sub-arcsecond, observations of CH$_3$CN in NGC1333-IRAS4A (hereafter IRAS4A).

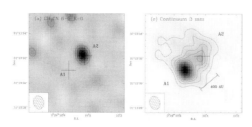

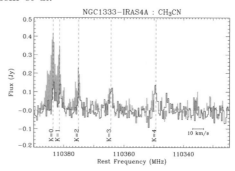

Figure 1. PdB maps of IRAS4A. — (a) CH_3CN $J = 6 - 5, K = 0$ line map with 2σ contour levels of 16 mJy; **(e)** 3 mm continuum emission in IRAS4A. The rms and contour levels are 1 and 10 mJy beam^{-1} respectively. — Crosses show the positions for A1 and A2. The beam size shown in the lower left corner is $1''1 \times 0''8$. (Adapted from Bottinelli *et al.* 2008)

Figure 2. CH_3CN spectra towards IRAS4A. — Shaded = spectrum obtained at the IRAM 30m Bottinelli *et al.* 2004; white = Plateau de Bure spectrum averaged over the emission region of IRAS4A2. Dashed lines indicate the frequencies of the CH_3CN transitions. (From Bottinelli *et al.* 2008)

2. Observations and implications

Observations. IRAS4A was observed with the IRAM Plateau de Bure Interferometer in its most extended (A) configuration. 3 mm and 1.3 mm continuum emission, and five CH_3CN transitions at 110.4 GHz were obtained simultaneously. The reader is referred to Bottinelli *et al.* (2008) for details of the data calibration and reduction. Figure 1-a shows the integrated line emission of CH_3CN averaged over the CH_3CN $J = 6 - 5, K = 0$ transition while the continuum at 3 mm is displayed in Figure 1e [see Bottinelli *et al.* (2008) for the $K = 1 - 3$ and 1.3 mm maps]. The line emission maps are unprecedented and show that only the north-west region (A2), the weakest component in the continuum, possesses CH_3CN emission. The spectrum averaged over A2 is compared to the spectrum obtained at the IRAM-30m (Figure 2), and shows that at least 50% (taking into account optical depth effects) of the emission from the low-energy lines ($K = 0, 1$) is filtered out by the PdB, indicating extended emission.

Implications. Using the density structure derived by Jørgensen *et al.* (2002), we can show that the CH_3CN abundance in the cold, outer regions is $\sim 2 \times 10^{-11}$, and that it jumps by about a factor 120 inside a radius of (85 ± 20) AU. From the temperature structure of Jørgensen *et al.* (2002), we deduce that the sublimation temperature of CH_3CN (where the abundance jump occurs) should be in the range 60 to 90 K, in agreement with laboratory experiments. This supports the hypothesis that CH_3CN is formed in the gas phase during the prestellar phase, and freezes out onto the grain mantles in the colder parts of the envelope.When the temperature reaches the CH_3CN sublimation temperature, those ices sublimate, injecting this molecule in the gas phase, which is equivalent to the presence of a hot corino, as seen in the IRAS4A data presented here.

References

Bottinelli, S., Ceccarelli, C., Neri, R., & Williams, J. P. 2008, *ApJ*, submitted

Bottinelli, S., Ceccarelli, C., Williams, J. P. & Lefloch, B. 2007, *A&A*, 463, 601

Bottinelli, S., Ceccarelli, C., Lefloch, B., Williams, J. P., *et al.* 2004a, *ApJ*, 615, 354

Bottinelli, S., Ceccarelli, C., Neri, R., Williams, J. P., *et al.* 2004b, *ApJ* (Letters), 617, L69

Cazaux, S., Tielens, A. G. G. M., Ceccarelli, C., Castets, A., *et al.* 2003, *ApJ* (Letters), 593, L51

Jørgensen, J. K., Schöier, F. L., & van Dishoeck, E. F. 2002, *A&A*, 389, 908

Organic Matter in Space
Proceedings IAU Symposium No. 251, 2008
S. Kwok & S. Sandford, eds.

© 2008 International Astronomical Union
doi:10.1017/S1743921308021315

Complex organic molecules in the intermediate mass protostar OMC2-FIR4

Nicolas Crimier, Cecilia Ceccarelli, Bertrand Lefloch, and Alex Faure

Laboratoire d'Astrophysique de Grenoble, Université Joseph Fourier, UMR 5571-CNRS,
France
email: `ncrimier@obs.ujf-grenoble.fr`

Abstract. We present the physical structure (density and temperature profiles) and the distribution of formaldehyde and methanol in intermediate mass protostar OMC2-FIR4 in the Orion molecular cloud complex.

Keywords. Astrochemistry, ISM: individual (OMC2-FIR4), line: formation

1. Introduction

The first stages of star evolution are characterized, among other things, by the formation Complex Organic Molecules (COMs; see e.g. Ceccarelli's contribution in this volume). Before the collapse sets in, in the pre-stellar condensations, at least some simple COMs like formaldehyde and methanol are synthesized on the grain surfaces by hydrogenation of CO. The birth of a protostar at the center warms up the surrounding matter up to 100 K or more, with two consequences: the previously formed grain ice mantles sublimate injecting into the gas phase their molecular content, and, second, new and more complex molecules are synthesized in the gas phase because of new formation routes opening up at the large gas temperatures and the availability of the molecules synthesized on the grain surfaces. What and how much COMs are formed at the end depend on the mass of the protostar, as suggested by some comparative studies between low and high mass protostars (Bottinelli *et al.* 2007). Surprising enough, low mass protostars synthesize larger quantities of COMs with respect to high mass protostars. This may be the result of both the COMs formation and destruction processes. Here we present new observations of methanol and formaldehyde, the simplest of but the most crucial COMs, towards the intermediate mass (IM) protostar OMC2-FIR4, the brightest submillimeter source of the Orion Molecular Cloud 2 with a bolometric luminosity of 1000 $L_\odot$.

2. Density, temperature, and CH$_3$OH and H$_2$CO abundance profiles

We derived the density and dust temperature profiles of the protostar OMC2-FIR4 envelope, by modeling the dust continuum emission by means of the radiative transfer code DUSTY (Ivezic & Elitzur 1999). We considered the Spectral Energy Density (SED) between 1.3 mm and 24 μm and the maps at 350, 450 and 850 μm respectively (see Crimier *et al.* 2008 for details). We assumed a power low density profile, $n(r) = n_0 (r/r_0)^{-\alpha}$, for an envelope of radius $R_{\max}$. n_0, α and $R_{\max}$ are then constrained by comparison between the observations and the model predictions. Following the claim by Jørgensen *et al.* 2006 that an Inter-Stellar Radiation Field (ISRF) with $G_0 = 10^4$ illuminates the OMC2-FIR4 envelope, we run models with G_0 between 1 and 10^4. We found that the continuum observations cannot constrain the G_0 value: all the ran models yield similar χ^2 values.

Deriving correct abundances from observations of molecular lines requires knowledge of the gas temperature profile. Very often, the approximation of $T_{\rm gas} = T_{\rm dust}$ is assumed. While this is a good enough approximation in low mass protostars, in IM mass protostars it may break down. Thus we modeled the thermal balance across the envelope, following the method described in Ceccarelli *et al.* (1996). To test the hypothesis of large G_0 illuminating OMC2-FIR4 we run 2 models with $G_0 = 1$ and 1000 respectively. Moreover, a crucial parameter in the thermal coupling between dust and gas is the water abundance, as water is a major coolant of the gas. Figure 1 presents the gas temperature and density (found by the modeling of the continuum discussed in §2) profiles in the two cases $G_0 = 1$ and 1000, and for different H_2O abundances in the inner part. The dust and gas are thermally decoupled in the inner region ($T_{\rm dust} \geqslant 100$ K), because of the water ices sublimation.

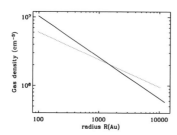 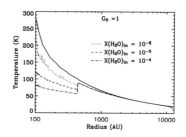

Figure 1. The gas density profile (left panel) for G_0 equal to 1 (plain curve) and 1000 (pointed curve) and temperature profile (right panel) for different values of the water abundance $X(H_2O)_{in}$.

We obtained multi-frequency observations of o-H_2CO, p-H_2CO and CH_3OH at the IRAM and JCMT telescopes (details in Crimier *et al.*, in prep.). Following the method by Maret *et al.* (2004) we derived the o-H_2CO, p-H_2CO and CH_3OH abundances in the inner and outer envelope of OMC2-FIR4 (Table 1). For a first analysis, we assumed LTE level distribution. From this preliminary analysis, the $\chi(H_2CO)/\chi(CH_3OH)$ ratio in the inner region is ~0.5, a value in between what found in Hot Corinos (~1) and Hot Cores (~0.1) (Bottinelli *et al.* 2007).

Table 1. o-H_2CO, p-H_2CO and CH_3OH column densities and abundances.

Molecules	T_{GAS}	N_x (cm^{-2})	Comments (opacities)	Abundances $G_0 = 1$	$G_0 = 10^3$
CH_3OH	20 K	~2.10^{15}	Moderately thick	~$6.7.10^{-9}$	~$1.3.10^{-8}$
	80 K	~8.10^{15}	thin	~$1.5.10^{-7}$	~$1.2.10^{-7}$
p-H_2CO	20 K	~5.10^{13}	thick	~$1.7.10^{-10}$	~$3.3.10^{-10}$
	80 K	~3.10^{14}	thin	~$5.5.10^{-9}$	~$4.6.10^{-9}$
o-H_2CO	20 K	~1.10^{14}	Moderately thick	~$3.3.10^{-10}$	~$6.7.10^{-10}$
	80 K	~8.10^{14}	thin	~$1.5.10^{-8}$	~$1.2.10^{-8}$

References

Bottinelli, S., Ceccarelli, C., Williams, J. P., & Lefloch, B. 2007, *A&A*, 463, 601
Ceccarelli, C., Hollenbach, D. J., & Tielens, A. G. G. M. 1996, *ApJ*, 471, 400
Crimier, N., Ceccarelli, C., Lefloch, B., & Faure, A. 2008, *A&A*, submitted
Ivezic, Z. & Elitzur, M. 1999, *MNRAS*, 303, 864
Jørgensen, J. K., Johnstone, D., van Dishoeck, E. F., & Doty, S. D. 2006, *A&A*, 449, 609
Maret, S., Ceccarelli, C., Caux, E. *et al.* 2004, *A&A*, 416, 577

Organic Matter in Space
Proceedings IAU Symposium No. 251, 2008
S. Kwok & S. Sandford, eds.

© 2008 International Astronomical Union
doi:10.1017/S1743921308021327

Methanol formation: A Monte Carlo study

Ankan Das[1], Kinsuk Acharyya[2], Sonali Chakrabarti[3,1], and Sandip K. Chakrabarti[2,1]

[1] Indian Centre for Space Physics,
Chalantika 43, Garia Station Road, Kolkata 700084, India
email: ankan@csp.res.in

[2] S. N. Bose National Centre for Basic Sciences
Salt Lake, Kolkata 700098, India
email: acharyya@bose.res.in, chakraba@bose.res.in

[3] Maharaja Manindra Chandra College
20 Ramakanta Bose Lane, Kolkata 700003, India
email: sonali@csp.res.in

Abstract. We carry out a Monte-Carlo simulation to study the formation of methanol on the grain surfaces. We found that the recombination efficiencies are strongly dependent on the extrinsic properties of the grain, such as the number of sites on the grain surface and the flux of the accreting matter. This uses the concept of effective grain surface (denoted through a factor α) area which changes as the grain is populated.

Keywords. ISM: abundances, ISM: molecules, methods: numerical, molecular processes

1. Introduction

It is now well known that the grain chemistry plays a vital role in the chemical evolutions of the molecular cloud. The rate equation method is very extensively used by several authors to study the grain surface chemistry (Hasegawa *et al.* 1992, Roberts & Herbst 2002, Acharyya *et al.* 2005). However, this method is only applicable when there are large number of reactants on the grain surface. Given the fact that the interstellar medium is very dilute, very often the criteria for using the rate equation are not fulfilled. Furthermore, it is assumed that the recombination efficiency is independent of the surface area of the grain. The only advantage of this method is that computationally it is faster and it can very easily be coupled with the gas phase reactions. More realistic method to handle the grain chemistry is the Monte-Carlo approach. This is very accurate method because here we can trace each and every species over the entire period of time. The major disadvantage of this is that it is computationally time consuming. Using Monte-Carlo approach Chakrabarti *et al.* (2006a, b) introduced a new concept of the effective grain surface area and showed that it is this effective area which must be used for the grain chemistry. It was argued that the surface area S in the usual rate equation is to be replaced by S^α where α is to be determined from Monte-Carlo method for a few cases and extrapolated value is to be used in the equation. Its usefulness was demostrated while computing the formation of H_2 molecules on the grains. In the present work, we carry out a similar analysis to study the formation of water and methanol on the grains.

2. Procedure

We study the chemical evolution on grains assuming H, O and CO as the sole accreting species. For the sake of simplicity, we assume that a grain surface is a square lattice S number of sites. We assume further that each site has four nearest neighbours, as in an fcc[100] lattice. In order to mimic the spherical grain structure, we assume a periodic

Robin T. Garrod *et al.*

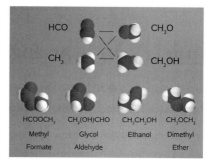

Figure 1. Four complex molecules, and the functional groups from which they may form on dust-grain surfaces during the hot-core warm-up phase.

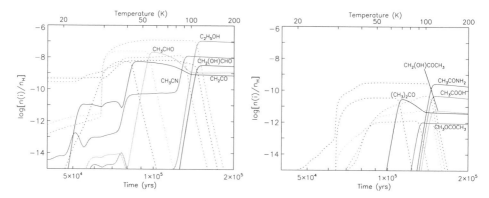

Figure 2. Modelled molecular abundances during the hot-core warm-up phase. Solid lines indicate gas-phase species; dotted lines of the same colour indicate the grain-surface species.

dimethyl ether, glycolaldehyde and ethanol, as well as other molecules like acetaldehyde (CH_3CHO), are formed at $\sim$30–40 K, when the radicals CH_3 and HCO become mobile on the grains. Dimethyl ether is still formed most strongly in the gas phase.

In summary, we find that complex organic molecules may be formed efficiently on grain-surfaces by the addition of heavy, functional-group radicals derived from the icy mantles. The various abundances of structural isomers may be explained by different formation routes. The evaporation of certain species at intermediate temperatures ($\sim$40–50 K) agrees well with observed rotational temperatures of such molecules, implying a strong interaction between grain-surface and gas-phase processes. While some species are formed on the grains, other species (including dimethyl ether, CH_3OCH_3) are still formed in the gas phase as a result of the evaporation of ices. This may occur at any temperature at which important species evaporate–not just at the canonical hot-core temperature of 100 K or so. Species of greater complexity, e. g. acetone, acetamide, are formed as a result of the break-down and re-construction of smaller complex molecules. The longer timescales required for these processes make high-mass star-forming regions more favourable to greater molecular complexity.

References

Garrod, R. T. & Herbst, E. 2006, *AA*, 457, 927
Garrod, R. T., Widicus Weaver, S. L., & Herbst, E. 2008, *ApJ*, 682, 283
Geppert, W. D., Hamberg, M., Thomas, R. D., *et al.* 2006, *Faraday Discuss.*, 133, 51
Horn, A., Møllendal, H., Sekiguchi, O., *et al.* 2004, *ApJ*, 611, 605
van Dishoeck, E. F. & Blake, G. A. 1998, *ARAA*, 36, 317

Organic Matter in Space
Proceedings IAU Symposium No. 251, 2008
S. Kwok & S. Sandford, eds.

© 2008 International Astronomical Union
doi:10.1017/S1743921308021340

Molecular gas dynamics in the Rosette Nebula

Henry E. Matthews[1] and William R. F. Dent[2]

[1] Herzberg Institute of Astrophysics, National Research Council of Canada, Victoria, BC
V2E 2E7, Canada
email: henry.matthews@nrc-cnrc.gc.ca

[2] Royal Observatory Edinburgh, Scotland, UK
email: :dent@roe.ac.uk

Abstract. We present observations of the Rosette Nebula and its near environment in the CO 3–2 transition obtained with an angular resolution of 20″. The gas dynamics of the region are complex; we find (1) a ring of gas expanding at about 20 km s^{-1}, (2) a number of collimated outflow sources, and (3) a chain of dust clumps having a velocity gradient along its length.

Keywords. ISM: molecules, ISM: kinematics and dynamics, instrumentation: detectors

1. Observations and Data Reduction

The Rosette Nebula, at a distance of 1400 pc, is a textbook example of the interaction between OB stars and surrounding gas and dust. NGC 2244, the central star cluster, contains six O-type stars with a total luminosity of 1000 L$_\odot$.

The data were obtained in 2006 and 2007 using a completely new suite of instrumentation at the James Clerk Maxwell Telescope (JCMT), on Mauna Kea, Hawaii, following major upgrades to the telescope and its observing capabilities. HARP, a 16-receptor (square 4×4) array receiver (Smith *et al.* 2003) operating in the 325–375 GHz (850 μm) region, was used with ACSIS, the spectral line correlator backend (Hovey *et al.* 2000) and a new telescope control system (OCS; Rees *et al.* 2002). The observations were carried out by continuously recording the data while the telescope was scanned across the target field. The scan rate used was 75″ per second, with a sample rate of 10 Hz. The entire region was scanned in both RA and Declination directions, ultimately resulting in a basket-weaved image solution with a well-determined and flat baselevel. The raw spectral resolution of these data was 0.42 km s^{-1}.

Data processing employed the SMURF package (Jenness *et al.* 2008) to first convert the data to spectral cubes. The Starlink KAPPA and CCDPACK routines were then used to remove a linear baseline, transform to a regular grid and combine the data into a single cube. Of these data, only 66 spectral channels were retained, covering the velocity interval of −3 through 25 km s^{-1} appropriate for the Rosette Nebula. Subsequent smoothing of the data to an angular resolution of 20″ resulted in a data cube in which the rms noise per point was 0.3 K T_A^* (corrected antenna temperature scale).

2. Results

The CO 3–2 image of the Rosette Nebula (see Figure 1, and Dent *et al.* 2008) obtained from these data shows that molecular material extends over a substantially larger area than the optical extent of the nebula. The structural features of the CO emission fall into four main types: compact flows, clumps, smooth extended regions, and elongated structures with narrow line velocity widths.

The data reveal a ring of gas expanding at 20 km s^{-1}, and show that compact, relatively high-velocity flows tend to be associated with the known IRAS sources, of which AFGL 961 is the most well-known, correlated with young star clusters. The CO 3–2/1–0 ratios indicate excitation temperatures of about 30–60 K in these regions. In a galactic context, these are relatively low energy outflows, having little influence on the Rosette Molecular Cloud.

Clumps of gas, typically 15″ to 3′ in size, are common. We estimate that in the present work we are able to detect such objects with masses as low as perhaps 30–100 Jupiter masses. Of particular note is a chain of clumps in the north-west quadrant of the Rosette which corresponds with a group of dust globules and elephant truck structures (Gahm et al. 2006) seen against the nebular background. The velocity information from these and other objects suggest that these objects are part of an expanding partial ring. The expansion rate of this structure, 16 km s^{-1}, suggests a dynamical age of 0.8 Myr, as compared with a systemic age of 2–3 Myr (Balog et al. 2007).

Kinetic temperatures derived from the CO 1–0 and 3–2 integrated line intensities are in the range 25–45 K. The clump mass distribution is estimated to be complete to about 0.1 M$_\odot$. For optically thin CO emission (A$_v \leqslant 1$) we find a mass-number slope of −0.6 (i.e., $dN(m)/dM \sim -1.6$) for the clump mass range 0.1–1.0 M$_\odot$.

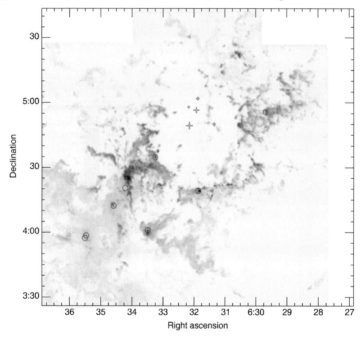

Figure 1. A molecular line image of the Rosette Nebula seen in emission from CO (3-2 transition). The image is about 2 degrees across, about twice the extent of the visible nebula. The positions of the luminous O-stars are indicated by crosses just above the centre of the image, and the circles show the locations of outflow sources.

References

Balog, Z., et al. 2007, *ApJ*, 660, 1532
Dent, W. R. F. D., et al. 2008, *MNRAS*, in preparation
Gahm, G. F., et al. 2006, *A&A*, 454, 201
Hovey, G., et al. 2000, *SPIE*, 4015, 114
Jenness, T., et al. 2008, *ADASS XVIII*, in press
Rees, N., et al. 2002, *ASPC*, 281, 500
Smith, H., et al. 2003, *SPIE*, 4855, 338

Organic Matter in Space
Proceedings IAU Symposium No. 251, 2008
S. Kwok & S. Sandford, eds.

Solid CH$_4$ toward low-mass protostars: How much is there to build complex organics?

Karin I. Öberg[1], A. C. Adwin Boogert[2], Klaus M. Pontoppidan[3], Geoffrey A. Blake[3], Neal J. Evans[4], Fred Lahuis[5], and Ewine F. van Dishoeck[1]

[1]Leiden Observatory, Leiden University, P.O. Box 9513, NL–2300 RA Leiden, the Netherlands
email: oberg@strw.leidenuniv.nl
[2]IPAC, NASA Herschel Science Center,Mail Code100-22, California Institute of Technology, Pasadena, CA 91125, USA
email: aboogert@ipac.caltech.edu
[3]Division of Geological and Planetary Sciences, California Institute of Technology Pasadena, CA 91125, USA
email: pontoppi@gps.caltech.edu
[4]Department of Astronomy, University of Texas at Austin, Austin, TX 78712-0259, USA
email: nje@bubba.as.utexas.edu
[5]SRON, PO Box 800, NL–9700 AV Groningen, The Netherlands
email: F.Lahuis@sron.nl

Abstract. We use Spitzer IRS spectra to determine the solid CH$_4$ abundance toward a large sample (52 sources) of low mass protostars. 50% of the sources have an absorption feature at 7.7 μm, attributed to solid CH$_4$. The solid CH$_4$/H$_2$O abundances are 2–13%, but toward sources with H$_2$O column densities above 2×10^{18} cm^{-2}, the CH$_4$ abundances (20 out of 25) are nearly constant at 4.7 ± 1.6%. Correlations with CO$_2$ and H$_2$O together with the inferred abundances are consistent with CH$_4$ formation through sequential hydrogenation of C on grain surfaces, but not with formation from CH$_3$OH and formation in gas phase with subsequent freeze-out.

Keywords. Astrochemistry, molecular processes, circumstellar matter, ISM: abundances, infrared: ISM

1. Introduction

CH$_4$ is believed to play a key role in the formation process of complex and prebiotic molecules (see, e.g., Markwick *et al.* 2000). In star forming regions most molecules are frozen out as ices and solid CH$_4$ has previously been detected towards mainly high mass protostars by, e.g., Boogert *et al.* (1996). Models predict solid CH$_4$ to form rapidly on cool grains through successive hydrogenation of atomic C, similarly to H$_2$O from O. Two other suggested formation pathways are photo-processing of CH$_3$OH and gas phase formation with subsequent freeze-out. Because formation pathway efficiency depends on environment, potential formation routes may be tested through exploring the distribution of CH$_4$ toward a large sample of objects of different ages, luminosities and ice column densities. In addition, correlations, or lack thereof, with other ice constituents may provide important clues to how the molecule is formed.

In this study we determine the CH$_4$ abundances and distribution pattern toward a sample of 52 low mass young stellar objects, in 11 different clouds and with a large spread in total ice column density. This is based on spectra acquired with the Spitzer Infrared Spectrometer (IRS) as part of our legacy program 'From molecular cores to protoplanetary disks' (*c2d*).

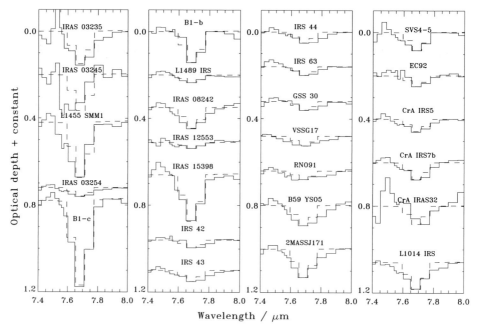

Figure 1. Continuum subtracted infrared spectra (solid line) of the CH_4 feature at 7.7 μm toward 25 low mass protostars, and laboratory spectra of CH_4 ice (dashed line).

2. Results and Discussion

Detailed results of this study are published in (Öberg *et al.*, ApJ in press). We detect solid CH_4 at 7.7 μm in 25 of our sources (Figure 1) and derive column densities by comparing the observed 7.7 μm features with a laboratory H_2O:CH_4 ice mixture. The calculated CH_4 abundances with respect to solid H_2O vary between 2 and 13%. Toward sources (20/25) with H_2O column densities above 3×10^{18} cm^{-2} all CH_4 abundances fall between 2 and 8%, however and the average is 4.7$\pm$1.6%. In the sources with no CH_4 detection, the average 3σ upper limit is 15%. These CH_4 abundances are comparable to what has been found towards high mass stars previously.

The nearly constant CH_4 abundance we found here can be contrasted with the large variations of the CH_3OH abundances in Boogert *et al.* (ApJ in press) for the same sources. CH_4 seems hence unrelated to CH_3OH. In general our results agree with model predictions where CH_4 is formed on the grain surface. Aikawa *et al.* (2005) predicts CH_4 ice abundances with respect to H_2O between $\approx$1–10%. In contrast gas phase models predict steady state CH_4/H_2 abundances of only $\sim10^{-7}$ (e.g., Woodall *et al.* 2007), compared with our inferred CH_4/H_2 abundances of $\sim 2 - 13 \times 10^{-6}$ (assuming a standard H_2O/H_2 ratio of 10^{-4}). The grain surface formation pathway is also supported by correlation studies between CH_4 and other observed ice molecules i.e. CH_4 correlates better with molecules formed on surfaces than those formed in the gas and subsequently frozen out.

References

Aikawa, Y., Herbst, E., Roberts, H., & Caselli, P. 2005, *ApJ*, 620, 330

Boogert, A. C. A., Schutte, W. A., Tielens, A. G. G. M., Whittet, D. C. B., Helmich, F. P., Ehrenfreund, P., Wesselius, P. R., de Graauw, T., & Prusti, T. 1996, *A&A* (Letters), 315, L377

Markwick, A. J., Millar, T. J., & Charnley, S. B. 2000, *ApJ*, 535, 256

Woodall, J., Agúndez, M., Markwick-Kemper, A. J., & Millar, T. J. 2007, *A&A*, 466, 1197

Organic Matter in Space
Proceedings IAU Symposium No. 251, 2008
S. Kwok & S. Sandford, eds.

Molecular evolution in star-forming cores: From prestellar cores to protostellar cores

Yuri Aikawa[1], Valentine Wakelam[2], Nami Sakai[3], R. T. Garrod[4], E. Herbst[5], and Satoshi Yamamoto[3]

[1] Department of Earth and Plantary Sciences, Kobe University,
Kobe 657-8501, Japan
email: aikawa@kobe-u.ac.jp

[2] Université Bordeaux I
[3] Department of Physics, University of Tokyo
[4] Max-Planck-Institute für Radioastronomie
[5] Departments of Physics, Chemistry, and Astronomy, The Ohio State University

Abstract. We investigate the molecular abundances in protostellar cores by solving the gas-grain chemical reaction network. As a physical model of the core, we adopt a result of one-dimensional radiation-hydrodynamics calculation, which follows the contraction of an initially hydrostatic prestellar core to form a protostellar core. Temporal variation of molecular abundances is solved in multiple infalling shells, which enable us to investigate the spatial distribution of molecules in the evolving core. The shells pass through the warm region of $T \sim 20$–100 K in several 10^4 yr and falls onto the central star in ~ 100 yr after they enter the region of $T > 100$ K. We found that the complex organic species such as $HCOOCH_3$ are formed mainly via grain-surface reactions at $T \sim 20$-40 K, and then sublimated to the gas phase when the shell temperature reaches their sublimation temperatures ($T \geqslant 100$ K). Carbon-chain species can be re-generated from sublimated CH_4 via gas-phase and grain-surface reactions. HCO_2^+, which is recently detected towards L1527, are abundant at $r = 100$–$2,000$ AU, and its column density reaches $\sim 10^{11}$ cm^{-2} in our model. If a core is isolated and irradiated directly by interstellar UV radiation, photo-dissociation of water ice produces OH, which reacts with CO to form CO_2 efficiently. Complex species then become less abundant compared with the case of embedded core in ambient clouds. Although a circumstellar (protoplanetary) disk is not included in our core model, we can expect similar chemical reactions (i.e., production of large organic species, carbon-chains and HCO_2^+) to proceed in disk regions with $T \sim 20$-100 K.

Keywords. Stars: formation, ISM: molecules

1. Introduction

Chemical reactions in molecular clouds are classified to two categories: gas-phase reactions and grain-surface reactions. Figure 1 schematically summarizes the various chemical processes in molecular clouds. In the cold clouds of $T \sim 10$ K, chemical reactions are triggered by cosmic-ray ionization and proceed mainly via ion-molecule reactions in the gas-phase. While neutral-neutral reactions often have activation barrier, some neutral-neutral reactions do not, and play an important role in forming molecules such as N_2. When atoms and molecules in the gas-phase collide with grains in the cold molecular clouds, they are efficiently adsorbed onto grain surfaces. The adsorbed species migrate on the grain surface, and react with each other when they meet. Grain-surface reactions are considered to contribute significantly to the formation of large organic species, because grain surfaces are relatively enriched with heavy-element species, and because the association reactions ($AB + C \rightarrow ABC$) are possible by depositing the excess energies on the grain surfaces. In the gas-phase reactions, on the other hand, products are mostly broken apart (AB + C → A+BC) to discard the excess energies as kinetic energies of the products.

Efficiency of each process in Figure 1 varies significantly with physical conditions. In prestellar cores with high density ($n_{\rm H} \geqslant 10^5$ cm^{-3}) and low temperature ($T \sim 10$ K), the adsorption overwhelms the desorption, although the non-thermal desorption by imping-ing cosmic-rays, UV, and reaction heat of grain-surface reactions is not negligible. On the grain-surface, hydrogenation (A + H → AH) should be the dominat reaction; since hydrogen atoms have relatively low mass and low binding energy onto the grain surface, they can migrate efficiently even under low temperatures.

In protostellar cores, on the other hand, hydrogen atoms are thermally sublimated to the gas phase. Heavy-element species, with higher binding energies (onto grains) than hydrogen atoms, are still on the grain surface, migrating to react with each other. As the core temperature rises further, the heavy element species are also sublimated according to their volatility. The sublimates undergo gas-phase reactions, which have been considered to be the dominant formation path of the large organic species observed in hot cores in high-mass star forming regions (e.g., Millar & Hatchell 1998). At the temperature of 200 K and density of $n_{\rm H} \sim 10^6$ cm^{-3}, for example, large organic species are formed from sublimated CH$_3$OH and H$_2$CO by the gas-phase reactions in $\sim 10^5$ yrs. But it is not clear if a similar process works in low-mass protostellar cores, in which large organic species are detected in recent years (Ceccarelli *et al.*, in this volume). In low-mass cores, the hot region ($T > 100$ K) with sublimated CH$_3$OH must be much smaller than in high-mass cores, and thus the infalling material would pass through such hot regions in a shorter timescale than 10^5 yrs. Furthermore, the recent theoretical studies and laboratory experiments on chemical reactions suggest that the gas-phase formation of large organic species would not be as efficient as previously assumed (Horn *et al.* 2004, Geppert *et al.* 2006). Garrod & Herbst (2006) showed grain-surface reaction at warm temperature ($\sim$several 10 K) can be an alternative or supplimental formation path of large organic species.

Molecular evolution in star-forming cores can be predicted quntitatively by solving the chemical reaction network, which includes the chemical processes mentioned above and are described mathematically as rate equations,

$$\frac{dx(i)}{dt} = \sum_{j,k} \alpha_{j,k} x(j) x(k) n_{\rm H} + \sum_{j} \beta_j x(j). \qquad (1.1)$$

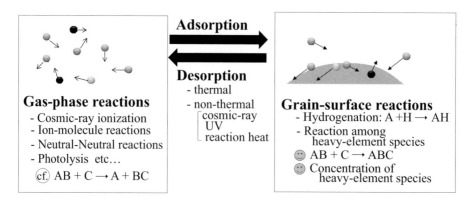

Figure 1. Chemical processes in molecular clouds.

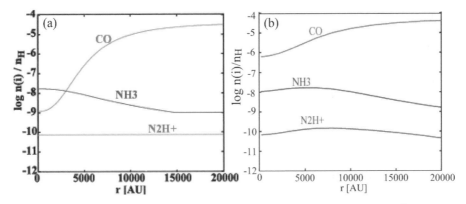

Figure 2. Radial distribution of molecular abundances in prestellar cores. Abundances derived from the observation of L1517B (Tafalla *et al.* 2002) (a) and our model (b).

The reaction rate coefficients, α and β, are in general a function of temperature, and $x(i)$ is the number density of species i relative to hydrogen nuclei (n_H). The solution of Eq. (1) then depends on density (n_H) and temperature, which are often assumed to be constant or a simple function of time. Garrod & Herbst (2006), for example, assumed that the density is constant and temperature rises as a square of the time ($T \propto t^2$). While such simple assumptions on the physical conditions are convenient in investigating the dependence of the chemical reaction network on physical parameters, it is also useful to adopt more realistic models of temperature and density evolution in order to evaluate the temporal and spatial variation of molecular abundances in star-forming cores. Masunaga & Inutsuka (2000), for example, constructed a one-dimensional (spherical) non-gray radiation-hydrodynamics code to calculate the physical evolution of a dense cloud core to form a protostar. Combining the chemical reaction network model with such radiation-hydrodynamics models, we can investigate temporal and spatial variation of molecular abundances in a star forming core (Aikawa *et al.* 2008).

In the following, we will present our model results on a collapsing prestellar core, which reproduces the observed chemical fractionation of C-bearing and N-bearing species (§2). In §3 we investigate the molecular evolution from a prestellar core to a protostellar core, and compare our model results with the observations of large organic species, carbon-chain species and HCO_2^+ in protostellar cores.

2. Prestellar Cores

Over the last several years, radio astronomers found chemical fractionation in prestellar cores. While the intensity map of dust continuum and nitrogen-bearing species (N_2H^+ and NH_3) are centrally peaked, the map of carbon-bearing species show a central hole. Comparison of these intensity maps revealed that C-bearing species are heavily depleted at the central regions, while N-beating species are not. Figure 2a shows the radial distribution of molecular abundances in L1517B (Tafalla *et al.* 2002).

In order to investigate the cause of this fractionation, we calculated molecular evolution in prestellar cores by combining the chemical reaction network model with the gravitational contraction of a spherical core. Since the radiation cooling is more efficient than the contraction heating at this evolutionary stage ($n_H \leqslant 10^7$ cm^{-3}), we adopt the isothermal collapse model to derive the temporal variation of density in infalling shells, in which the cheical reaction network is solved (Aikawa *et al.* 2005). Our present model includes two important updates from Aikawa *et al.* (2005). Firstly, we set the adsorption energy of CO and N_2 to be $E_{ads}(CO)=1180$ K and $E_{ads}(N_2)=1060$ K. While previous model works often assumed significantly lower adsorption energy of N_2 than that of CO, e.g. $E_{ads}(CO)=1780$ K and $E_{ads}(N_2)=750$ K, Öberg *et al.* (2005) found via laboratory experiments that the adsorption energy E_{ads} of CO and N_2 are similar. Secondly, we assume N_2H^+ recombination produces NH + H (10%) and N_2+H (90%), following the recent laboratory experiment (Geppert priv. com.). Previously, the branching ratio was claimed to be 65% for NH + H and 35% for N_2+H (Geppert *et al.* 2004).

Figure 2b shows radial distributions of molecular abundances in our model when the central density of the core is similar to that of L1517B ($\sim 10^5$ cm^{-3}). The model shows a reasonable agreement with the observation; in the central region, NH_3 abundance is slightly enhanced, and N_2H^+ abundance is almost constant, while CO is depleted. In our model, CO is adsorbed onto grains. Since CO is the main reactant of N_2H^+, the rate of N_2H^+ destruction decreases as CO depletes. Although N_2 can freeze-out onto grains with a similar adsorption energy of CO, slow (i.e. long-lasting) N_2 formation also helps to keep the abundance of N-bearing species undepleted (Aikawa *et al.* 2005, Maret *et al.* 2006).

3. Protostellar cores

3.1. *Physical and chemical evolution*

As the core contracts further, the contraction heating overwhelms the radiation cooling, and core temperature rises. Eventually, the protostar is formed, which further heats the envelope (i.e. protostellar core). In order to investigate the molecular evolution in protostellar cores, we adopt a core model of Masunaga & Inutsuka (2000). The model starts from a hydrostatic cloud core with the central density of $n_H \sim 6 \times 10^4$ cm^{-3}, and follows the core evolution until 9.3×10^4 yrs after the protostar is born ($\equiv t_{final}$). Although the original paper by Masunaga & Inutsuka (2000) mainly discusses the formation of a protostar, the model gives the radial distribution of density, temperature and infall velocity in the entire core as a function of time. Figure 3 shows the structure of the core model at assorted evolutionary stages. It should be noted that CO and large organic species such as CH_3OH are sublimated to the gas phase at ~ 20 K and ~ 100 K, respectively. At the moment of protostar birth ($t = 0$), for example, CO is sublimated inside ~ 100 AU.

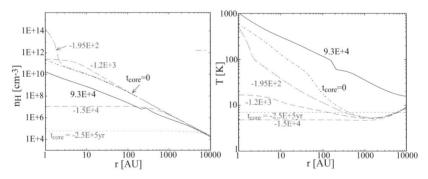

Figure 3. Radial distribution of density and temperature in a star-forming core at assorted evolutionary stages. The protostar is born at $t = 0$; $t < 0$ (gray lines) corresponds to the prestellar stage, while $t > 0$ (black lines) corresponds to the protostellar stage. The core is initially almost isothermal, but the central region start to warm up as the contraction heating overwhelms the radiation cooling ($t \sim -1.2 \times 10^3$ yr). The first core, i. e. an AU-sized hydrostatic core, appears at $t \sim -$several 100 yr, which then collapses to form a protostar.

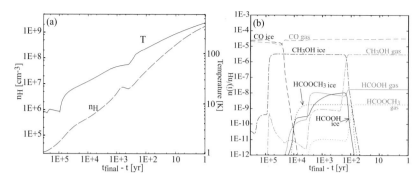

Figure 4. (a) Temporal variation of density and temperature in the infalling shell which reaches $r = 2.5$ AU at t_{final}. (b) Evolution of molecular abundances in the shell. Black lines represent the species adsorbed onto grain surfaces (i.e. ice), while gray lines represent gaseous species.

Based on the core model described above, we derive the temporal variation of density and temperature in infalling shells; Figure 4a shows such variation in the shell which reaches $r = 2.5$ AU at t_{final}. The temporal variation accelerates as the shell migrates towards the inner radius, where the infall velocity, temperature, and density steeply increases inwards. In order to highlight the rapid temporal variation near and at the final stage, the horizontal axis of Figure 4 is set to be the logarithm of $t_{\text{final}} - t$. We can see that once the shell enters the region of $T \geqslant 100$ K, it falls onto the central star within ~ 100 yrs, which is much shorter than the timescale ($\sim 10^5$ yr) needed to form the large organic species from sublimated CH_3OH via gas-phase reactions (§1).

Figure 4b shows the molecular evolution in the infalling shell. As initial molecular abundances, we adopt the "low-metal" elemental abundances and assumed that the species are in the form of atoms or atomic ions except for hydrogen, which is in its molecular form. Assuming that the initial cloud core is stable for 1×10^6 yrs, we solved the chemical reaction network with the fixed density and temperature over this period, which sets the initial molecular abundances for the collapse stage shown in Figure 4. We can see that large organic species such as $HCOOCH_3$ are mainly formed on the grain surfaces at $T \sim 20$–40 K, and then sublimate to the gas-phase when the shell temperature reaches their sublimation temperatures. One exception is $HCOOH$, which is formed by the gas-phase reaction of sublimated H_2CO with OH.

Following the molecular evolution in multiple infalling shells, we obtain spatial distribution of molecules in a protostellar core. Figure 5 shows the radial distribution of molecular abundances at t_{final}. Gaseous organic species are abundant inside about 100 AU, where the temperature exceeds their sublimation temperatures. Formic acid is again an exception; it extends beyond its sublimation radius because of the gas-phase formation. The organic species on grain surfaces are abundant at 100 AU–1,000 AU.

3.2. *Effect of UV radiation: embedded core vs isolated core*

Although we are interested in the central region of protostellar cores, the shells are initially (in the prestellar core stage) at outer radii $r \sim 10^4$ AU. In the model described in the previous subsection, we assumed that the model core is embedded in ambient clouds, and set the visual extinction of $A_v = 3$ mag at the outer boundary of the core. But this assumption is arbitrary; Bok Globules, for example, are not surrounded by ambient clouds and thus can be exposed directly to interstellar UV radiation. In order

Table 1. Gas-phase molecular abundances in IRAS 16293-2422 and model results.

Species	IRAS 16293-2422	model embedded	isolated[a]
H_2CO	$1.0(-7)^b$, $1.1(-7)^c$	$2.8(-6)$	$1.3(-11)$
CH_3OH	$1.0(-7)^d$, $9.4(-8)^c$	$3.0(-6)$	$8.5(-8)$
$HCOOCH_3$	$2.5-5.5(-7)^e$, $2.6-4.3(-9)^f$, $> 1.2(-8)^g$	$1.8(-9)$	$3.2(-11)$
$HCOOH$	$6.2(-8)^e$, $2.5(-9)^g$	$1.7(-8)$	$2.1(-8)$
CH_3OCH_3	$2.4(-7)^e$, $7.6(-8)^c$	$3.5(-10)$	$5.8(-12)$
CH_3CN	$1.0(-8)^e$, $7.5(-9)^h$	$3.0(-8)$	$1.3(-8)$

Notes:
[a] Gas-phase abundances at $r = 30.6$ AU are listed, because the abundances are mostly constant at $r \lesssim 100$ AU.
[b] Maret *et al.* (2004), [c] Chandler *et al.* (2005), [d] Maret *et al.* (2005), [e] Cazaux *et al.* (2003), [f] Kuan *et al.* (2004),
[g] Remijian & Hollis (2006), [h] Schöier *et al.* (2002).

to investigate the effect of UV radiation on core chemistry, we calculated a model with $A_v = 0$ mag at the outer core edge.

We found that the isolated core model evolves to be a protostelolar core with higher abundance of CO_2 and lower abundance of large organic molecules compared with the embedded core model (Aikawa *et al.* 2008). While the shells are at outer radii ($\sim 10^4$ AU) of the prestellar core, CO_2 are efficiently formed by the reactions of

$$H_2O + h\nu \to H + OH$$
$$CO + OH \to CO_2 + H$$

on the grain surfaces. In the observation of protostellar cores, molecular abundances are known to vary among objects. Our results indicate that such variation could, at least partially, result from the incident UV radiation at the prestellar stage.

3.3. *Comparison with observation*

Table 1 compares molecular abundances in our embedded and isolated core models and those estimated from the observations of a class 0 object IRAS 16293-2522. It should be noted that the estimated abundances vary significantly depending on the assumptions on temperature and density distribution in the protostellar core. Considering such uncertainties, our models are in reasonable agreement with the observations.

In order to understand the chemical conditions in protostellar cores, it is also important to observe species other than large organic species. Recently, interesting chemical species are detected towards L1527, which is considered to be in the transition phase from class 0 to class I. Firstly, Sakai *et al.* (2008a) detected carbon chains (e.g., C_4H, C_4H_2, and C_3H_2); C_4H emission extends over the 40" scale, and the line width toward the core center is broader than those toward the surrounding positions, indicating that C_4H emission

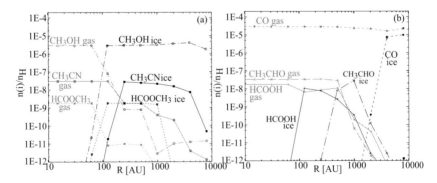

Figure 5. Radial distribution of molecules in our protostellar core model at t_{final}.

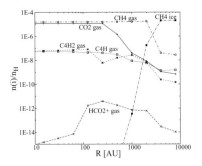

Figure 6. Radial distribution of carbon chains and HCO_2^+ in our protostellar core model at t_{final}.

is continuously distributed from the infalling envelope to the inner part of the core. Existence of carbon-chain species towards L1527 was a surprise, because carbon-chain species are usually associated with the early stages of cloud cores when the dominant form of carbon changes from atomic carbon to CO. Sakai *et al.* (2008a) proposed that carbon chains could regenerate from sublimated CH_4 in the warm central region. Indeed, carbon chains are abundant in our model (Figure 6). Methane is produced in the prestellar core phase and sublimate as the infalling shell temperature rises to 25 K. Then a fraction of sublimated CH_4 is transformed to carbon chains via both gas-phase and grain-surface reactions.

Secondly, Sakai *et al.* (2008b) detected HCO_2^+ towards L1517. While CO_2 ice is known to be abundant in protostellar cores (ex. 22% relative to H_2O ice towards Elias 29)(Ehrenfreund & Shutte 2000), the abundance of gaseous CO_2 has not been well-constrained. Since HCO_2^+ is formed by the reaction of $H_3^+ + CO_2 \rightarrow HCO_2^+ + H_2$, it can be a tracer of gaseous CO_2. Although the spatial distribution of HCO_2^+ is not well-constrained from the observation, Sakai *et al.* (2008b) argue, with a help of simple chemical analysis, that the HCO_2^+ seems to extend upto $\sim$2000 AU. The column density of HCO_2^+ is estimated to about 7.6×10^{10} cm^{-2}.

In comparison, distribution of HCO_2^+ in our model is shown in Figure 6; it is abundant at radius of 100 AU $< r <$ 2000 AU. Column density of HCO_2^+ reaches 10^{10}–10^{11} cm^{-2} towards the core center (impact parameter $\leqslant$ 2000 AU), which is consistent with the observation. In our model, gaseous CO_2 and HCO_2^+ increase at the region of $\sim$25 K, where some carbon-chains react with O_2 to form CO_2 in the gas-phase. CO_2 ice is abundantly formed by grain-surface reactions, but they do not desorb efficiently until the shell temperature reaches $\sim$70 K. In the central region of $r \leqslant$ 100 AU, on the other hand, CO_2 gas is abundant but HCO_2^+ is not, because NH_3 and H_2O, which have larger proton affinity than CO_2, are sublimated to the gas phase.

References

Aikawa, Y., Herbst, E., Roberts, H., & Caselli, P. 2005, *ApJ*, 620, 330

Aikawa, Y., Wakelam, V., Herbst, E., & Garrod, R. T. 2008, *ApJ*, 674, 984

Chandler, C. J., Brogen, C. L., Shirley, Y. L., & Loinard, L. 2005, *ApJ*, 632, 371

Cazaux, E., Tielens, A. G. G. M., Ceccarelli, C., Castets, A., Wakelam, V., Caux, E., Parise, B., & Teyssier, D. 2003, *ApJ* (Letter), 593, L51

Ehrenfreund, P. & Shutte, W. A. 2000, *in Astrochemistry: From Molecular Clouds to Planetary Systems*, (Chelsea, MI; Sheridan Books; Astronomical Society of the Pacific), p. 135

Garrod, R. T. & Herbst, E. 2006, *A&A*, 457, 927

Geppert, W. D., *et al.* 2004, *ApJ*, 609, 459

Geppert, W. D., Thomas, R. D., Ehlerding, A., *et al.* 2006, *Faraday Discuss.*, 133, 177

Horn, A., Møllendal, H., Sekiguchi, O., *et al.* 2004, *ApJ*, 611, 605

Kuan, Y.-J., Juang, H.-C., Charnley, S. B., Hirano, N., Takakuwa, S., Wilner, D. J., Liu, S.-Y., Ohashi, N., Bourke, T. L., Qi, C., & Zhang, Q. 2004, *ApJ* (Letter), 616, L27

Maret, S., Bergin, E. A., & Lada, C. J. 2006, *Nature*, 442, 425

Maret, S., Ceccarelli, C., Caux, E., Tielens, A. G. G. M., Jørgensen, J. K., van Dishoeck, E. F., Bacmann, A., Castets, A., Lefloch, B., Loinard, L., Parise, B., & Schöier, F. L. 2004, *A&A*, 416, 577

Maret, S., Ceccarelli, C., Tielens, A. G. G. M., Caux, E., Lefloch, B., Faure, A., Castet, A., & Flower, D. R. 2005, *A&A*, 442, 527

Masunaga, H. & Inutsuka, S. 2000, *ApJ*, 531, 350

Millar, T. J. & Hatchell J. 1998, *Faraday Discuss.*, 109, 15

Öberg, K., van Broekhuizen, F., Fraser, H. J., Bisschop, S. E., van Dishoeck, E. F., & Schlemmer, S. 2005, *ApJ* (Letter), 621, L33

Remijian, A. J. & Hollis, J. M. 2006, *ApJ*, 640, 842

Sakai, N., Sakai, T., Aikawa, Y., & Yamamoto, S. 2008b, *ApJ* (Letter), 675, L89

Sakai, N., Sakai, T., Hirota, T., & Yamamoto, S. 2008a, *ApJ*, 672, 371

Schöier, F. L., Jørgensen, J. K., van Dishoeck, E. F., & Blake, G. A. 2002, *A&A*, 391, 1001

Tafalla, M., Myers, P. C., Caselli, P., Walmsley, C. M., & Comito, C. 2002, *ApJ*, 569, 815

Discussion

BOTTINELLI: I was wondering what abundance of ammonia you had in your model and whether you can see difference in the chemistry between ammonia rich and ammonia poor ice.

AIKAWA: In our model the ammonia abundance is not assumed. It is determined by the chemical reaction network, and is about 10% of H_2O ice in outer radius.

CUPPEN: I have a question about the accretion model, your updated version. You show that ammonia is in agreement with the observations. Can you comment on the CO?

AIKAWA: CO in the gas phase is determined by the efficiency of cosmic ray desorption. This is one of the assumptions in the model. Observationally, we do not have a sensitivity to distinguish between 10^{-7} abundance to 10^{-6} abundance. Observers can only say that it is lower than 10^{-6}.

CECCARELLI: I saw that your model prediction of the abundance of large organic molecules is around 100 times lower than we observed.

AIKAWA: In our model we need methanol to produce these molecues. I want to emphasize that there is no assumptions on initial conditions or ice abundance. I just started with a very simple protostellar core and solved detailed but still simple chemical reaction network, to see how far we can go and how much organic species we can make. If you need more, I have to work on the chemical network to make more organics from methanol. Another solution could be axisymetric model, rather than spherical. Then the fluid parcels can stay longer in warm regions (of circumstellar disk) in which large organic molecules are formed.

Organic Matter in Space
Proceedings IAU Symposium No. 251, 2008
S. Kwok & S. Sandford, eds.

© 2008 International Astronomical Union
doi:10.1017/S1743921308021376

Collisional excitation of complex organic molecules

Alexandre Faure[1], Eric Josselin[2], Laurent Wiesenfeld[1], and Cecilia Ceccarelli[1]

[1]Laboratoire d'Astrophysique de Grenoble, Université Joseph Fourier/CNRS UMR 5571, France
email: afaure@obs.ujf-grenoble.fr

[2]G.R.A.A.L, Université de Montpellier II/CNRS UMR 5024, France

Abstract. A major difficulty in modelling the infrared and (sub)millimeter spectra of gas-phase complex organic molecules is the lack of state-to-state collisional rate coefficients. Accurate quantum or classical scattering calculations for large polyatomic species are indeed computationally highly challenging, particularly when both rotation and low frequency vibrations such as bending and torsional modes are involved. We briefly present here an approximate approach to estimate and/or extrapolate rotational and rovibrational rates for polyatomic molecules with many degrees of freedom.

Keywords. Molecular data, molecular processes, ISM: molecules

1. Introduction

In the past few years, observations of star forming regions have revealed unexpected high abundances of complex organic molecules such as ketene (H_2CCO), methyl formate ($HCOOCH_3$), dimethyl ether (CH_3OCH_3), etc. (e.g. Ceccarelli *et al.* 2007). The detailed modelling of these observations is however currently hampered by the lack of state-to-state collisional rates. A major problem relates to the fact that quantum scattering calculations for such complex species are impractical owing to the excessively large number of coupled channels to be included in the solution of the Schrödinger equation. Moreover, classical mechanics has other severe intrinsic limitations (e.g. Faure & Wiesenfeld 2004). As a result, the largest system for which absolute rovibrational state-to-state cross sections are available is E-type methanol colliding with He (Pottage *et al.* 2004a).

2. Rotational excitation and the information theory

The information theory was introduced in the early 1970's as a procedure to compact and correlate the large amounts of data, both theoretical and experimental, on molecular collisions (Levine 1978 and references therein). In this approach the decrease of (de)excitation rates with increasing inelasticity is described through purely statistical factors and an exponential gap law. Despite its simplicity, this theory was shown to properly account for the underlying qualitative trends of energy transfer processes. Based on this theory, we have derived a simple analytic formula to describe pure rotational rate coefficients. This formula, supplemented with collisional propensity rules, has been calibrated and checked against the accurate quantum calculations of Pottage *et al.* (2004b) for E-type methanol. The results are presented in Figure 1.

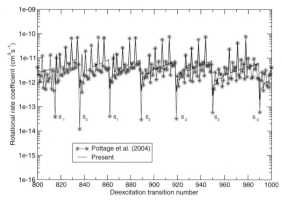

Figure 1. Sample of rotational rates for E-type methanol colliding with para-H_2 at 100 K. The present calculations are based on an exponential gap law supplemented with the quantum propensity rules $\Delta K = 0, \Delta J < 6$. The agreement with the results of Pottage *et al.* (2004b) is typically within a factor of 3, except for very low rates.

3. Rovibrational excitation and critical densities

There is to our knowledge no state-to-state rate coefficients available for rovibrational processes involving polyatomic molecules. Based on data available in the literature, however, it appears that these rates can be assumed, *to first order*, to be proportional to the pure rotational rates. We have derived a formula where the proportionality coefficient i) is calibrated over theoretical or experimental data, ii) is defined so that rates obey the detailed balance principle and iii) depends on temperature. For example, for the torsional mode of E-type methanol, this coefficient is about 10^{-2} at 700 K (Pottage *et al.* 2004a).

If we now consider the critical densities of rovibrational levels of complex molecules, we can summarize the situation as follows:

- For levels in the ground vibrational state, critical densities are typically $n < 10^5$ cm^{-3}. Non-LTE rotational populations are expected in regions with density $n < 10^7$ cm^{-3}.
- For levels in excited vibrational states, critical densities are $n > 10^9$ cm^{-3}. Non-LTE vibrational populations are expected in, e.g., circumstellar envelopes with density $10^7 < n < 10^{11}$ cm^{-3}.

4. Conclusion

The present method allows to estimate and/or extrapolate collisional rates for complex organic molecules, including possibly vibrational modes. It should provide guidance at the order-of-magnitude level. It is also expected to become more accurate at high temperatures (>300 K) where standard quantum methods are impractical. More details and examples will be published elsewhere.

References

Ceccarelli, C., Caselli, P., Herbst, E., Tielens, A. G. G. M., Caux, E. 2007 in: V. B. Reipurth, D. Jewitt, and K. Keil (eds.), *Protostars and Planets V* (University of Arizona Press, Tucson), p. 47
Faure, A. & Wiesenfeld, L. 2004, *J. Chem. Phys.*, 121, 6771
Levine, R. D. 1978, *Ann. Rev. Phys. Chem.*, 29, 59
Pottage, J. T., Flower, D. R., & Davis, S. L. 2004a, *J. Phys. B*, 37, 165
Pottage, J. T., Flower, D. R., & Davis, S. L. 2004b, *MNRAS*, 352, 39

Organic Matter in Space
Proceedings IAU Symposium No. 251, 2008
S. Kwok & S. Sandford, eds.

© 2008 International Astronomical Union
doi:10.1017/S1743921308021388

Ion chemistry in the interstellar medium

Oscar Martinez Jr., Theodore P. Snow, and Veronica M. Bierbaum

Department of Chemistry and Biochemistry, Center for Astrophysics and Space Astronomy,
Department of Astrophysical and Planetary Sciences, University of Colorado,
Boulder, CO 80309-0215, U.S.A.
email: oscar.martinez@colorado.edu, theodore.snow@colorado.edu,
veronica.bierbaum@colorado.edu

Abstract. Without accurate data on reaction rates and branching ratios, models of interstellar chemistry are unreliable. Recent research has identified a number of reactions of unusual importance because the rates and branching ratios are unknown or poorly known. Efforts to expand and improve on current databases are underway using a flowing afterglow-selected ion flow tube (FA-SIFT) coupled to a quadrupole mass spectrometer. Our current focus is on the reactions of C^+, a major cation in the interstellar medium, with the neutrals O_2, H_2O, CH_4, NH_3 and C_2H_2. Future planned work includes studies of polycyclic aromatic hydrocarbons (PAHs), developing comprehensive pathways for their formation, and identification of those PAHs important to interstellar chemistry. The recent discovery of ISM anions has highlighted the importance of examining mechanisms of anionic chemistry in the interstellar medium, and we plan to obtain data relevant to the formation and destruction processes of molecular anions in space.

Keywords. Astrobiology, astrochemistry, astronomical data bases: miscellaneous, ISM: evolution, ISM: kinematics and dynamics, ISM: molecules

1. Introduction

There exist numerous discrepancies in rate constants and branching ratios compiled into chemical databases available to the astrophysical community. A study by Markwick-Kemper (2005) ranked the importance of reactions used in astrochemical models by examining the effect of uncertainties in rate constants and branching ratios. Of the ~4000 reactions in the UMIST 99 catalog, only ~1200 reactions have less than 25% uncertainty and ~2700 reactions have certainty only "within a factor of 2".

Of particular importance to the evolution and chemistry of the interstellar medium (ISM) are reactions involving the carbon cation, C^+. Found throughout the ISM, its chemistry with interstellar neutrals is prevalent in dense molecular clouds. Unlike diffuse clouds, where carbon is photoionized, in dense molecular clouds C^+ is formed via the reaction $He^+ + CO \longrightarrow C^+ + O + He$ with supplemental ionization by cosmic rays (Herbst & Klemperer 1973, Watson 1974). C^+ reactions with NH_3, CH_4, O_2, H_2O and C_2H_2 were among the top 100 in Markwick-Kemper's study; although prior measurements for the reactions have been made, our focus was to eliminate the large uncertainty associated with the prior measurements and to clear the discrepancies between values reported in available databases.

2. Experimental

The flowing afterglow selected ion flow tube (FA-SIFT) is ideal for studies of ion-neutral reactions relevant to the interstellar medium. A variety of ionization methods and chemical versatility exist for the FA-SIFT method. The FA-SIFT, with its high ion

ZIURYS: If we go from AGB circumstellar envelopes to planetary nebula to the diffuse ISM, molecular abundances slowly go down, maybe in order of magnitude per step.

SARRE: So the carbon content may be increasingly in forms such as large molecules or dust.

ZIURYS: The molecules are not confined to the gas phase. The large molecules may be also be fragmenting and there is undoubtedly gas-grain interactions.

Lucy Ziurys asking one of her many questions (photo by Sze-Leung Cheung).

Organic Matter in Space
Proceedings IAU Symposium No. 251, 2008
S. Kwok & S. Sandford, eds.

Organic molecular anions in interstellar and circumstellar environments

M. A. Cordiner[1], T. J. Millar[1], C. Walsh[1], E. Herbst[2], D. C. Lis[3], T. A. Bell[3], and E. Roueff[4]

[1] Astrophysics Research Centre, School of Mathematics and Physics, Queen's University, Belfast, BT7 1NN, U.K.
email: m.cordiner@qub.ac.uk

[2] Departments of Physics, Astronomy and Chemistry, The Ohio State University, Columbus, OH 43210, U.S.A.
email: herbst@mps.ohio-state.edu

[3] California Institute of Technology, MC 320-47, Pasadena, CA 91125, U.S.A.
email: dcl@caltech.edu

[4] LUTh & UMR 8102 du CNRS, Observatoire de Paris, 5, Place J. Janssen, F-92190, Meudon, France
email: Evelyne.Roueff@obspm.fr

Abstract. The synthesis of organic molecular anions in TMC-1 and IRC+10216 is investigated. Modelled C_2H^-, CN^-, C_3N^-, C_5N^- and C_7N^- column densities are sufficiently great that these species might be observable in IRC+10216. Density-enhanced shells in the outer envelope of IRC+10216 are found to enhance the C_2H^- and CN^- column densities by shielding these anions from destruction by UV radiation. From a newly-derived upper column density limit of 6.6×10^{10} cm^{-2} for C_2H^- in IRC+10216 we deduce the primary production mechanism for this anion to be $C_2H_2 + H^- \longrightarrow C_2H^- + H_2$. In TMC-1, due to the low radiative electron attachment rates calculated for C_2H^-, CN^- and CH_2CN^-, these species have modelled column densities below the detection threshold. They could, however, be produced in reactions we have not yet considered.

Keywords. Astrochemistry, ISM: molecules, ISM: abundances, ISM: clouds, stars: carbon

1. Introduction

The molecular anions C_4H^-, C_6H^- and C_8H^- have recently been detected in large quantities in the envelope of the carbon-rich AGB star IRC+10216 (McCarthy *et al.* 2006, Cernicharo *et al.* 2007, Remijan *et al.* 2007). C_6H^- and C_8H^- have also been detected in the cold, dense interstellar cloud TMC-1 (McCarthy *et al.* 2006, Brüenken *et al.* 2007), and C_4H^- and C_6H^- in the protostellar source L1527 (Sakai *et al.* 2007, Agúndez *et al.* 2008). Fractional abundance ratios of the anions relative to their neutral parents of up to 0.1 indicate that anions may play a more significant role in interstellar physics and chemistry than previously believed.

The possibility that a relatively large fraction of interstellar molecular material might be in the form of anions was first suggested by Herbst (1981) who pointed out that carbon chain molecules and other radicals with large electron affinities may have high radiative attachment rates, leading to interstellar anion-to-neutral ratios of the order of a few percent. Using detailed chemical models (Millar *et al.* 2007) we have previously considered the formation of hydrocarbon anions in IRC+10216 and TMC-1 and were able to reproduce the large observed anion abundances. In these proceedings, results

Table 1. Calculated anion and neutral molecular column densities (N/cm^{-2}) from chemical models of TMC-1 and IRC+10216. Anion-to-neutral ratios $R = N(\text{X}^-)/N(\text{X})$ are also given.

Species	TMC-1		IRC+10216	
	N	R	N	R
C_2H^-	1.4×10^9	2.9×10^{-5}	6.1×10^{10}	1.2×10^{-5}
C_2H	4.9×10^{13}		5.2×10^{15}	
C_3H^-	3.0×10^{10}	0.0021	2.7×10^6	1.9×10^{-8}
C_3H	1.8×10^{13}		1.4×10^{14}	
C_4H^-	8.1×10^{10}	0.0077	2.4×10^{13}	0.015
C_4H	1.1×10^{13}		1.6×10^{15}	
C_5H^-	1.9×10^{11}	0.045	1.5×10^{13}	0.065
C_5H	4.1×10^{12}		2.3×10^{14}	
C_6H^-	1.9×10^{11}	0.059	9.0×10^{13}	0.069
C_6H	3.2×10^{12}		1.3×10^{15}	
C_7H^-	3.7×10^{11}	0.19	5.6×10^{13}	0.14
C_7H	1.9×10^{12}		4.1×10^{14}	
C_8H^-	2.6×10^{11}	0.047	2.0×10^{13}	0.045
C_8H	5.4×10^{12}		4.4×10^{14}	
C_9H^-	4.5×10^{11}	0.14	1.8×10^{13}	0.11
C_9H	3.2×10^{12}		1.6×10^{14}	
$C_{10}H^-$	1.0×10^{11}	0.041	5.6×10^{12}	0.0035
$C_{10}H$	2.5×10^{12}		1.6×10^{14}	
CN^-	3.3×10^5	2.9×10^{-8}	1.4×10^{10}	5.8×10^{-6}
CN	1.1×10^{13}		2.4×10^{15}	
C_3N^-	5.5×10^8	7.0×10^{-4}	8.5×10^9	2.1×10^{-4}
C_3N	7.9×10^{11}		4.1×10^{13}	
C_5N^-	3.4×10^9	0.053	1.3×10^{13}	0.059
C_5N	6.5×10^{10}		2.2×10^{14}	
C_7N^-	3.4×10^9	0.053	2.9×10^{12}	0.051
C_7N	6.5×10^{10}		5.7×10^{13}	
CH_2CN^-	7.0×10^7	2.6×10^{-5}	7.2×10^6	3.0×10^{-7}
CH_2CN	2.7×10^{12}		2.4×10^{13}	

are presented from updated models of these environments, which include the additional anions C_2H^-, C_3H^-, CN^-, C_3N^-, C_5N^-, C_7N^- and CH_2CN^-.

2. TMC-1

We modelled the chemistry of TMC-1 using the time-dependent model described by Woodall *et al.* (2007), augmented with the C_nH^- anion chemistry from Millar *et al.* (2007) extended down to $n = 2$. Electron radiative attachment rates for C_nH species are from Herbst & Osamura (2008). Chemical networks for $C_{2n+1}N^-$ species are from the model by Millar *et al.* (2000), with an additional C_3N^- formation channel involving dissociative electron attachment to HNC_3 (the chemistry for which has been taken from the OSU database (Harada & Herbst 2008). CH_2CN^- is assumed to be formed by radiative electron attachment to CH_2CN, the rate for which was calculated using the phase-space approach discussed by Herbst & Osamura (2008). The electron affinity of CH_2CN was taken to be 1.55 eV (Lykke *et al.* 1987), and the necessary vibrational frequencies were estimated to be the same as those of the isoelectronic NH_2CN. The G value, expressing the ratio of the electronic degeneracy of the anion to that of the reactants was set to 1/4. The rate coefficient k_r for radiative stabilization of the anion was estimated to be 500 s^{-1} based on calculated values of analogous ions, while the rate coefficient k_{-1} for dissociation of the anionic complex was calculated to be 1.79×10^8 s^{-1}. Since k_{-1} is much greater than k_r, radiative attachment to CH_2CN is relatively inefficient.

Anions in the TMC-1 model are destroyed predominantly in reactions with atomic H, C, N and O. Results are shown in Table 1.

3. IRC+10216

We have updated the chemical model of IRC+10216 (Millar *et al.* 2000, Millar *et al.* 2007) to be consistent with the chemical reaction rates in the (dipole-enhanced) RATE06 database (Woodall *et al.* 2007). Initial C_2H_2 and HCN abundances are from Fonfria *et al.* (2008). Anion chemistry has been updated to include the additional species mentioned in Section 2. Further, the following CN^- and C_2H^- formation reactions have been included (with rate coefficients from Prasad & Huntress 1980, Mackay *et al.* 1977):

$$HCN + H^- \longrightarrow CN^- + H_2 \tag{3.1}$$

$$C_2H_2 + H^- \longrightarrow C_2H^- + H_2 \tag{3.2}$$

These processes are important in IRC+10216 (and not in TMC-1) due to the large abundances of HCN and C_2H_2 in the stellar outflow.† Dominant anion destruction mechanisms are by reaction with H and photodetachment by the interstellar radiation field. Photodetachment rates were calculated according to Eq. 2 of Millar *et al.* (2007).‡ The model (which will be explained in detail by Cordiner & Millar 2008), includes density-enhanced shells of gas and dust which have been observed in the stellar outflow (Mauron & Huggins 2000, Dinh-V-Trung & Lim 2008). Seven concentric shells are included; each is 2″ thick with a gas and dust density 5 times greater than the underlying density distribution and an inter-shell spacing of 8″. The presence of shells increases the shielding of the inner envelope from interstellar UV radiation, which allows CN^- and C_2H^- to survive photodetachment for longer and raises their column densities (by about a factor of two) compared to models without shells.

Results of the IRC+10216 model are shown in Table 1. Notable new species included in this model that have sufficiently high column densities to be detectable with current instruments include $C_{2n-1}N^-$ (for $n = 1$ to 4). C_2H^- is predicted to be detectable, and during the preparation of this article new data were recorded at the Caltech Submillimeter Observatory which place a 3σ upper limit of 6.6×10^{10} cm^{-2} on the column density of C_2H^- in IRC+10216. This value was obtained from a preliminary analysis of a $J = 3-2$ spectrum, using an estimated Einstein A coefficient of 7.44×10^{-4} s^{-1} and an assumed (LTE) rotational temperature of 20 K for the molecule. This upper limit is consistent with our modelled column density of 6.1×10^{10} cm^{-2}.

To produce C_2H^-, we have also considered reactions of the kind (Mackay *et al.* 1977)

$$C_2H_2 + X^- \longrightarrow C_2H^- + XH \tag{3.3}$$

for the various anions X^- in our model. However, such reactions cannot be rapid (with the exception of Eq. (3.2), i.e. $X = H$), because they would result in a modelled C_2H^- column density ~ 100 times greater than the observed upper limit. It is therefore hypothesised that reaction 3.3 is endothermic for anions X^- with high electron binding energies such as C_nH^- and C_n^-.

4. CH_2CN^- and dipole-bound electronic states

CH_2CN^- has been included in these models due to the suggestion by Cordiner & Sarre (2007) that this molecule could be the carrier of the 8037 Å diffuse interstellar band. In both TMC-1 and IRC+10216 the modelled CH_2CN^- abundance is very low due to the small rate of anion formation by radiative electron attachment. Unless an alternative, more rapid, production mechanism exists for this anion, its low calculated

---

† H^- is produced in the model mainly by cosmic-ray dissociation: $H_2 + CR \longrightarrow H^+ + H^-$.

‡ The CN^- photodetachment rate used here is ~ 100 times less than the value in the RATE06 database.

abundances suggest that the oscillator strength of the responsible transition must be exceptionally high for CH_2CN^- to be the carrier of any DIBs. It has been suggested (see, e.g., Sommerfeld 2005, Cordiner & Sarre 2007) that dipole-bound electronic states (possessed by strongly dipolar molecules) may assist in the formation of some anions by acting as a doorway through which rapid radiative stabilisation can occur. This effect might increase the abundance of potential DIB-carrying anions such as CH_2CN^- in the diffuse ISM, and may also be important in radiative electron attachment to C_nH species with dipole moments $\gtrsim 2.5$ D (i.e., $n \neq 1,2,4$), perhaps reconciling some of the discrepancies (see Agúndez *et al.* 2008) between current observational and theoretical anion-to-neutral ratios.

5. Summary

Our new chemical models calculate abundances of CN^-, C_2H^- and C_3N^- that suggest these species may be observable in IRC+10216 but not in TMC-1. C_5N^- and C_7N^- might be observable in both sources due to their high calculated anion-to-neutral ratios of $\sim 5\%$. There is some uncertainty in these predictions due to the uncertain nature of the adopted rate coefficients. Further experimental and theoretical work is needed to constrain these rate coefficients, together with further observational studies of anions in a variety of extraterrestrial regions.

References

Agúndez, M., Cernicharo, J., Guélin, M. *et al.* 2008, *A&A*, 478, 19
Brünken, S., Gupta, H., Gottlieb, C. A. *et al.* 2007, *ApJ* (Letters), 664, L43
Cernicharo, J., Guélin, M., Agúndez, M. *et al.* 2007, *A&A* (Letters), 467, L37
Cordiner, M. A. & Millar, T. J. 2008, *in preparation*
Cordiner, M. A. & Sarre, P. J. 2007, *A&A*, 472, 537
Dinh-V-Trung & Lim, J. 2008, *ApJ*, in press
Fonfria, J. P., Cernicharo, J., Richter, M. J., & Lacy, J. H. 2008, *ApJ*, 673, 445
Harada, N. & Herbst, E. 2008, *in preparation*
Herbst, E. 1981, *Nature*, 289, 656
Herbst, E. & Osamura, Y. 2008, *ApJ*, in press
Lykke, K. R., Neumark, D. M., Andersen, T., Trapa, V J., & Lineberger, W. C. 1987, *J. Chem. Phys.*, 87, 12, 6842
Mauron, N. & Huggins, P. J. 2000, *A&A*, 359, 707
Mackay, G. I., Tanaka, K., & Bohme, D. K. 1977, *Int. J. Mass Spec.*, 24, 125
McCarthy, M. C., Gottlieb, C. A., Gupta, H. C., & Thaddeus, P. 2006, *ApJ* (Letters), 652, L141
Millar, T. J., Herbst, E., & Bettens, R. P. A. 2000, *MNRAS*, 316, 195
Millar, T. J., Walsh, C., Cordiner, M. A., Ní Chuimín, R., & Herbst, E. 2007, *ApJ* (Letters), 662, L87
Prasad, S. S. & Huntress Jr., W. T. 1980, *ApJSS*, 43, 1
Remijan, Anthony J., Hollis, J. M., Lovas, F. J., Cordiner, M. A., Millar, T. J., Markwick-Kemper, A. J., & Jewell, P. R. 2007, *ApJ* (Letters), 664, L47
Sakai, N., Sakai, T., Osamura, Y., & Yamamoto, S. 2007, *ApJ* (Letters), 667, L65
Sommerfeld, T. 2005, *J. Phys.:Conf.Ser*, 4, 245
Woodall, J., Agúndez, M., Markwick-Kemper, A. J., & Millar, T. J. 2007, *A&A*, 466, 1197

Discussion

ZIURYS: You said that having anions helps you create cations as well. No one has really seen many cations in IRC+10216. There was a tentative detection of HCO^+ and that was it. Did you actually predict the abundance of the cations like HCO^+ from your model of IRC+10216?

CORDINER: Yes. Incorporating anions does give us a better fit between the model and the observations for HCO^+ by reducing its abundance.

Organic Matter in Space
Proceedings IAU Symposium No. 251, 2008
S. Kwok & S. Sandford, eds.

Ethylene in the circumstellar envelope of IRC+10216

K. H. Hinkle[1], L. Wallace[1], M. J. Richter[2], and J. Cernicharo[3]

[1]National Optical Astronomy Observatory
P.O. Box 26732, Tucson, Arizona 85726, U.S.A.
email: [hinkle;wallace]@noao.edu

[2]University of California at Davis
Physics Department, 1 Shields Avenue, Davis, CA 95616, U.S.A.
email: richter@physics.ucdavis.edu

[3]Departamento de Astrofísica Molecular e Infrarroja
Instituto de Estructura de la Materia, CSIC, Serrano 121, 28006 Madrid, Spain
email: cerni@damir.iem.csic.es

Abstract. Ethylene (C_2H_4) is a symmetric molecule that is best detected using mid-infrared transitions. We report on observations of the 10.5 μm ν_7 band using the cryogenic grating spectrograph TEXES. These confirm the previous ethylene detection in the IRC+10216 circumstellar shell. We detect 18 ethylene lines. The lines are both narrow and weak with depths of no more than $\sim$2%. The ethylene lines suggest an excitation temperature of $\sim$80 K.

Keywords. Techniques: spectroscopic, stars: carbon, circumstellar matter, ISM: molecules, infrared: stars

---

Due to symmetry ethylene does not have a permanent dipole moment and there are no strong microwave rotational transitions. Astronomically ethylene is best detected through the ν_7 out-of-plane bending mode at 10.5 μm. Ethylene (C_2H_4) was first detected in space by observing the ν_7 lines in the circumstellar shell of the dust-obscured carbon-rich AGB star IRC+10216. Infrared heterodyne spectroscopy and a 1.5 m telescope were employed (Betz 1981, Goldhaber *et al.* 1987).

Heterodyne spectroscopy is a novel mid-infrared technique useful for molecules with line frequencies in near coincidence with available reference lines. The detected ethylene lines were shallow, at most a few percent deep. The observations consisted of individual ethylene line profiles that were spectrally highly resolved but rather noisy. We felt that inspection of the entire spectral region might lead to (1) insight into possible blends and (2) detection of additional ethylene features. This in turn would lead to much improved information on the creation and destruction of ethylene in the circumstellar environment.

The observations were again of IRC+10216, a carbon star obscured in a dust shell of material shed from the star in mass loss. IRC+10216 is relatively close by and very bright in the thermal infrared, allowing highly sensitive infrared and microwave searches for molecular lines. The carbon-rich mass outflow from IRC+10216 results in a rich selection of organic circumstellar molecules. A recent review by Olofsson (2005) lists more than 30 species known in the IRC+10216 circumstellar envelope.

We observed the 10.4 μm (958 cm^{-1}) to 10.8 μm (924 cm^{-1}) spectral region to conduct a full search for the ν_7 ethylene lines. The new observations were carried out using the cryogenic grating spectrograph TEXES on the NASA IRTF. TEXES is a sensitive high-resolution cryogenic grating spectrograph of innovative design which works in the mid-IR (Lacy *et al.* 2002). The observed spectral resolution ($\lambda/\Delta\lambda$) was $\sim$10^5. A single exposure covered $\sim$5 cm^{-1} in segments of $\sim$0.7 cm^{-1}. Multiple grating settings were used to span the observed spectral range. Exposures were taken so the spectra overlapped eliminating gaps and systematics.

The observed spectrum confirms the previous work of Betz (1981) and Goldhaber *et al.* (1987); ethylene lines previously reported in these papers were detected as were considerably stronger circumstellar lines of ammonia and silane. A total of 18 C_2H_4 lines were identified. The ethylene lines are indeed weak with central depths of at most a few percent (Figure 1). The lines are also narrow, with full widths dominated by the TEXES instrumental resolution of ~ 3 km s^{-1}. The line profiles observed with the heterodyne equipment, at > 10 times the TEXES resolution, have FWHM ~ 2 km s^{-1}. Ethylene has an outflow velocity of ~ 14 km s^{-1}.

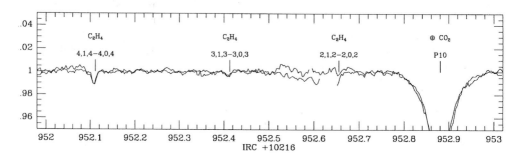

Figure 1. A section of the spectrum of IRC+10216 as observed by TEXES. The spectrum has been continuum normalized with only the 5% of intensity next to the continuum displayed. Three ethylene lines are identified and one telluric line of CO_2. This spectral region was observed twice. The overlap and offset of the spectral sections is visible (see text).

The current observations show that the ethylene lines are not blended by any other molecular bands. The signal-to-noise of the new measurements exceeds that of the measurements done in the 1980s. However, the ethylene lines are very weak with typical equivalent width $\sim 1 \times 10^{-4}$ cm^{-1}. As a result considerable uncertainty remains in the equivalent widths. A preliminary Boltzmann plot gives a rotational excitation temperature of ~ 80 K.

The excitation temperature and velocity of the lines are indications of where the lines are formed in the circumstellar shell. In the near- and mid-infrared line profiles from molecules existing throughout the circumstellar shell, such as CO, HCN, and C_2H_2, have complex profiles. Semi-empirical circumstellar models can be derived from this information (e.g., Keady *et al.* 1988, Fonfría *et al.* 2008) and can be applied to modeling the ethylene spectrum. Detailed modeling of the ethylene lines is underway. With an excitation temperature of ~ 80 K ethylene lines could arise from the external shell of IRC+10216 where many radicals have been found.

References

Betz, A. L. 1981, *ApJ*, 244, L103
Fonfría, J. P., Cernicharo, J., Richter, M. J., & Lacy, J. H. 2008, *ApJ*, 673, 445
Goldhaber, D. M., Betz, A. L., & Ottusch, J. J. 1987, *ApJ*, 314, 356
Keady, J. J., Hall, D. N. B., & Ridgway, S. T. 1988, *ApJ*, 326, 832
Lacy, J. H., Richter, M. J., Greathouse, T. K., Jaffe, D. T., & Zhu, Q. 2002, *PASP*, 114, 153
Olofsson, H. 2005, in: A. Wilson (ed.), *Proceedings of the Dusty and Molecular Universe: a Prelude to Herschel and ALMA*, ESA SP-577, (Noordwijk: ESA), p. 223

Organic Matter in Space
Proceedings IAU Symposium No. 251, 2008
S. Kwok & S. Sandford, eds.

© 2008 International Astronomical Union
doi:10.1017/S1743921308021455

Circumstellar H$_2$O in M-type AGB stars

Matthias Maercker[1], Fredrik L. Schöier[2], and Hans Olofsson[2]†

[1]Stockholm Observatory, AlbaNova University Center, 106 91 Stockholm, Sweden
email: `maercker@astro.su.se`

[2]Onsala Space Observatory, 439 92 Onsala, Sweden
email: `schoier@chalmers.se`, `hans.olofsson@chalmers.se`

Abstract. Surprisingly high amounts of H$_2$O have recently been reported in the circumstellar envelope around the M-type AGB star W Hya. However, substantial uncertainties remain, as the required radiative transfer modelling is difficult due to high optical depths, sub-thermal excitation and the sensitivity to the combined radiation field from the central star and dust grains.

Keywords. Stars: AGB and post-AGB, circumstellar matter, mass loss

1. Introduction

We perform a detailed radiative transfer analysis and determine abundances of circumstellar H$_2$O in the envelopes around six M-type AGB stars (Maercker *et al.* 2008a). ISO LWS spectra are used to constrain the circumstellar abundance distribution of ortho-H$_2$O. We also make predictions for H$_2$O emission lines in the range of the upcoming Herschel/HIFI mission. The HIFI data will resolve the line profiles, providing valuable additional information. To investigate the constraints set by also fitting the line profile of spectrally resolved lines, we include spectrally resolved Odin data of the 557 GHz ($1_{10} - 1_{01}$) ortho-H$_2$O line for R Cas, R Dor and W Hya (Maercker *et al.* 2008b). Finally, the new models are used to adjust the predictions made for the HIFI lines.

2. Radiative transfer modelling

A 'standard' model with a spherically symmetric circumstellar envelope (CSE), formed by a constant mass-loss rate and expanding at a constant velocity is assumed, resulting in a density structure where $\rho \propto r^{-2}$. Amorphous silicate dust (Justtanont & Tielens 1992) is used, with a single grain size ($a_d = 0.05\,\mu$m) and mass density ($\rho_d = 2.0\,\mathrm{g\,cm^{-3}}$). The dust and CSE parameters were determined separately and used as an input for the water vapour models. The dust density profile, the dust temperature profile, and the dust optical depth are determined using Dusty (Ivezić *et al.* 1999), fitting the SEDs to 2MASS and IRAS fluxes. The mass-loss rates, gas expansion velocities, and kinteic temperature distributions of the CSEs are determined in a circumstellar CO excitation analysis using a non-LTE radiative transfer code based on the Monte Carlo method (Schöier & Olofsson 2001). The modelling of the circumstellar H$_2$O emission lines is done using the ALI (accelerated lambda iteration) method (Bergman, OSO internal report). The code is a detailed non-LTE, and non-local, radiative transfer code, including 45 energy levels in the ground state and 45 energy levels in the first excited vibrational state of ortho-H$_2$O. Collision rates within the ground state, the first excited vibrational state and between the two states are included. The radiation from the central star, the

† On behalf of the Odin team

Table 1. Results of the best-fit models from fitting the ISO data. The difference to the observed intensities is given in % for the ISO lines (average difference of all lines), and for the Odin line where data is available. For WX Psc and IK Tau the theoretical radius from Netzer & Knapp (1987) is used, as no constraints could be set on the radius in the models.

Source	$\dot{M}$ $[10^{-6}\,M_\odot\,\mathrm{yr}^{-1}]$	$v_{\exp}$ $[\mathrm{km\,s}^{-1}]$	r_e $[10^{15}\mathrm{cm}]$	f_0 $[10^{-4}]$	δ_{Odin} %	δ_{ISO} %	χ^2_{red}
TX Cam	7.0	18.5	20.0	3.0	-	−7	0.6
R Cas	0.9	10.5	3.0	3.5	+3	−5	0.3
R Dor	0.2	6.0	1.1	3.0	+46	−1	0.6
W Hya	0.1	7.2	2.2	15.0	+67	−1	0.5
WX Psc	40	19.3	43.0	0.02	-	−7	2.7
IK Tau	10	19.0	17.0	3.5	-	−9	1.2

circumstellar dust, and the cosmic microwave background are included. The adjustable parameters in the modelling are the abundance of ortho-H_2O (relative to H_2 and assuming a Gaussian abundance distribution), and the size of the H_2O envelope. The outer radius of the H_2O envelope is defined by the e-folding radius (abundance decrease to 37%). A comparison to the radius given by theoretical models of photodissociation of water (Netzer & Knapp 1987) is done.

3. Results and conclusions

The results of the H_2O line modelling based on the ISO observations are presented in Table 1. For all sources the H_2O abundances are high with respect to expectations based on stellar atmosphere equilibrium calculations. An exception is WX Psc, possibly indicating processes that reduce the amount of observed water vapour (such as adsorption onto dust grains) in this source. The H_2O abundances are accurate to approx. ±50% within the adopted model (the absolute accuracy is within a factor of approximately 5). We find that the lines generally are subthermally excited and the emitting region to be excitation limited. The predictions for HIFI show that these lines are readily observable, with self-absorption and P-Cygni profiles apparent in the model spectra. In order to fit the spectrally resolved Odin line in R Cas, R Dor and W Hya, a reduction in expansion velocity compared to the CO models is needed (−19%, −33%, and −25%, respectively). Inclusion of the Odin line sets considerably tighter constraints on the envelope size for all three objects, and requires a reduction of the outer radius for W Hya by 32%. The fits to the spectrally resolved lines give information on the velocity field of the CSEs. In addition, the low-energy Odin line sets significantly stronger constraints on the size of the CSEs, as the line is excited throughout the envelope. The origin of the HIFI lines lies inside the region where the Odin line is formed, the resolved line profiles from these observations will likely set important constraints on the velocity structure of, and abundance distribution within the CSEs.

References

Ivezić, Ž, Nenkova, M., & Elitzur, M. 1999, User Manual for DUSTY, (Univ. Kentucky Internal Rep.)
Justtanont, K. & Thielens, A. G. G. M. 1992, *ApJ*, 389, 400
Maercker, M., Schöier, F. L., Olofsson, H., Bergman, P., & Ramstedt, S. 2008a, *A&A*, 479, 779
Maercker, M., Schöier, F. L., Olofsson, H., *et al.* 2008b, *A&A*, submitted
Netzer, N. & Knapp, G. R. 1987, *ApJ*, 323, 734
Schöier, F. L. & Olofsson, H. 2001, *A&A*, 368, 969

Organic Matter in Space
Proceedings IAU Symposium No. 251, 2008
S. Kwok & S. Sandford, eds.

© 2008 International Astronomical Union
doi:10.1017/S1743921308021467

Molecular line observations of the SiO maser source IRAS 19312+1950

J. Nakashima[1], S. Deguchi[2], H. Imai[3], and A. Kemball[4]

[1]Department of Physics, University of Hong Kong, Pokfulam Rd., Hong Kong
email: junichi@hku.hk

[2]Nobeyama Radio Observatory, National Astronomical Observatory of Japan

[3]Department of Physics, Kagoshima University

[4]NCSA, University of Illinois at Urbana-Champaign

Abstract. IRAS 19312+1950 is a unique SiO maser source, exhibiting a rich set of molecular radio lines, although SiO maser sources are usually identified as oxygen-rich evolved stars, in which chemistry is relatively simple comparing with carbon-rich environments. The rich chemistry of IRAS 19312+1950 has raised a problem in circumstellar chemistry if this object is really an oxygen-rich evolved star, but its evolutional status is still controversial. In this paper, we briefly review the previous observations of IRAS 19312+1950, as well as presenting preliminary results of recent VLBI observations in maser lines. PDF file of the poster is available from http://www.geocities.jp/nakashima_junichi/

Keywords. Circumstellar matter, ISM: jet and outflow, masers, stars: imaging, stars: individual (IRAS 19312+1950)

1. Introduction

SiO maser sources are usually identified as oxygen-rich (O-rich) evolved stars, and it has been long thought that O-rich evolved stars do not significantly contribute to the enrichment of organic matter in the universe unlike carbon stars do. However, recent observations of IRAS 19312+1950, which is an SiO maser source (Nakashima & Deguchi 2000), suggest that O-rich stars are non-negligible in the chemical evolution of the universe, because a rich set of molecular lines are detectable toward IRAS 19312+1950. If IRAS 19312+1950 is really an O-rich evolved star, its chemical properties are quite suggestive in terms of the interstellar chemistry, but we must securely identify its evolved-star status before going to further astrophysical/astrochemical interpretation, because some observational properties of this object are not consistent with the evolved-star status.

2. Previous Observations

The extended infrared nebulosity of IRAS 19312+1950 was first realized in the 2MASS images by Nakashima & Deguchi (2000), and soon after that a more fine near-infrared image was obtained by UH 2.2 m/SIRIUS. The SIRIUS imaging revealed that the envelope of IRAS 19312+1950 exhibits a point symmetric structure.

Deguchi *et al.* (2004) and Nakashima *et al.* (2004) searched molecular millimeter lines toward IRAS 19312+1950 using the NRO 45 m telescope. Even though the molecular lines search was not unbiased survey (i.e., selected lines were observed), they detected 22 different molecular species. Nakashima & Deguchi (2004, 2005) made interferometric observations with BIMA in the bright molecular lines. They concluded that the properties

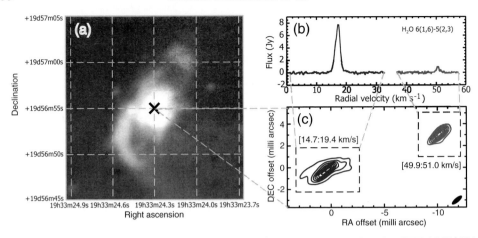

Figure 1. (a): Near-infrared H-band image of IRAS-19312+1950 taken by SUBARU/CIAO (Murakawa *et al.* 2007). (b): VLBA total flux profile of the H_2O maser line ($6_{1,6}$–$5_{2,3}$). (c): VLBA total intensity map of the red- and blue-shifted components of the H_2O maser line.

of the broad component seen in the line profiles are explained by an expanding sphere model.

3. VLBA & MERLIN Observations in Maser Lines

Maser emission detected from IRAS 19312+1950 enables us to investigate the kinematics in the vicinity to the central star if we use the VLBI technique. We have recently made VLBA observations in the SiO ($J = 1$–0, $v = 1$ and 2) and H_2O ($6_{1,6}$–$5_{2,3}$) maser lines. In addition, we have made MERLIN observations in the H_2O (22 GHz) and OH (1612 MHz) maser lines. In the H_2O observations, we clearly detected a double peak in the total flux line profile, and confirmed that the red- and blue-shifted components are seated in spatially separated regions (Figure 1).

4. Discussion

Our current interpretation of IRAS 19312+1950 is a rare AGB or post-AGB star embedded in a small dark cloud, even though we have no clear idea about why an AGB/post-AGB star is embedded in a small dark cloud. A recent model calculation by Murakawa *et al.* (2007) have also suggested that the mass loss rate of the central star is consistent with that of an AGB star, but they have also withheld the discussion about the origin of the ambient material. To securely confirm the evolutional status of this object, presumably, observations listed below will be a key: (1) monitoring observations in maser lines with VLBI technique, (2) measuring a $^{12}C/^{13}C$ ratio through, for example, radio observations in the $H^{12}CO^+$ and $H^{13}CO^+$ lines.

References

Deguchi, S., Nakashima, J., & Takano, S. 2004, *PASJ*, 56, 1083
Murakawa, K., Nakashima, J., Ohnaka, K., & Deguchi, S. 2007, *A&A*, 470, 957
Nakashima, J. & Deguchi, S. 2000, *PASJ* (Letters), 52, L43
Nakashima, J. & Deguchi, S. 2004, *ApJ* (Letters), 610, L41
Nakashima, J. & Deguchi, S. 2005, *ApJ*, 633, 282
Nakashima, J., Deguchi, S., & Kuno, N. 2004, *PASJ*, 56, 193

Organic Matter in Space
Proceedings IAU Symposium No. 251, 2008
S. Kwok & S. Sandford, eds.

© 2008 International Astronomical Union
doi:10.1017/S1743921308021479

The physics and chemistry of circumstellar envelopes of S-stars on the AGB

Sofia Ramstedt[1], Fredrik L. Schöier[2], and Hans Olofsson[1,2]

[1]Dept. of Astronomy, Stockholm University,
Roslagstullsbacken 21, SE-10691, Stockholm, Sweden
email: sofia@astro.su.se

[2]Onsala Space Observatory

Abstract. Presented here are the preliminary results of a long-term study of S-stars on the AGB. S-stars are important as possible transition objects between oxygen-rich M-stars and carbon stars. The aim of the study is to compare results from our newly gathered observational database for the S-stars with those already obtained for the M- and carbon stars. We can thus follow the changes as the stars evolve along the AGB and more firmly establish the suggested M-MS-S-SC-C evolutionary sequence. It will also allow us to determine the relative importance of processes such as non-equilibrium chemistry, grain formation, and photodissociation in regulating the chemistry in circumstellar envelopes of AGB stars.

Keywords. Stars: AGB and post-AGB, (stars:) circumstellar matter, stars: mass loss, stars: abundances

1. Observational database

The sample consists of 42 S-type AGB stars, and it is flux-limited. The CO data has been gathered using the Onsala 20 m telescope ($J = 1 \rightarrow 0$), the IRAM 30 m telescope ($J = 1 \rightarrow 0$, $2 \rightarrow 1$), the JCMT ($J = 3 \rightarrow 2$), and APEX ($J = 3 \rightarrow 2$). The SiO data has been gathered using the Onsala 20 m ($J = 2 \rightarrow 1$, v=0, v=1), the IRAM 30 m ($J = 2 \rightarrow 1$, $5 \rightarrow 4$), the JCMT ($J = 6 \rightarrow 5$, $8 \rightarrow 7$), and APEX ($J = 8 \rightarrow 7$). Further HCN data has been gathered using the IRAM 30 m ($J = 1 \rightarrow 0$, $3 \rightarrow 2$), the JCMT ($J = 3 \rightarrow 2$), and APEX ($J = 4 \rightarrow 3$). Emission from other molecules (CS, SiS) has been searched for, and detected in some of the sample sources.

2. Radiative transfer modelling

The circumstellar envelopes (CSEs) are assumed to be spherically symmetric and formed by constant mass-loss rates. Mass-loss rates and physical properties of the CSEs are estimated from the CO data, using a non-LTE, non-local radiative transfer code based on the Monte Carlo method (Schöier & Olofsson 2001, Olofsson *et al.* 2002, Ramstedt *et al.* 2006). The energy balance is solved self-consistently and the effects of dust on the radiation field and the thermal balance are included. The molecular excitation analysis also includes the radiation field generated by the central star and reprocessed by thermal dust grains in the envelope. Once the physical properties are determined, abundances of the other molecules observed in the CSEs can be estimated using the same radiative transfer code.

3. Results and conclusions

• We have estimated the fractional abundance of SiO in 27 S-stars and find that the abundances can be more than an order of magnitude larger than predicted by thermal equilibrium chemistry (see Figure 1). For a certain mass-loss rate the circumstellar SiO abundance seems independent of the C/O-ratio.

• The previous analysis of M-stars (González Delagado *et al.* 2003) and carbon stars (Schöier *et al.* 2006) shows a clear trend that the SiO abundance decreases as the density of the wind increases. Eventhough the high-mass-loss-rate S-stars are sparse, we see indications of the same trend in this analysis. The trend is indicative of adsorption of SiO onto grains.

• We have started the modelling of the HCN emission to determine fractional abundances in the S-stars. The results will be compared to the results for the M- and carbon stars (see contribution by Schöier & Olofsson in these proceedings).

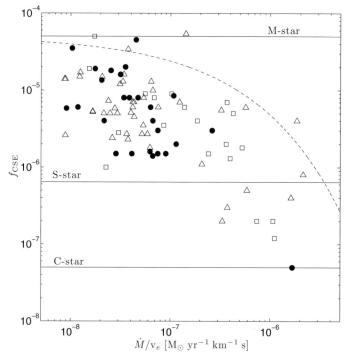

Figure 1. SiO fractional abundances (f_{CSE}), obtained from the radiative transfer modelling, as a function of a density measure for M-stars (open triangles), carbon stars (open squares) and S-stars (filled dots). The horizontal lines mark the abundances predicted by equilibrium chemistry calculations. The dashed line shows the expected $f(\infty)$ (scaled to 5×10^{-5}, roughly the expected fractional abundance from stellar equilibrium chemistry when C/O<1) for a model including adsorption of SiO onto dust grains.

Acknowledgements

The authors acknowledge support from the Swedish Research Council. This work has benifitted from research funding from the European Community's sixth Framework Programme under RadioNet R113CT 2003 5058187.

References

González Delgado, D., Olofsson, H., & Kerschbaum, F. 2003, *A&A*, 411, 123
Olofsson, H., González Delgado, D., Kerschbaum, F., & Schöier, F. L. 2002, *A&A*, 391, 1053
Ramstedt, S., Schöier, F. L., Olofsson, H., & Lundgren, A. A. 2006, *A&A* (Letters), 454, L103
Schöier, F. L. & Olofsson, H. 2001, *A&A*, 368, 969
Schöier, F. L, Olofsson, H., & Lundgren, A. A. 2006, *A&A*, 454, 247

Organic Matter in Space
Proceedings IAU Symposium No. 251, 2008
S. Kwok & S. Sandford, eds.

© 2008 International Astronomical Union
doi:10.1017/S1743921308021480

Molecular lines in the envelopes of evolved stars

Yong Zhang[1], Sun Kwok[1], and Dinh-V-Trung[2]

[1]Department of Physics, University of Hong Kong, Pokfulam Road, Hong Kong
email: zhangy96@hku.hk; sunkwok@hku.hk

[2]Institute of Astronomy and Astrophysics, Academia Sinica, P.O Box 23-141,
Taipei 106, Taiwan
email: trung@asiaa.sinica.edu.tw

Abstract. We report a spectral line survey of the circumstellar envelopes of evolved stars at millimeter wavelengths. The data allow us to investigate the chemical processes in different physical environments and evolutionary stages. A total of more than 500 emission features (mostly rotational transitions of molecules) are detected in the survey. Our observations show that the sources in different evolutionary stages have remarkably different chemical composition. As a star evolves from AGB stage to proto-planetary nebula, the abundances of Si-bearing molecules (SiO, SiCC, and SiS) decrease, while the abundances of some long-chain molecules, such as CH_3CN, C_4H, and HC_3N, increase. After further evolution to planetary nebula, the abundances of neutral molecules dramatically decrease, and the emission from molecular ions becomes more intense. These differences can be attributed to the changes of the role that dust, stellar winds, shock waves, and UV/X-rays from the central star play in different evolutionary stages. These results will provide significant constraints on models of circumstellar chemistry.

Keywords. Stars: AGB and post-AGB, line: identification, ISM: molecules, radio lines: ISM

1. Introduction

Since 1970, more than 60 molecular species, including many organic molecules, have been detected in the circumstellar envelopes of evolved stars. Asymptotic giant branch (AGB) stars, their descendant planetary nebulae (PNs), and the transition objects between the two phases, proto-planetary nebulae (PPNs), therefore represent major sites of molecular synthesis. Since the evolutionary time scales of these phases are very short (10^4–10^5 yr, $< 10^3$ yr, 10^3–10^4 yr for AGB stars, PPNs, and PNs, respectively), chemical reaction time scales are very well constrained by these time scales. The study of the changing chemical composition and molecular abundance between objects in consecutive phases of evolution provides useful information on the chemical pathways of molecular synthesis. Furthermore, these comparisons can lead us to a better understanding of the roles of dust, shock waves, and UV and X-ray radiation on the chemical processes. For this purpose, we present a molecular line survey in a sample of evolved stars.

2. Observations

Our sample consists of three AGB stars (IRC+10216, CRL 3068, and CIT 6), one PPN (CRL 2688), and one young PN (NGC 7027). The spectral survey was carried out between 2005 April and 2008 January using the Arizona Radio Observatory 12 m and 10 m telescopes, covering the frequency ranges from 71–161 GHz and 218–268 GHz, respectively. The typical sensitivity is $T_R < 10$ mK at a spectral resolution of 1 MHz. The spectra are presented in Figure 1. The temperature scales at the ARO 12 m and 10 m are T_R^* and T_A^*, respectively. The main beam brightness temperatures were derived through

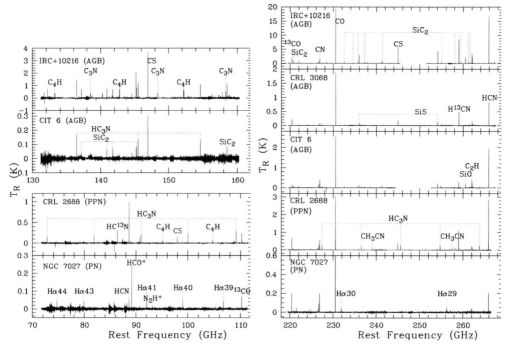

Figure 1. *Left panel*: The 12 m data; *Right panel*: The 10 m data.

$T_R = T_R^*/\eta_m^*$ and $T_R = T_A^*/\eta_{mb}$, where η_m^* is the corrected beam efficiency and η_{mb} the beam efficiency.

3. Conclusions

A systematic comparison of the molecular emission in the five objects has revealed the following spectral differences:

a) Refractory metal- and Si-bearing species, which are plentifully present in IRC+10216, show weak or no emission in the other objects. These molecules may be depleted onto dust grains with stellar evolution.

b) Compared to the other objects, CRL 2688 has stronger C_4H, HC_3N and CH_3CN emission, suggesting that these species are rapidly reprocessed during the evolution from AGB to PPN.

c) Compared to other objects, CIT 6 has a large CN/HCN intensity ratio due to photodissociation of HCN into CN. However, the CN/HCN ratios in CRL 2688 and NGC 7027 are relatively low, indicating that CN and/or HNC may be efficiently transferred into HCN in hot environments.

d) Emissions from neutral molecules in NGC 7027 are fainter compared to those in other objects. This is partly due to the destruction of neutral molecules caused by strong shocks and UV radiation in the young PN.

e) For our observations, NGC 7027 is the only source in which we have detected ionized species (HCO^+, HCS^+, and N_2H^+) and recombination lines. This is attributed to photoionization by the much stronger UV radiation from the central star.

Further observations are underway to expand the evolutionary coverage in order to better constrain the relation between chemical abundance and physical conditions.

Organic Matter in Space
Proceedings IAU Symposium No. 251, 2008
S. Kwok & S. Sandford, eds.

© 2008 International Astronomical Union
doi:10.1017/S1743921308021492

The 1 mm spectrum of VY Canis Majoris: Chemistry in an O-rich envelope

Emily D. Tenenbaum[1], Stefanie N. Milam[2], Aldo J. Apponi[1],
Neville J. Woolf[1], Lucy M. Ziurys[1], and Fredrik L. Schöier[3]

[1]Steward Observatory, Dept. of Chemsitry, University of Arizona
933 N. Cherry Ave. Tucson, AZ 85721 USA
email: `emilyt@as.arizona.edu`

[2]NASA Ames Research Center and SETI Institute
MS 245-6, Moffett Field, CA 94035-1000, USA

[3]Onsala Space Observatory, SE-439 92, Onsala, Sweden

Abstract. We present preliminary results of an unbiased spectral survey at 1 mm of the oxygen-rich supergiant, VY CMa. A number of exotic molecules have been detected, including NaCl and PO, and a relatively rich organic chemistry is observed. Results of the survey will be compared with carbon-rich stars.

Keywords. Stars: individual (VY Canis Majoris, IRC +10216), circumstellar matter, astrochemistry, submillimeter

1. Introduction

Although numerous spectral band surveys have been carried out for carbon-rich circumstellar shells, an oxygen-rich counterpart has never been studied in equivalent detail. In 2006, with the goal of more deeply investigating oxygen-rich circumstellar chemistry, we began a spectral survey of the oxygen-rich supergiant VY CMa. This star is one of the brightest objects in the near-infrared sky, with an unusually large mass loss rate of $\sim 5 \times 10^{-4}$ solar masses per year (Muller *et al.* 2007). Unlike less massive AGB stars such as IRC + 10216, which have uniformly expanding shells, the mass loss in this object is very sporadic, resulting in a clumpy, structured envelope consisting of arcs, knots and jets (Humphreys *et al.* 2007). The main goal of our survey is to observe a continuous spectrum from 215 to 280 GHz towards VY CMa using the Arizona Radio Observatory 10 m Sub-millimeter Telescope (SMT) on Mount Graham, Arizona. The SMT is equipped with a new side-band separating 1 mm receiver, giving system temperatures ranging from 150 to 400 K over the survey band. In addition, a simultaneous survey of IRC +10216 is being conducted in the same 1 mm band.

2. Results

Currently, the data-taking phase of the survey is 60% complete and it should be finished by November 2008. Seventy-five emission lines have been detected so far, including 12 unidentified lines. Eighteen molecules have been detected in VY CMa and of these, eight were first detected in VY CMa by our survey. In particular, the species NaCl, NS, HCO$^+$, and PN were detected for the first time in an oxygen-rich shell (Ziurys *et al.* 2007), and PO has been identified - the first interstellar molecule with a P-O bond (Tenenbaum *et al.* 2007).

Six carbon-containing molecules have been observed towards VY CMa. Emission from CO, CS, HCO$^+$, HCN, and HNC has been modeled using the non-LTE large velocity gradient code developed by Bieging & Tafalla (1993). Sample spectra of these molecules are shown in Figure 1. Emission from CS and HCN arises from regions close to the star, suggesting that shock-induced chemistry plays a role in the formation of these species. Our findings are in agreement with theoretical work by Cherchneff (2006) which predicts that non-TE shocked regions above the photosphere are areas of active carbon-chemistry.

3. Future Work

After the data-taking phase of the survey is complete, the observed emission will be modeled using a 3–D Monte Carlo radiative transfer code for circumstellar envelopes (Schöier & Olofsson 2001). From the modeling we will gain additional information about molecular abundances and envelope kinematics. When both the VY CMa and IRC +10216 surveys are complete, we will compare the chemistry in C-rich versus O-rich circumstellar envelopes. Interferometry observations of molecular emission from VY CMa will be critical to understanding the morphology of the circumstellar envelope. SMA observations of CO and SO emission have recently been published (Muller *et al.* 2007) and the envelope of VY CMa will be an optimal target for observations with ALMA.

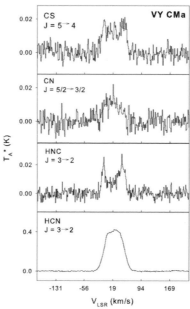

Figure 1. Emission from carbon-bearing molecules in VY CMa observed with the SMT.

References

Bieging, J. H. & Tafalla, M. 1993, *AJ*, 105, 576
Cherchneff, I. 2006, *A&A*, 456, 1001
Humphreys, R. M., Helton, L. A., & Jones, T. J. 2007, *AJ*, 133, 2716
Muller, S., Trung, D. V., Lim, J., Hirano, N., Muthu, C., & Kwok, S. 2007, *ApJ*, 656, 1109
Schöier, F. L. & Olofsson, H. 2001, *A&A*, 368, 969
Tenenbaum, E. D., Woolf, N. J., & Ziurys, L. M. 2007, *ApJ*, 666, L29
Ziurys, L. M., Milam, S. N., Apponi, A. J. & Woolf, N. J. 2007, *Nature*, 447, 1094

Organic Matter in Space
Proceedings IAU Symposium No. 251, 2008
S. Kwok & S. Sandford, eds.

HCO$^+$ emission possibly related with a shielding mechanism that protects water molecules in the young PN K 3-35

Y. Gómez[1], D. Tafoya[1,2], G. Anglada[3], L. Loinard[1], J. M. Torrelles[4],
L. F. Miranda[3], M. Osorio[3], R. Franco-Hernández[1,2], L. Nyman[5],
J. Nakashima[6], and S. Deguchi[7]

[1]Centro de Radioastronomía y astrofísica, UNAM, México
email: y.gomez@astrosmo.unam.mx

[2]Harvard-Smithsonian Center for Astrophysics, Cambridge, MA, USA

[3]Instituto de Astrofísica de Andalucía, CSIC, Granada, Spain

[4]Instituto de Ciencias del Espacio, and IEE de Catalunya, Barcelona, Spain

[5]European Southern Observatory, Santiago, Chile

[6]Department of Physics, University of Hong Kong, Hong Kong

[7]Nobeyama Radio Observatory, NAO, Nagano, Japan

Abstract. Water maser emission has been detected only toward three planetary nebulae (PNe). In particular, in K3-35, the first PN where water vapor maser emission was detected, the components are located in a torus-like structure with a radius of 85 AU and also at the surprisingly large distance of 5000 AU from the star, in the tips of the bipolar lobes. The existence of these water molecules in PNe is puzzling, probably related to some unknown mechanism shielding them against the ionizing radiation. We report the detection of HCO$^+$ $(J = 1 - 0)$ emission toward K 3-35, that not only suggests that dense molecular gas ($\sim 10^5$ cm^{-3}) is present in this PN, but also that this kind of PN can enrich their surroundings with organic molecules.

Keywords. Planetary Nebula, stars: individual (K 3-35), radio lines: molecules, stars: circumstellar matter

1. Introduction

The presence of H$_2$O maser emission in planetary nebulae (PNe) is opening a new field in the evolutionary study of intermediate mass stars (Gómez 2007, Gómez *et al.* 2008). Even when water maser emission is typically found in the envelopes of asymptotic giant branch (AGB) stars, it is not expected to persist in the PNe phase, where the envelope not only begins to be rarefied but also becomes ionized. Water maser emission has been detected toward three PNe (K 3-35, Miranda *et al.* 2001; IRAS 17347-3139, de Gregorio-Monsalvo *et al.* 2004; IRAS 18061-2505, Gómez *et al.* 2008). In the particular case of K 3-35, the radio continuum emission exhibits a point-symmetric morphology that has been modelled with a precessing jet evolving in a dense AGB circumstellar medium (Velázquez *et al.* 2007). OH and H$_2$O maser emission has been reported toward this PN (Miranda *et al.* 2001), with the H$_2$O masers located at the tips of the jets and toward the core in a torus-like structure (Gómez 2007, Uscanga *et al.* 2008 in preparation). In order to understand which is the mechanism that is maintaining the presence of water molecules, we made a search for molecular gas toward K 3-35 under the hypotesis that dense material can protect the water molecules from the UV radiation of the central star.

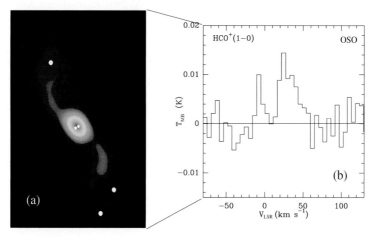

Figure 1. (a): VLA (3.6 cm) radio continuum image of K 3-35, white dots mark the position of the water masers detected by Miranda *et al.* (2001). (b): HCO$^+$ ($J = 1 - 0$) spectra detected with the Onsala telescope (Tafoya *et al.* 2007).

2. Summary

The survey for molecular lines toward the young planetary nebula K 3-35 was made using the 20-m telescope of Onsala Space Observatory, the 45-m radio telescope of Nobeyama and the IRAM 30-m telescope. This survey included the molecules SiO, HCO$^+$, H^{13}CO$^+$, HNC, HCN, HC$_5$N, HC$_3$N, CS, CN, CH$_3$OH, H$_2$O, CO and ^{13}CO. The details of these observations are published in Tafoya *et al.* (2007).

We discuss here the detection of HCO$^+$ ($J = 1 - 0$) emission toward K 3-35 (see Figure 1). The HCO$^+$ peak emission is centered at $\sim$28 km s^{-1} and the line has a FWHM of $\sim$20 km s^{-1}. We have found that the HCO$^+$ abundance in K 3-35 is $\sim$6 $\times$ 10^{-7}, which is similar to values found in other young PNe. Broad CO (2-1) and (1-0) emission were also detected toward K 3-35. Using the CO (2-1;1-0) emission we have estimated an excitation temperature T_{ex} $\sim$20 K. Assuming LTE and that the CO emission is optically thin, a molecular mass for the envelope of $\sim$0.017 M$_\odot$ is derived. The ratio of molecular to ionized mass is $\sim$1.9. All these results support the presence of a massive molecular envelope in K 3-35 that could be responsible for the shielding mechanism that protects water molecules from being destroyed by the stellar radiation. High angular observations are needed to image the dense molecular gas envelope in this PN.

Acknowledgements

YG acknowledges financial support from PAPIIT-UNAM (IN100407) and CONACyT (49947), México.

References

de Gregorio-Monsalvo, I., Gómez, Y., Anglada, G., Cesaroni, R., Miranda, L. F., Gómez, J. F., & Torrelles, J. M. 2004, *ApJ* 601, 921

Gómez, Y. 2007, *Astrophysical masers and their environments, IAU Proceedings* 242, 292

Gómez, J. F., Suárez, O., Gómez, Y., Miranda, L. F., Torrelles, J. M., Anglada, G., & Morata, O. 2008, *AJ* in press.

Miranda, L. F., Gómez, Y., Anglada, G., & Torrelles, J. M. 2001, *Nature* 414, 284

Tafoya, D., *et al.* 2007, *ApJ* 133, 364

Velázquez, P. F., Gómez, Y., Esquivel, A., & Raga, A. C. 2007, *MNRAS* 382, 1965

Organic Matter in Space
Proceedings IAU Symposium No. 251, 2008
S. Kwok & S. Sandford, eds.

© 2008 International Astronomical Union
doi:10.1017/S1743921308021510

Synthesis of organic compounds in the circumstellar environment

Sun Kwok

Department of Physics, University of Hong Kong, Hong Kong, China
email: sunkwok@hku.hk

Abstract. Through the techniques of millimeter-wave and infrared spectroscopy, over 60 species of gas-phase molecules and a variety of inorganic and organic solids have been detected in the short phase of stellar evolution between the asymptotic giant branch and planetary nebulae. The chemical pathways that lead to the synthesis of complex organic compounds in such low-density environments are therefore important topics of astrochemistry. In this review, we summarize the observational evidence for the existence of complex aliphatic and aromatic compounds in these circumstellar environments, and discuss the nature of their possible carriers. Also discussed are a number of unidentified emission features which may also have an organic origin. The possible relations between these circumstellar organic matter with Solar System organic matter are explored.

Keywords. Stars: AGB and post-AGB, Planetary nebulae: general, ISM: molecules, ISM: lines and bands, infrared: ISM

1. Introduction

Nucleosynthesis of the element carbon and its dredge up to the photosphere represents the beginning of organic synthesis in the late stages of stellar evolution. The low surface temperature of aymptotic giant branch (AGB) stars allow the formation of gas-phase molecules such as CO, CN and C_2 in the photosphere. The onset of mass loss in the late AGB stage leads to the large-scale formation of a variety of gas-phase molecules in the stellar wind, including many organic molecules (Ziurys, these proceedings). Through mm, submm, and infrared spectroscopy, over 60 different molecular species have been identified in the stellar winds of AGB stars. These molecules, and the solid particles that condense out of them, represent a major source of organic compounds in the Galaxy. The detection of presolar grains in meteorites originating from AGB stars establishes a direct link between stars and the Solar System, and AGB stars are a potential source of organic enrichment of the early Solar System (Zinner, these proceedings).

In this review, we summarize the observational evidence for the presence of organic solids in AGB stars, as well as their descendents proto-planetary nebulae (PPNe) and planetary nebulae (PNe). The rich and complex organic compounds synthesized in such a short (10^4 yr) phase of stellar evolution establishes circumstellar environment as an important laboratory for the study of astrochemistry.

2. Stellar synthesis of inorganic compounds

The detection of the 9.7 and 18 μm Si$-$O stretching and Si$-$O$-$Si bending modes of amorphous silicates in oxygen-rich AGB stars was the first realization that solids can efficiently form in the circumstellar environment of evolved stars (Woolf & Ney 1969). In the IRAS sky survey, over 4000 O-rich and over 700 C-rich AGB stars have been found to possess the 9.7 μm silicate feature and the 11.3 μm SiC feature, respectively (Kwok *et al.* 1997). *ISO* observations have detected a rich family of emission features attributed to cystalline silicates (olivine and pyroxene) in AGB stars and PNe (Jäger *et al.* 1998). A variety of refractory oxides (e.g., corunum, spinel) have also been seen in AGB stars (Posch *et al.* 1999). The detection of these substances in the circumstellar environment represents the beginning of the field astromineralogy (Henning 2003).

3. Stellar synthesis of organic compounds

The family of infrared emission features first discovered in the PN NGC 7027 by Russell *et al.* (1977) was identified as originating from stretching and bending modes of aromatic compounds (Duley & Williams 1981). These features are not seen in AGB stars, suggesting that the carriers are either synthesized or only excited into emission after the end of the AGB phase. To understand the chemical synthesis history of these aromatic compounds, it would be useful to observe objects in transition between the AGB and PN stages. After the *IRAS* mission, an increasing number of such transition objects (often called "proto-PN" or "PPN") have been identified (Kwok 1993). *ISO* observations of these objects have found that they show the typical aromatic infrared bands (AIB), including the C−H stretch at 3.3 μm, C−C stretches at 6.2 and 7.7 μm, C−H in-plane bending at 8.6 μm, and the C-H out-of-plane bending mode at 11.3 μm. In addition, features attributed to aliphatic side chains attached to the rings, such as the 3.4 μm C-H stretches and the 6.9 μm C−H in-plane bending modes arising from the methyl ($-CH_3$) and methylene ($-CH_2$) groups attached to the aromatic rings (Geballe *et al.* 1992, Hrivnak *et al.* 2007, Figure 2). Also observed in the PPN spectra are broad emission plateau features around 8 and 12 μm, which can be interpreted as arising from

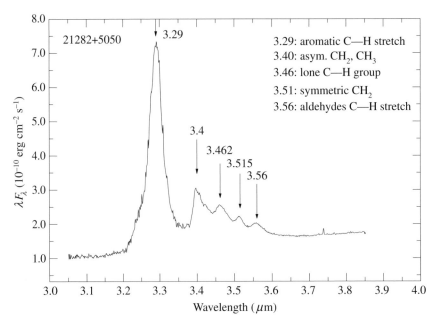

Figure 1. A spectrum of IRAS 21282+5050 taken at the Keck Telescope showing the 3.4 μm aliphatic features. The emission feature at 3.56 μm could be due to an aldehyde group attached to the aromatic rings.

the respective in-plane and out-of-plane bending modes of a mixture of aliphatic side groups (Kwok *et al.* 2001).

Another spectral distinction between PN and PPN is the presence of the other C–H out-of-plane bending modes at 12.1, 12.4, and 13.3 μm from aromatic rings with various degree of exposed edges (Witteborn *et al.* 1989, Kwok *et al.* 1999, Hony *et al.* 2001). In general, these features are seen in PPNe, and weak in PNe (Figure 2). These C–H bending modes probably arise from a mixture of aromatic units in different configurations, examples of which can be seen in Hu & Duley (2007).

The aromatic units can also undergo in-plane and out-of-plane distortion mode C–C–C vibrations which can lead to emission plateaus in the 15-20 μm region (Van Kerkhoven *et al.* 2000). Such emission plateaus are seen in the spectrum of the young PN NGC 7027 (Figure 3).

A broad ($\sim 1~\mu$m) emission feature at 15.8 μm is seen in the *Spitzer IRS* spectra of the PPNe IRAS 06530-0213 and 23304+6147 (Hrivnak *et al.*, these proceedings). Although quite prominent in these objects, this feature is not seen in PNe.

4. Nature of the carriers of the infrared bands

Unlike rotational transitions, the infrared features only tell us about the nature of the chemical bonds and the vibrational modes, but not the exact chemical structure of the carriers. All we can say is that the carrier is a carbonaceous compound with a mixture of aromatic and aliphatic structures. Example schematic of such structures can be found in Pendleton & Allamandola (2002) and Kwok *et al.* (2001). The closest natural analogs of such structure is probably kerogen, which are random arrays of aromatic rings and

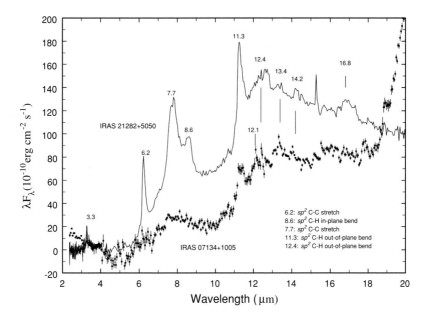

Figure 2. The *ISO SWS01* spectra of the young PN IRAS 21282+5050 and the PPN IRAS 07134+1005, showing various aromatic C-H and C-C stretching and bending modes. The PPN spectra are characterized by the 12.1, 12.4, 13.3 μm out-of-plane bending mode features from small aromatic units.

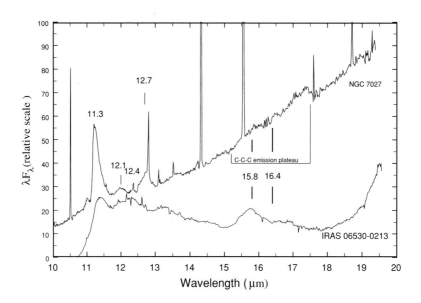

Figure 3. The *Spitzer IRS* spectra of young PN NGC 7027 and the PPN IRAS 06530−0213 showing 11.3, 12.1 μm AIB features and the unidentified 15.8 μm feature. The 15-17.5 μm emission plateau in NGC 7027 could be due to C−C−C vibrational modes. The narrow feature in the NGC 7027 are atomic lines.

aliphatic chains with functional groups made up of H, O, N, and S attached (Papoular *et al.* 1989, Papoular *et al.* 1996). In particular, the relative strengths of the 3.3 and 3.4 μm feature and the 8 and 12 μm emission plateaus seen in PPNe exhibit close resemblence to that of kerogen (Guillois *et al.* 1996). Soot, which is formed by combustion of hydrocarbon molecules in a flame, has also been suggested as a possible natural analog of circumstellar carbonaceous dust. Since the chemical composition of soot is highly dependent on the initial mix of the burning gas, its structure is too indefinite to be tested for spectral comparison.

Artificial substances created in laboratory simulations offer several possible counterparts to the circumstellar carbonaceous materials. Examples include hydrogenated amorphous carbon (HAC) produced from laser ablation of graphite in a hydrogen atmosphere (Scott *et al.* 1997, Duley, these proceedings), quenched carbonaceous composites (QCC) made from the quenching of plasma of methane (Sakata *et al.* 1987, Wada, these proceedings), and carbon-based nanoparticles created by laser pyrolysis of gas-phase C_2H_4, C_4H_6 molecules (Herlin *et al.* 1998).

We should remember that circumstellar AIB emission features lie on top of a strong infrared continuum (see, e.g., the continuum in the spectrum of NGC 7027 in Figure 3). This featureless dust continuum is likely to be the remnants of the circumstellar dust left over from the AGB stage, as evolved carbon stars such as IRC+20216, AFGL 3068 and IRAS 21318+5631 show such strong dust continuum. The carrier of this continuum has often been attributed to amorphous carbon because of its featureless properties. Less evolved AGB stars show the 11.3 μm SiC feature, but this gives away to a featureless dust emission in highly evolved carbon stars. Complex and disordered organic compounds such as kerogen and tholins also show a strong infrared continuum and it is conceivable that this continuum is provided by complex organic particles of $\sim 10^3$ C atoms. Tholins is an artificial substance made from the irradiation of gaseous mixtures of nitrogen and methane (Khare, these proceedings). Since nitrogen is a common element in circumstellar gas molecules (including cyanopolyynes), it is possible that N is incorporated into the solid component of circumstellar grains.

5. Unidentified features

In addition to the aromatic and aliphatic infrared features, AGB stars, PPNe, and PNe show other strong emission features which origins are not currently understood.

The 21-μm feature. The discovery of this very strong emission feature (Figure 4) in carbon-rich PPNe was a surprise (Kwok *et al.* 1989). Now over a dozen PPNe (but not AGB stars and PNe) have been seen to possess this feature (Hrivnak *et al.*, these proceedings). Many different candidates have been proposed as the carrier of this feature, including the recent proposals of SiC (Speck & Hofmeister 2004) and FeO (Posch *et al.* 2004).

The 30-μm feature. A broad emission feature peaking around 30 μm is commonly observed in carbon-rich AGB stars, PPNe, and PNe (Figure 5). A significant fraction of the total energy output of PPNe and PNe can be emitted through this feature (Hrivnak *et al.* 2000).

Extended red emission. The extended red emission (ERE) is a broad emission band first detected in HD 44179 (the Red Rectangle), and is also seen in PNe (e.g., NGC 7027). It is believed to arise from photoluminescence of a semiconductor. A variety of inorganic (silicon nanoparticles, Witt *et al.* 1998; nanodiamonds, Chang *et al.* 2006) and organic (HAC, Duley 1985; QCC, Sakata *et al.* 1992) solids have been suggested as possible carriers.

The exact relationship between the above features and the AIB phenomenon is not clear, although carbon-rich association is assumed to be a common factor. All 21-μm

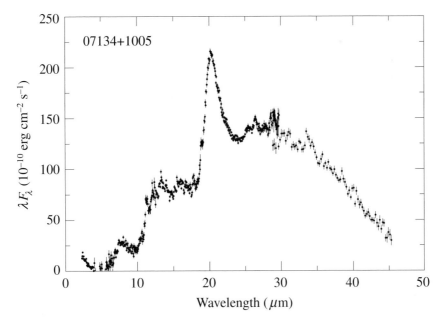

Figure 4. *ISO* SWS spectrum of the PPN IRAS 07134+1005 showing the strong unidentified 21 μm emission feature.

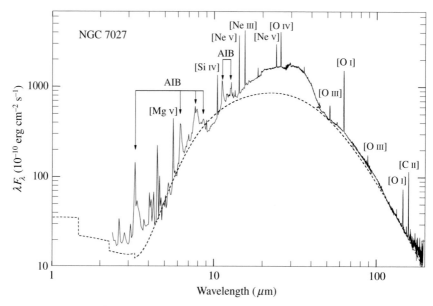

Figure 5. The *ISO SWS/LWS* spectrum of NGC 7027 between 1-200 μm. Some of the stronger atomic lines and AIB features are marked. The dashed line is a model fit composed of a sum of dust emission, *f-b* and *f-f* gas emission. The strong 30 μm feature can be clearly seen above the dust continuum.

sources show the 30 μm feature, but not vice versa. The ERE, although widely seen in the diffuse interstellar medium, is not commonly seen in evolved stars. This can be attributed to the need for an UV background for photoluminescence. Our difficulty in the identification of these features can be traced to our lack of understanding of spectral behavior of nanoparticles, as the mobility of electrons in a small cluster is reduced in a nanoparticle in comparison to bulk materials. Because of boundary effects and the resulting loss in symmetry, some vibrational modes can be more infrared active than in large crystals of the same chemical composition (Li 2004).

6. Chemical evolution and the role of photochemistry

The detection of organic compounds in the ejecta of evolved stars gives us important information on how these species are formed. The evolution from AGB to PPN to PN is very short, and this gives us precise knowledge on the time scale of chemical synthesis. Since the AIB features first emerge in the PPN phase, what are the steps leading to the formation of ring molecules? Acetylene (C_2H_2), believed to be the first building block of benzene, is commonly detected in evolved carbon stars through its ν_5 fundamental band at 13.7 μm. Polymerization of C_2H_2 leads to the formation of diacetylene (C_4H_2) and triacetylene (C_6H_2) in PPN, cumulating in the formation of benzene (Cernicharo *et al.*, 2001). The weakening of the 3.4 and 6.9 μm from PPNe to PNe suggests a change from aliphatic to aromatic structures. This could be the result of photochemistry where the onset of UV radiation modify the aliphatic side groups through isomerizations, bond migrations, cyclization and ring closures and transform them into ring systems (Kwok *et al.* 2001). Hydrogen loss can also result in fully aromatic rings that are more stable than alkanes or alkenes. Evidence for such H loss can also be found in the weakening of the 12.1, 12.4, and 13.3-μm features and the strengthening of the 11.3-μm feature from PPNe to PNe. The weakening of the aliphatic component due to photo-processing is also supported by an observed correlation between the central wavelengths of the 7.7 and 11.3 μm features with the temperature of the central star (Sloan, these proceedings).

7. Connection with the Solar System

The connection between inorganic solids produced in AGB stars and the Solar System was established by the discovery of presolar grains in meteorites (Zinner, these proceedings; Nittler, these proceedings). It is clear that SiC and other refractory oxide grains produced by AGB stars have been able to find their way across the Galaxy and be incorporated into the constituents of the early Solar System. The 3.4 μm aliphatic feature seen in PPNe are also seen in the spectra of interplanetary dust and comets (Flynn *et al.* 2003). The insoluble organic matter (IOM) in meteorites, comets, and interplanetary dust is mostly macromolecular and is similar to kerogen in structure (Kerridge 1999, Cruikshank, these proceedings). The D/H and $^{15}N/^{14}N$ ratios seen in IOM are closer to interstellar values than Solar System values (Messenger 2000, Alexander, these proceedings). Is it possible that IOM is also stellar in origin?

In Section 4, we have noted the spectral similarities between the infrared spectra of PPN and kerogen. Terrestrial kerogen is biological in origin and produced by prolonged geological processing. If the carrier of the AIB features in PPNe is indeed similar to kerogen, then chemical synthesis of such complex organic compound is much more efficient in the circumstellar environment, which accomplishes the synthesis in thousands of years under extremely low density conditions.

8. Conclusions

We have shown that stars in the late stages of evolution are prolific producers of complex organic compounds. Through stellar winds, these organic materials are ejected to the interstellar medium and spread throughout the Galaxy (Kwok 2004). Although we are uncertain how much of these materials are destroyed or further processed in the interstellar medium, we do know at least some of the inorganic grains made to the early Solar System and were incorporated in meteorites. It is possible that stellar organic matter also contributed to the organic content in asteroids, comets, meteorites, and interplanetary dust particles.

The unidentified emission features seen in evolved stars may provide valuable clues to possible new organic compounds that can be synthesized under unusual physical conditions. These materials have the potential of broadening our understanding of organic chemistry beyond the terrestrial environment.

Acknowledgements

The work was supported by a grant from the Research Grants Council of the Hong Kong Special Administrative Region, China (Project No. HKU 7028/07P).

References

Cernicharo, J., Heras, A. M., Tielens, A. G. G. M., Pardo, J. R., Herpin, F., Guélin, M., & Waters, L. B. F. M. 2001, *ApJ* (Letters), 546, L123
Chang, H.-C., Chen, K., & Kwok, S. 2006, *ApJ* (Letters), 639, L63
Duley, W. W. 1985, *MNRAS*, 230, 1P
Duley, W. W. & Williams, D. A. 1981, *MNRAS*, 196, 269
Flynn, G. J., Keller, L. P., Feser, M., Wirick, S., & Jacobsen, C. 2003, *Geochim. Cosmochim. Acta*, 67, 4791

Geballe, T. R., Tielens, A. G. G. M., Kwok, S., & Hrivnak, B. J. 1992 *ApJ* (Letters), 387, L89

Guillois, O., Nenner, I., Papoular, R., & Reynaud, C. 1996, *ApJ*, 464, 810

Henning, Th. (ed.), 2003, Astromineralogy (Springer)

Herlin, N., Bohn, I., Reynaud, C., Cauchetier, M., Galvez, A., & Rouzaud, J.-N. 1998, *A&A*, 330, 1127

Hony, S. *et al.* 2001, *A&A*, 370, 1030

Hrivnak, B. J., Volk, K., & Kwok, S. 2000, *ApJ*, 535, 275

Hrivnak, B. J., Geballe, T. R., & Kwok, S. 2007, *ApJ*, 662, 1059

Hu, A. & Duley, W. W. 2007, *ApJ* (Letters), 660, L137

Jäger, C. *et al.* 1998, *A&A*, 339, 904

Kerridge, J. F. 1999, *Space Sc. Rev.*, 90, 275

Kwok, S. 1993, *Ann. Rev. Astr. Ap.*, 31, 63

Kwok, S. 2004, *Nature*, 430, 985

Kwok, S., Volk, K., & Bernath, P. 2001, *ApJ* (Letters), 554, L87

Kwok, S., Volk, K., & Bidelman, W. P. 1997, *ApJS*, 112, 557

Kwok, S., Volk, K., & Hrivnak, B. J. 1989, *ApJ* (Letters), 345, L51

Kwok, S., Volk, K., & Hrivnak, B. J. 1999, *A&A* (Letters), 350, L35

Li, A. 2004, in: A. N. Witt, G. C. Clayton, & B. T. Draine (eds.), *Astrophysics of Dust*, Vol. 309, p. 417

Messenger, S. 2000, *Nature*, 404, 968

Papoular, R., Conard, J., Giuliano, M., Kister, J., & Mille, G. 1989, *A&A*, 217, 204

Papoular, R., Conard, J., Guillois, O., Nenner, I., Reynaud, C., & Rouzaud, J.-N. 1996, *A&A*, 315, 222

Pendleton, Y. J. & Allamandola, L. J. 2002, *ApJS*, 138, 75

Posch, Th., Mutschke, H., & Andersen, A. 2004, *ApJ*, 616, 1167

Posch, T., Kerschbaum, F., Mutschke, H., Fabian, D., Dorschner, J. & Hron, J. 1999, *A&A*, 352, 609

Russell, R. W., Soifer, B. T., & Willner, S. P. 1977, *ApJ* (Letters), 217, L149

Sakata, A., Wada, S., Onaka, T., & Tokunaga A.T. 1987, *ApJ* (Letters), 320, L63

Sakata, A., Wada, S., Narisawa, T. *et al.* 1992, *ApJ* (Letters), 393, L83

Scott, A. D., Duley, W. W., & Jahani, H. R. 1997, *ApJ* (Letters), 490, L175

Speck, A. K. & Hofmeister, A. N. 2004, *ApJ*, 600, 986

Van Kerckhoven, C., *et al.* 2000, *A&A*, 357, 1013

Witt, A. NJ., Gordon, K. D., & Furton, D. G. 1998, *ApJ* (Letters), 501, L111

Woolf, N. J. & Ney, E. P. 1969. *ApJ* (Letters), 155, L181

Discussion

PENDLETON: Do you think that stars made aliphatic molecules first and then they become aromatic? Some people said that it might go the other way that you might have aromatic molecules and they break apart to shorter chains.

KWOK: The first organic compounds that are synthesized are probably very amorphous, dirty, and complicated. Benzene is probably the first to form, with further rings gradually added. In the process, all kinds of side chains got attached to these basic aromatic units. This is just not a clean chemistry. By the time that the peripheral materials got cleaned up by photo processes, the aromatic clusters then link together and become larger and larger.

ZIURYS: What fraction of these organic compounds do you think contain atoms other than carbon and hydrogen, e. g., oxygen and nitrogen?

KWOK: One thing I am certain is that these compounds will very likely contain impurities. For example, one carbon atoms in the aromatic rings may be substituted by nitrogen. This will result in a change in the frequency of the vibrational modes. However, at present we do not have the sensitivity nor spectral resolution to have confirmed detection of such species.

ATHENA: Can you say a little bit about the interpretation of the ERE?

KWOK: The ERE was discovered in the Red Rectangle and has been seen in planetary nebulae, for example NGC 7027. They have been suggested to originate from HAC, QQC, C_{60}, silicon nanoparticles, etc., as well as from nanodiamonds as we proposed in a paper in 2006. The only thing we are certain is that it is a photoluminescence phenomenon from a semiconductor, most likely a carbonaceous one. ERE may be one of the manifestation of carbonaceous organic compounds that we are discussing here.

SANDFORD: I would like to make a comment about the issue of aliphatic versus aromatic. Most of the discussions on aliphatics have been about side groups hanging off the edges of aromatic domains, but in fact there is the additional possibility that the aromatic grains can have extra hydrogen atoms on some of the peripheral carbons. In this case, you have a hybrid molecule which spectrum looks like aromatic plus aliphatic. With UV photolysis you don't have to build new rings – you just need to knock a hydrogen off one of the aliphatic rings and it starts to become more aromatic.

MATHEWS: Going back to the question about the aromatic and aliphatic being together, I think what we have are HCN polymers and they are not as easy to identify as aromatics. Since we have ammonia, methane and nitrogen which can give rise to HCN and acetylene. Acetylene will give you hydrocarbons and HCN will give you HCN polymers. I predict you always find the two together. I am glad you emphasize that further observational techniques are necessary before we actually see the HCN polymers in the circumstellar envelopes.

KWOK: I think the presence of HCN polymers is very likely. However, my understanding is that they don't have strong spectroscopic signatures which make astronomical identification difficult. They may be widely present but we can't be sure that they are there.

HRIVNAK: We did a study to look at the strength of the 3.4 μm feature compared to the 3.3 μm feature and in general the 3.4 gets weaker as you go from PPNe to PNe. But there are some counterexamples.

KWOK: Qualitatively we know when a star evolves from low temperature to high temperature which corresponds to the changes in spectral type. But in practice since the progenitors of planetary nebulae have a variety of central stars mass they evolve at highly different rates, there may not be a strict one-to-one correspondence between chemcal evolution and spectral types.

Sun Kwok answering questions after his talk (photo by Sze-Leung Cheung).

Organic Matter in Space
Proceedings IAU Symposium No. 251, 2008
S. Kwok & S. Sandford, eds.

© 2008 International Astronomical Union
doi:10.1017/S1743921308021522

A Spitzer Space Telescope study of dust features in planetary nebula and HII regions

Jeronimo Bernard-Salas[1], **Els Peeters**[2,3], **Vianney Lebouteiller**[1]
Gregory C. Sloan[1], **Bernhard R. Brandl**[4], and **James R. Houck**[1]

[1] Cornell University, Space Sciences Building
Ithaca, NY 14853-6801, USA
email: jbs@isc.astro.cornell.edu

[2] University of Western Ontario, PAB 213, Canada

[3] SETI Institute, Mountain View, CA 94043, USA

[4] Leiden University
P.O. Box 9513, 2300 RA Leiden, The Netherlands

Abstract. One of the key questions of infrared astronomy is how the characteristics of dust depend on the physical properties of the surrounding medium. To address this question, we present results from the Spitzer Space Telescope on two projects designed to study the dust properties of a sample of 25 Planetary Nebulae (PNe) in the Magellanic Clouds, and three well-known Giant HII regions (NGC 3603, 30 Doradus and N 66/NGC 346). Most PNe show emission from polycyclic aromatic hydrocarbons (PAHs) and only two of them show amorphous silicates. Eleven PNe display a strong broad feature around 11 μm which is attributed to silicon carbide and 8 of them show magnesium sulfide. One PNe, SMP LMC 11, shows spectacular absorption bands due to molecules which are the precursors from which more complex hydrocarbons are formed. The Spitzer spectra of the HII regions, NGC 3603, 30 Doradus, and NGC 346 are very rich, displaying a wealth of spectral features within each region. This not only allows us to compare the dust at different metallicities but also to study the spatial variations of many features across a given region and correlate it with the distance to the ionizing cluster(s) and other parameters.

Keywords. Infrared: general, dust, Magellanic Clouds, planetary nebulae: general, HII regions, ISM: individual (30 Doradus), ISM: lines and bands

1. Introduction

The planetary nebula phase is one of the latest phases of evolution of intermediate mass stars. Formation of dust occurs during the transition between the asymptotic giant branch (AGB) and the PN phase. Besides the strong fine-structure emission lines characteristic of PNe, the infrared spectra of PNe display a variety of dust features, which provide information on how dust is formed. In contrast, the complex structures of giant HII regions give the opportunity to study the dust in different physical environments (distance to central cluster, shocks, etc...). In addition, it is thought that metallicity has an impact on some dust properties (Galliano *et al.* 2008). With this in mind, we have observed a sample of PNe in the Magellanic Clouds (MCs), and 3 giant HII regions in the Milky Way (MW), LMC, and SMC using the Spitzer Space Telescope (Werner *et al.* 2004). For the PNe project we observed 18 LMC PNe and 7 PNe in the SMC using the low-resolution modules (SL, LL, 5.2–37 μm) of the Infrared Spectrograph (IRS, Houck *et al.* 2004). In the HII region program we used the SL module (5.2–14 μm) to observed the 3 largest HII

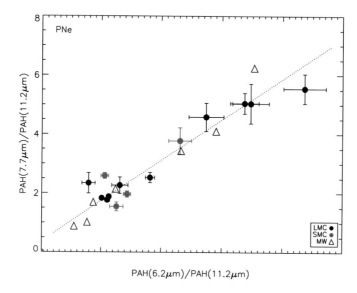

Figure 1. PAH ratio for the PNe. The dotted line is the best fit to the data.

regions in each of the following galaxies: the MW (NGC3603), the LMC (30 Doradus), and the SMC (N66).

2. Magellanic Clouds PNe

We have observed 25 PNe in the MCs, 17 in the LMC and 8 in the SMC. Gas phase abundances have been derived by Bernard-Salas *et al.* (2008b) and the dust properties will be presented in a complementary paper (Bernard-Salas *et al.* 2008a). Polycyclic Aromatic Hydrocarbons (PAHs), typical of carbon-rich environments, are detected in 14 of the PNe. In addition to the 6.2, 7.7, 8.6 and 11.2 μm PAH bands, in some objects with better S/N weaker bands such as the 12.7, 16.4 and 17.4 μm bands are detected.

Ionized PAHs emit more in the 6.2 and 7.7 μm bands while the 11.2 μm band is stronger in neutral PAHs, therefore the ratio of the 6.2 or 7.7 μm PAH over the 11.2 μm feature can be used as a measure of the ionization fraction of the PAHs. This is shown in Figure 1 and it can be seen that there is a good correlation between these band ratios. For comparison several Galactic PNe are also plotted in Figure 1. It is interesting to note that the range of band ratios is independent of metallicity. Given that at low metallicities the hardness of the radiation field is stronger, one could have expected that the ratios of the LMC and SMC PNe would be higher (more ionised). The correlation in Figure 1 is the same of that seen in HII regions where Galliano *et al.* (2008) conclude that the ionization fraction is responsible for their trend.

PAHs are known to vary from object to object and also within a class of objects. Peeters *et al.* (2002) proposed a classification scheme of the PAHs according to their peak position and profile. Within this classification most Galactic PNe (studied by ISO) were class B and a couple of them class A. The resolution of the SL and LL modules in the IRS is low (60–120). This makes it difficult to study variations in detail, but we can conclude that most MCs PNe are class B with a couple of them probably belonging to class A. Summarizing, the ratios of MCs PNe PAH fluxes and PAH profiles are similar to their Galactic counterparts, and independent on metallicity (at least within the metallicity range that we are probing).

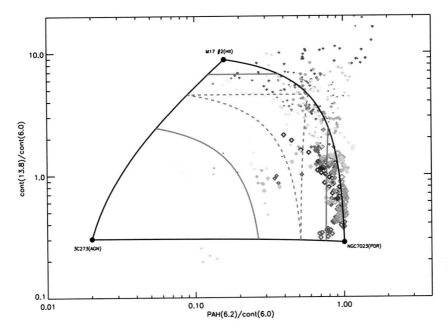

Figure 2. Laurent diagram for the different positions in 30 Doradus. The black solid lines anchor the templates for AGN, Hɪɪ, and PDR dominated spectra. The grey lines indicate the contribution in percentage of each system (see Lebouteiller *et al.* 2007 and Bernard-Salas *et al.* 2008b for details).

SMP LMC 11 shows a broad absorption band in the 12-17 μm region in which narrower absorptions features are superimposed (like those of AFGL 618, Cernicharo *et al.* 2001), and the IR-spectrum shows that this object is in the pre-planetary nebula phase. These absorptions features correspond to acetylene, poly-acetylenic chains, benzene, and some yet un-identified features (Bernard-Salas *et al.* 2006), and constitute the basic molecules from which PAHs are formed. SMC SMC 24 may show hydrogenated hydrocarbons (HACs) and low excitation lines and must be a young PN. According to Sloan *et al.* (2007), and Sloan this volume further processing of these HACs will eventually expose the PAHs.

Of the sample of PNe showing PAHs, eight of them show silicon carbide (SiC) and eleven magnesium sulfide (MgS). These features are common of carbon-rich environments. MgS is also seen in MW PNe, but the presence of SiC in 8 of the 24 PNe in the MCs (or out of 14 carbon-rich PNe) is not expected. This feature is seen in the AGB phase but it is rarely seen in Galactic PNe: only 2 MW PNe (maybe 3) show SiC. The reason for this remains unclear but it seems that the presence of this feature in the PN phase is metallicity dependent, only occurring at low metallicities. Additionally, the strength of the feature decreases with the hardness of the radiation field, either it is destroyed or processed.

Only two of the PN in the sample show amorphous silicates which is typical of oxygen-rich dust. The low proportion of oxygen-rich PNe compared to the Galactic PNe is expected and it is linked to the higher efficiency of the third dredge-up (which brings out carbon) at low metallicities.

3. Giant Hɪɪ regions

NGC3603, 30 Doradus, and N66 are the largest Hɪɪ regions in the MW, the LMC, and SMC, respectively. Results on NGC3603 are presented in Lebouteiller *et al.* (2007), and those of 30 Doradus and N66 by Bernard-Salas *et al.* (2008c) and Whelan *et al.* (2008), respectively. A comparison of the properties of the 3 objects is given in Lebouteiller *et al.* (2008).

For each object several (9 to 17) positions were observed. These were centered in different locations to represent different conditions of the gas (red knots, shock fronts, stellar cluster etc.). The variety of spectra is striking, and goes from positions dominated by PAHs (e.g., red nebulosities), to fine-structure line emission with weak PAHs (e.g., close to the central cluster), to only line emission dominated spectra with no PAHs (shock front where the PAHs have been destroyed), and spectra with strong silicates absorption (embedded young star).

Taking advantage of the long slit in SL we can obtain spatial information along all the positions which were observed, this results in hundreds of spectra for each HII region. Figure 2 displays a Laurent diagram for these positions within 30 Doradus which traces very well the photo-dissociation region (PDR) and HII region interface. The ratio of ionized to neutral PAHs shows little variation with the distance to the (main) stellar cluster in NGC3603 and 30 Doradus. This indicates that the PAH mixture is very constant over the whole region.

References

Bernard-Salas, J., Peeters, E., Sloan, G. C., *et al.* 2006, *ApJL*, 652, 29
Bernard-Salas, J., Peeters, E., Sloan, G. C., *et al.* 2008a, in prep
Bernard-Salas, J., Pottasch, S. R., Gutenkunst, S., *et al.* 2008b, *ApJ*, 672, 274
Bernard-Salas, J., *et al.* 2008c, in prep
Cernicharo, J., Heras, A. M., Tielens, A. G. G. M., *et al.* 2001, *ApJ*, 546, L123
Galliano, F., Madden, S. C., Tielens, A. G. G. M., Peeters, E., & Jones, A. P. 2008, *ApJ*, in press
Houck, J. R., Roellig, T. L., van Cleve, J., *et al.*, 2004, *ApJS*, 154, 18
Laurent, O., Mirable, I. F., Charmandaris, V., *et al.* 2000, *A&A*, 359, 887
Lebouteiller, V., *et al.* 2007, *ApJ*, 655, 390
Lebouteiller, V., *et al.* 2008, *ApJ*, in prep
Peeters, E., Van Kerckhoven, C., Tielens, A. G. G. M., *et al.* 2002, *A&A*, 390, 1089
Sloan, G. C., Duley, W. W., Kraemer, K. E., *et al.* 2007, *ApJ*, 664, 1144
Werner, M. W., Roellig, T. L., Low, F. J., *et al.*, 2004, *ApJS*, 154, 1
Whelan, D., *et al.* 2008, in prep

Discussion

MULAS: A comment about the band ratios that you use to determine ionization of PAHs; while is it true that ionization has a large effect, it is not the only parameter. It is a multidimensional space. Especially the ratio of the 11.3 μm PAH band with shorter wavelength bands. The 11.3 μm band is not so weak in ionized big PAHs, so you should be a bit careful with the interpretation.

BERNARD-SALAS: Indeed, there are other processes besides ionization that could a priori explain the correlation we see in the PAH ratios, the main ones being dehydrogenation, extinction and to a lesser extent molecular size and structure. Dehydrogenation is only relevant for molecules with less than 25 C-atoms, and the PAHs we believe are present in space are larger than that. We can rule out extinction due to silicates which affects the 11.3 μm band because these PNe have no silicates. Additionally, in the 30 Dor spectra the positions with silicate absorption and the ones without it show the same trend. Also, in a recent study by Galliano *et al.* (2008) they study the same PAH ratio for a great variety of sources. Their ratios differ by a factor of 10 and they conclude that only ionization can produce such a large effect. The PNe we present here have ratios differing by a factor of 3 but they fall right in the trend seen by Galliano and collaborators, so while we cannot rule out other options it is most likely that ionization drives the relation that is shown. Concerning your last comment, Bauschlicher *et al.* (2008) found that PAH cations in large PAHs do not have a major contribution to the 11.3 μm band, but I will be very interested if this is not the case.

From left to right: Mikako Matsurra, Greg Sloan, Angela Speck

Angela Speck talking about pre-solar grains.

Organic Matter in Space
Proceedings IAU Symposium No. 251, 2008
S. Kwok & S. Sandford, eds.

Spitzer spectroscopy of unusual hydrocarbons in cool radiative environments

G. C. Sloan

Cornell University, Space Sciences 108, Ithaca, NY 14853-6801
email: `sloan@isc.astro.cornell.edu`

Abstract. The *Spitzer Space Telescope* has discovered several objects with unusual spectra, where the emission features from polycyclic aromatic hydrocarbons (PAHs) are shifted to longer wavelengths than normally observed. Previously, only two of these class C PAH spectra had been identified. The new and larger sample reveals that PAHs emit at longer wavelengths when processed by cooler radiation fields. Limited laboratory data show that samples with mixtures of aromatic and aliphatic hydrocarbons produce emission features at longer wavelengths than purely aromatic samples. The aliphatic bonds are more fragile and would only survive in cooler radiation fields. In harsher radiation fields, the aliphatics attached to the aromatic hydrocarbons are destroyed.

Keywords. Infrared: ISM, (ISM:) dust, extinction

1. Introduction

Peeters *et al.* (2002) studied 57 spectra with emission features from polycyclic aromatic hydrocarbons (PAHs) observed with the Short Wavelength Spectrometer on the *Infrared Space Observatory*. They classified the majority as class A or B, depending on whether the 7–9 μm PAH complex peaked at 7.6 or 7.8 μm, respectively. The spectra of two sources, AFGL 2688 and IRAS 13416−6243, both in transition from the asymptotic giant branch (AGB) to planetary nebula (PN), showed a peak at longer wavelengths, ∼8.2 μm. They were the only two class C PAH spectra in the sample.

2. New spectra

The Infrared Spectrograph (IRS, Houck *et al.* 2004) on the *Spitzer Space Telescope* (Werner *et al.* 2004) has added several more sources to the sample of unusual class C PAH spectra. Sloan *et al.* (2005) studied four Herbig AeBe (HAeBe) stars with emission from PAHs, but not silicates. They noticed that the spectra were intermediate between class B and C, with the 7.6–7.8 μm feature shifted to 7.9–8.0 μm. The coolest HAeBe star in their sample, HD 135344, had the most redshifted feature.

Peeters *et al.* (2002) included isolated HAeBe stars in class B, but they are generally distinct from the rest of the class; we will refer to them as class "B′" here, with the exception of HD 135344, which is class "B/C". Figure 1 illustrates these different classes of spectra.

Another cool object, HD 233517, a K2 giant associated with an infrared excess, also shows a class C PAH spectrum (Jura *et al.* 2006), as does MSX SMC 029, a post-AGB object in the Small Magellanic Cloud (Kraemer *et al.* 2006). The intermediate-mass T Tauri star SU Aur also shows class C PAHs in its spectrum (Furlan *et al.* 2006).

HD 100764 is the second cool red giant with PAHs detected in its spectrum. Sloan *et al.* (2007) examined this spectrum, along with all other known class C PAH spectra. Instead of measuring the peak wavelength of the PAH features, they measured the *central* wavelength, defined to be the wavelength with half the emitted flux to either side. The

It follows that the scatter apparent in Figure 2 may not be statistical but may indicate the relative degrees of processing history in the different environments sampled.

5. A Final Note

The introduction of the PAH model to explain the unidentified infrared (UIR) emission features by Leger & Puget (1984) and Allamandola *et al.* (1985) led to a great deal of controversy that took more than a decade to sort out. Many other models have competed with the PAH model, including hydrogenated amorphous carbon (HAC, Duley & Williams 1979), quenched carbonaceous composite (QCC, Sakata *et al.* 1984), and even coal-like material (Papoular *et al.* 1989).

The community may by reaching a consensus that the UIR features as typically seen in PDRs arise from relatively simple PAHs (e.g., Allamandola *et al.* 1999, Hony et al. 2001, Peeters *et al.* 2002, Li & Draine 2002). However, most of the alternatives to the simple PAH model represent similar mixtures of hydrocarbons with aromatic and aliphatic bonds (e.g., Jones *et al.* 1990, Pendleton & Allamandola 2002, Kwok 2004). Evidence is mounting that in radiative environments less harsh than the typical PDR, it is these mixtures that we are observing.

References

Allamandola, L. J., Hudgins, D. M., & Sandford, S. A. 1999, *ApJ* (Letter), 511, L115

Allamandola, L. J., Tielens, A. G. G. M., & Barker, J. R. 1985, *ApJ* (Letter), 290, L25

Duley, W. W. & Williams, D. A. 1979, *Nature*, 277, 40

Furlan, E., *et al.* 2006, *ApJS*, 165, 568

Geballe, T. R., Joblin, C., D'Hendecourt, L. B., Jourdain de Muizon, M., Tielens, A. G. G. M., & Leger, A. 1994, *ApJ* (Letter), 434, L15

Geballe, T. R., Tielens, A. G. G. M., Allamandola, L. J., Moorhouse, A., & Brand, P. W. J. L. 1989, *ApJ*, 341, 278

Geballe, T. R., Tielens, A. G. G. M., Kwok, S., & Hrivnak, B. J. 1992, *ApJ* (Letter), 387, L89

Geballe, T. R., & van der Veen, W. E. C. J. 1990, *A&A*, 235, 9

Hony, S., Van Kerckhoven, C., Peeters, E., Tielens, A. G. G. M., Hudgins, D. M., Allamandola, L. J. 2001 *A&A*, 370, 1030

Houck, J. R., *et al.* 2004, *ApJS*, 154, 18

Jones, A. P., Duley, W. W., Williams, D. A. 1990, *QJRAS*, 31, 567

Jura, M., *et al.* 2006, *ApJ* (Letter), 637, L45

Keller, L. D., *et al.* 2008 *ApJ*, in press

Kraemer, K. E., Sloan, G. C., Bernard-Salas, J., Price, S. D., Egan, M. P., & Wood, P. R. 2006, *ApJ* (Letter), 652, L25

Kwok, S. 2004, *Nature*, 430, 985

Leger, A. & Puget, J. L. 1984, *A&A* (Letter), 137, L5

Li, A. & Draine, B. T. 2002, *ApJ*, 572, 232

Papoular, R., Conrad, J., Giuliano, M., Kister, J., & Mille, G. 1989, *A&A*, 217, 204

Peeters, E., Hony, S., Van Kerckhoven, C., Tielens, A. G. G. M., Allamandola, L. J., Hudgins, D. M., & Bauschlicher, C. W. 2002, *A&A*, 390, 1089

Pendleton, Y. J. & Allamandola, L. J. 2002, *ApJS*, 138, 75

Sakata, A., Wada, S., Tanabe, T., & Onaka, T. 1984, *ApJ* (Letter), 287, L51

Sloan, G. C., Bregman, J. D., Geballe, T. R., Allamandola, L. J., & Woodward, C. E. 1997, *ApJ*, 474, 735

Sloan, G. C., *et al.* 2005, *ApJ*, 632, 956

Sloan, G. C., *et al.* 2007, *ApJ*, 664, 1144

Werner, M. W., *et al.* 2004, *ApJS*, 154, 1

Organic Matter in Space
Proceedings IAU Symposium No. 251, 2008
S. Kwok & S. Sandford, eds.

Unidentified infrared bands and the formation of PAHs around carbon stars

Angela Speck[1], Mike Barlow[2], Roger Wesson[2], Geoff Clayton[3], and Kevin Volk[4]

[1] Physics & Astronomy Department, University of Missouri- Columbia
email: speckan@missouri.edu
[2] Physics & Astronomy Department, University College London
[3] Dept. of Physics & Astronomy, Louisiana State University
[4] Gemini Observatory

Abstract. Although unidentified infrared bands (UIBs) have been observed in many astrophysical environments, there is one notable exception: carbon (C) stars. Only a handful of C stars have been shown to emit UIBs and most have hot companions. This makes C stars with hot companions an ideal location to investigate the emitters of the UIBs. PAHs are excited by absorption of single photons whose energy is then distributed over the whole molecule. These molecules then emit the energy at the characteristic wavelengths, but the precise wavelengths and strength ratios depend on the size, composition and charge state of the individual PAHs. Furthermore, the wavelength of photons needed to excite PAHs depends on their size and charge state. While small PAHs undoubtedly need higher energy (UV) photons, it has been suggested that large or ionized PAHS (>100 C atoms) can be excited by visible or even near-IR photons. The lack of PAH emission from single carbon stars suggests that either PAHs do not form around C stars or that only small neutral grains form, which cannot be excited by a C star's radiation field.

There are two competing formation mechanisms for PAHs around C stars: (1) "bottom-up" where acetylene molecules react to form aromatic rings, building up to PAHs; or (2) "top-down", where small carbon grains react with H atoms and desorb PAHs

Using spatially resolved spectroscopic observations from Gemini/Michelle, of five carbon stars with hot companions, we investigate the circumstance under which PAH emission occurs and try to discriminate between formation mechanisms.

Discussion

ZINNER: Average carbon-rich AGB stars produce silicon carbide, not graphite. If we take a look at isotropic distribution you, find in silicon carbide grains in meteorites have ^{12}C to ^{13}C ratios of between 40 and 80, and this agrees with what you see in average carbon stars. But the isotropic composition of the meteoritic graphite grains which are from AGB stars have much higher ^{12}C to ^{13}C ratios, up to 300, 500 or so. The reason is that together with the high ^{12}C to ^{13}C ratio you also have to have high C/O ratio. So I think these graphite grains are preferentially formed from AGB stars with a very high C/O ratio.

SPECK: I think this fits in with what I was showing about extreme carbon stars. Regardless of the C/O ratio, because of the high pressure and the mass loss rates, you can form graphite easily there. Since extreme carbon stars are optically obscured, the measurement of C/O ratio is hard.

ALAXANDER: In order to form benzene and PAHs in circumstellar envelopes by acetylene addition, you have to make them under a narrow temperature range about 900 to 1000 K and it takes 10^6 years which is much too long for AGB evolution.

SPECK: Something that I am slightly disturbed about is the issue of the temperature. It does seem that there is a peak in that zone around 1000 K which suggests that the reactions are happening in stars of low mass loss rates. I think that the work Isabelle did 15 years ago may be under-estimating the formation rate because some of the conditions they chose were not really appropriate to carbon stars.

ZIURYS: You see these UIR PAH bands in young planetary nebula, but do you see them in old planetary nebulae, like the Helix?

SPECK: The problem with the Helix is that you're running into other issues such as clumping. I don't know the answer to the question in terms of exactly at what age the UIR bands start to diminish.

HENNING: In the paper by Goto, she finds that in carbon-rich post-AGB stars, the extent of the aliphatic zone is decreasing as you go to later stages of evolution. So spatial information is available through adaptive optics observations, at least for this phase of stellar evolution.

SPECK: I think by looking at the range of stars where they have a similar radiation field but naturally different separations, we can really start to get a feeling for the conditions on how these things form and evolve.

HENNING: Is there any evidence with 3.4 micron absorption feature in some of the carbon stars or AGB stars? At some point you would expect to see at least some of the objects with absorption features.

SPECK: I would say 'no' even in the extreme carbon stars where we have got ISO spectra and should be able to detect any absorption feature. I don't think we see anything that suggests aliphatic stuff. We see acetylene and HCN, that's it.

Organic Matter in Space
Proceedings IAU Symposium No. 251, 2008
S. Kwok & S. Sandford, eds.

Carbon-rich AGB stars in our Galaxy and nearby galaxies as possible sources of PAHs

M. Matsuura[1], G.C. Sloan[2], J. Bernard-Salas[2], A. A. Zijlstra[3],
P. R. Wood[4], P. A. Whitelock[5,6,7], J. W. Menzies[5], M. Feast[6],
E. Lagadec[3], M. A. T. Groenewegen[8],
M. R. Cioni[9], J. Th. van Loon[10], and G. Harris[11]

[1]National Astronomical Observatory of Japan, Osawa 2-21-1, Mitaka, Tokyo 181-8588, Japan
email: m.matsuura@nao.ac.jp

[2]Astronomy Department, Cornell University, 610 Space Sciences Building, Ithaca, NY
14853-6801, USA

[3]Jodrell Bank Centre for Astrophysics, School of Physics and Astronomy, The University of
Manchester, Oxford Street, Manchester M13 9PL, UK

[4]Research School of Astronomy & Astrophysics, Mount Stromlo Observatory, Australian
National University, Cotter Road, Weston ACT 2611, Australia

[5]South African Astronomical Observatory, P.O. Box 9, 7935 Observatory, South Africa

[6]Astronomy Department, University of Cape Town, 7701 Rondebosch, South Africa

[7]NASSP, Department of Mathematics and Applied Mathematics, University of Cape Town,
7701 Rondebosch, South Africa

[8]Instituut voor Sterrenkunde, KU Leuven, Celestijnenlaan 200B, 3001 Leuven, Belgium

[9]Centre for Astrophysics Research, University of Hertfordshire, Hatfield AL10 9AB, UK

[10]Astrophysics Group, School of Physical and Geographical Sciences, Keele University,
Staffordshire ST5 5BG, UK

[11]Department of Physics and Astronomy, University College London, Gower Street, London
WC1E 6BT, UK

Abstract. We have obtained infrared spectra of carbon-rich AGB stars in three nearby galaxies
– the Large and Small Magellanic Clouds, and the Fornax dwarf spheroidal galaxy. Our primary
aim is to investigate gas compositions and mass-loss rate of these stars as a function of metallicity,
by comparing AGB stars in several galaxies with different metallicities. C_2H_2 are detectable
from AGB stars, and possibly PAHs are subsequently formed from C_2H_2. Thus, it is worth
investigating chemical processes at low metallicity. These stars were observed using the Infrared
Spectrometer (IRS) onboard the *Spitzer Space Telescope* which covers 5–35 μm region, and the
Infrared Spectrometer And Array Camera (ISAAC) on the Very Large Telescope which covers the
2.9–4.1 μm region. HCN, CH and C_2H_2 molecular bands, as well as SiC and MgS dust features
are identified in the spectra. The equivalent width of C_2H_2 molecular bands is larger at lower
metallicity, thus PAHs might be abundant in AGB stars at low metallicity. We find no evidence
that mass-loss rates depend on metallicity. Chemistry of carbon stars is affected by carbon
production during the AGB phase rather than the metallicities. We argue that lower detection
rate of PAHs from the interstellar medium of lower metal galaxies is caused by destruction of
PAHs in the ISM by stronger UV radiation field.

Keywords. Stars: AGB and post-AGB, ISM: molecules, stars: late-type, (stars:) circumstellar
matter, (galaxies:) Magellanic Clouds, (galaxies:) Local Group

197

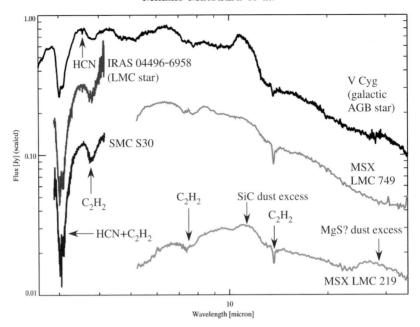

Figure 1. Infrared spectra of Galactic, LMC and SMC carbon-rich AGB stars

1. Introduction

Most polycyclic aromatic hydrocarbons (PAHs) are probably formed in carbon-rich asymptotic giant branch (AGB) stars (Allamandola *et al.* 1989). However, PAHs are undetected from AGB stars, because of the lack of UV radiation from these low temperature stars (effective temperature of 2000–3000 K). Instead, C_2H_2, which is a parent molecule to form PAHs (Allamandola *et al.* 1989), is detectable in AGB stars. This motivates us to study C_2H_2 in AGB stars.

AGB stars distribute their chemical products through mass loss process. The mass loss from AGB stars is intense; approximately 50–80% of the stellar mass is lost during the AGB phase for Galactic stars. Understanding this process and the composition of the gas lost from AGB stars is important for understanding both stellar and galactic evolution. In particular, carbon-rich molecules are primarily formed in carbon-rich AGB stars. Theoretical work has suggested that AGB mass-loss rates depend on metallicity (Bowen & Willson 1991, Willson 2006), because the stellar wind is triggered by radiation pressure on dust grains and the dust is made up of astronomical metals. Therefore a study of mass-loss rate of extra-galactic AGB stars is vital, because it allows to study chemical enrichment process at low metallicity.

2. Observations and results

We have obtained infrared spectra, using the *Very Large Telescope* (VLT) and the *Spitzer Space Telescope* (SST). This paper summarises our series of studies (Matsuura *et al.* 2002, 2005, 2006, 2007, Sloan *et al.* 2006, van Loon *et al.* 2006, Zijlstra *et al.* 2006, Lagadec *et al.* 2007, 2008) of the 3–35 μm spectra of extragalactic carbon stars. Our targets are located in three nearby galaxies, namely the Large Magellanic Cloud (LMC; [Fe/H]~ -0.3), the Small Magellanic Cloud (SMC; [Fe/H]~ -0.6), and the Fornax dSph galaxy ([Fe/H]~ -1.0).

Figure 1 shows the spectrum of a star in our Galaxy obtained by the Infrared Space Observatory (ISO) and spectra of LMC and SMC stars observed with VLT and Spitzer.

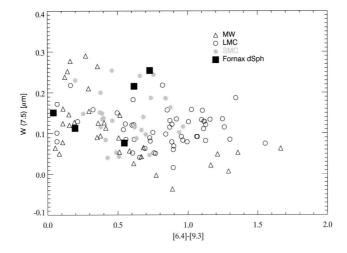

Figure 2. The equivalent width of 7.5 μm C$_2$H$_2$ as a function of infrared colour [6.4]−[9.3]. Symbols show the host galaxies, MW representing the Milky Way.

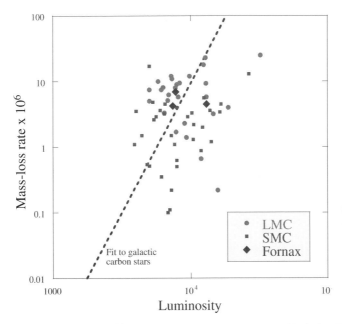

Figure 3. Mass-loss rate (M$_\odot$ yr^{-1}) as a function of luminosity (L$_\odot$) for Formax sample (Matsuura *et al.* 2007) and for the LMC and SMC samples (Groenewegen *et al.* 2007). Fornax stars show high mass-loss rates for their luminosities. The dotted line is the fit to the luminosity vs mass-loss rate relation for Galactic carbon stars (Groenewegen *et al.* 1998).

Major molecular bands and dust excess are indicated in the figure. C$_2$H$_2$ bands are found at 3.8, 7, and 13 μm. The 3.1 μm absorption is due to HCN and C$_2$H$_2$. There are two dust features found, SiC at 11.3 μm and MgS at $\sim$ 30 μm.

First we measured the equivalent widths of the molecular bands, so as to evaluate the metallicity dependence of these features. The equivalent width of the C$_2$H$_2$ molecular bands are measured following the method of Zijlstra *et al.* (2006). Figure 2 shows the

7.5 μm C_2H_2 equivalent width as a function of infrared colour [6.4]−[9.3]. The [6.4] and [9.3] values are calculated from the Spitzer/IRS spectra, and the definition of these magnitudes is also given by Zijlstra *et al.* (2006). Within the 0.6< [6.4]−[9.3] <1.0 range, a high value of W(7.5), with respect to infrared colour, is found for stars in Fornax and the SMC while a low W(7.5) is found in our Galaxy. This infrared colour measures effective temperature of stars for blue stars, but also circumstellar dust excess for red stars. C_2H_2 equivalent width increases towards lower metallicity.

Figure 3 shows the derived mass-loss rate as a function of luminosity. This is also confirmed by the further LMC study of Sloan *et al.* (2008). Low metallicity dependence of mass-loss rates is found for carbon-rich stars. The stars in Fornax dSph galaxy are at the upper end of the SMC mass-loss rates at a given luminosity. LMC stars appear to reach a higher mass-loss rate than the SMC and Fornax stars at a given luminosity.

3. Discussion

We found that C_2H_2 is more abundant at low metallicity. C_2H_2 formation relies on excess carbon atoms after all oxygen atoms are locked into carbon monoxide. Thus the higher abundance of C_2H_2 at lower metallicity is due to the higher C/O ratio (Matsuura *et al.* 2005) caused by carbon synthesised in AGB stars. C_2H_2 is thought to be a parent molecule in the formation of PAHs, as indicated by chemical models (Allamandola *et al.* 1989). It is still unknown whether PAHs are formed during the AGB phase or afterwards. If PAHs are indeed formed during the AGB phase, an over-abundance of C_2H_2 in a low metal environment will affect the growth of these important molecules. This implies that weaker PAH bands from the interstellar medium of lower metal galaxies could be caused by destruction of strong UV radiation fields at lower metallicities.

References

Allamandola, L. J., Tielens, G. G. M., & Barker, J. R. 1989, *ApJS*, 71, 733
Bowen, G. H. & Willson, L. A. 1991, *ApJ* (Letter), 375, L53
Groenewegen, M. A. T., Whitelock, P. A., Smith, C. H., & Kerschbaum, F. 1998, *MNRAS*, 293, 18
Groenewegen, M. A. T., Wood, P. R., Sloan, G. C., *et al.* 2007, *MNRAS*, 376, 313
Lagadec, E., Zijlstra, A. A., Matsuura, M., Menzies, J. W., van Loon, J. Th., & Whitelock, P. A. 2008, *MNRAS*, 383, 399
Lagadec, E., Zijlstra, A. A., Sloan, G. C., *et al.* 2007, *MNRAS*, 376, 1270
Matsuura, M., Wood, P. R., Sloan, G. C., *et al.* 2006, *MNRAS*, 371, 415
Matsuura, M., Zijlstra, A. A., Bernard-Salas, J., *et al.* 2007, *MNRAS*, 382, 1889
Matsuura, M., Zijlstra, A. A., van Loon, J. Th., *et al.* 2002, *ApJ* (Letter), 580, L133
Matsuura, M., Zijlstra, A. A., van Loon, J. Th., *et al.* 2005, *A&A*, 434, 691
Sloan, G. C., Kraemer, K. E., Matsuura, M., *et al.* 2006, *ApJ* 645, 1118
Sloan, G. C., Kraemer, K. E., Wood, P. R., *et al.* 2008, *ApJ*, in press, astro-ph/0807.2998
van Loon, J. Th., Marshall, J. R., Cohen, M., *et al.* 2006, *A&A*, 447, 971
Willson, L. A. 2006, in: Letizia Stanghellini, J. R. Walsh, and N.G. Douglas (eds.), *Planetary nebulae beyond the Milky Way, Proceedings of the ESO workshop* (Berlin: Springer), p. 99
Zijlstra, A. A., Matsuura, M., van Loon, J. Th., *et al.* 2006, *MNRAS*, 370, 1961

Organic Matter in Space
Proceedings IAU Symposium No. 251, 2008
S. Kwok & S. Sandford, eds.

Probing chemical processes in AGB stars

Fredrik L. Schöier and Hans Olofsson

Onsala Space Observatory, SE-439 92 Onsala, Sweden
email: schoier@chalmers.se, hans.olofsson@chalmers.se

Abstract. We are conducting multi-transition observations of circumstellar line emission from common molecules such as HCN, SiO, CS, SiS and CN for a large sample of AGB stars with varying photospheric C/O-ratios and mass-loss charachteristics. Our recently published results for SiO and SiS clearly show that major constraints on the relative roles of non-equilibrium chemistry, dust condensation, and photodissociation can be obtained from the study of circumstellar molecular line emission. Presented here are also preliminary results based on detailed radiative transfer modelling of HCN line emission.

Keywords. Stars: AGB and post-AGB, circumstellar matter, mass loss

1. Introduction

Stellar winds from evolved stars carry the products of internal nuclear processes, activated during the evolution along the asymptotic giant branch (AGB), and thus contributes to the enrichment of heavy elements in the ISM and the chemical evolution of galaxies (e.g., Gustafsson 2004). Of paramount importance is gaining knowledge of the fraction of these products that are accreted onto dust grains and the fraction that remains in the gas phase. Moreover, the mass loss will also have a profound effect on the future evolution of the star, and eventually terminate its life as a star, as well as their descendents, the planetary nebulae. The study of the circumstellar envelopes (CSEs) produced by the stellar winds is crucial to our understanding of the late stages of stellar evolution and its implications on the cosmic gas cycle.

Molecules can easily form in large abundance in the cool atmospheres of AGB stars and initiate a relatively complex chemistry that is further enhanced by photodissociation in the CSEs. The molecular and grain type setups in the CSEs are to a large extent determined by the C/O-ratio of the central star but also the thickness of the envelope and the ambient ultraviolet field are of importance. AGB stars divide in two chemically distinct groups; carbon stars (C/O>1) which show a more rich chemistry than the M-type AGB stars (C/O<1). There is also a class of putative transition objects, the S-stars, where the photospheric C/O-ratio is ≈ 1. For objects with high mass-loss rates where the photosphere can not be directly observed the study of the CSE is the only way to follow the future evolution of the star.

We have already obtained multi-transition CO observations for a large sample of some 150 AGB stars with varying C/O-ratios. The characteristics of their stellar winds such as mass-loss rate, expansion velocity, and kinetic temperature structure has been determined through detailed non-LTE, and non-local, radiative transfer modelling of the CO data (Schöier & Olofsson 2001, Olofsson *et al.* 2002, Gonzalez Delgado *et al.* 2003, Ramstedt *et al.* 2006). These models form the basis for the excitation analysis of other molecular species.

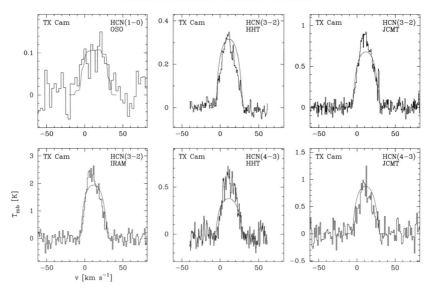

Figure 1. Multi-transition spectra (histograms) of HCN line emission toward the M-type AGB star TX Cam (HHT data from Bieging *et al.* 2000). Spectra from the best-fit single-dish model (assuming a Guassian abundance distribution) using a fractional HCN abundance $f_0 = 3.5 \times 10^{-7}$ and a envelope size $r_e = 2.0 \times 10^{16}$ cm are also shown (solid lines). The hyperfine splitting of the $J = 1 \rightarrow 0$ transition significantly broadens the line. This effect is explicitly taken into account in the modelling. The calibration uncertainty in the observed spectra is $\pm 20\%$. (Schöier *et al.*, in prep.)

2. Circumstellar chemistry

Most of the published abundance estimates are carried out for limited samples, based on rather simple methods, and are typically order of magnitude estimates. At this level, chemical modeling does a reasonable job in explaining many of the observed abundances (Millar 2003). However, there are some notable exceptions such as the detection of H_2O, H_2CO and NH_3 toward the carbon star IRC +10216 (Melnick *et al.* 2001, Ford *et al.* 2004, Hasegawa *et al.* 2006) in amounts significantly higher than predicted by stellar equilibrium chemistry. Processing by shocks does not seem to increase these abundances any further. A possible explanation for these high abundances could be catalytic processes (Willacy 2004) or, in the case of H_2O, evaporation of Kuiper-belt like objects (Melnick *et al.* 2001). The detection of carbon bearing species such as CN, CS and HCN in envelopes of M-type AGB stars is also somewhat surprising. In the atmospheres of M-type AGB stars the lack of any free carbon prevents them to form in equilibrium chemistry. However, as shown by Duari *et al.* (1999) non-equilibrium gas-phase chemistry can potentially produce significant amounts of carbon bearing molecules such as HCN and CO_2.

3. Results and discussion

3.1. *SiO and SiS*

The first more detailed studies of circumstellar abundances in larger samples of sources have been performed by Gonzalez-Delgado *et al.* (2003), Schöier *et al.* (2006a) and Schöier *et al.* (2007). Average SiO and SiS fractional abundances were obtained from a detailed radiative transfer analysis of multi-transition single-dish observations. In the case of SiO (see also contribution by Ramtedt *et al.* in this proceeding) the M-type AGB stars have much lower fractional abundances than expected from photospheric equilibrium chemistry. For the carbon stars the derived abundances can be up to two orders of magnitude higher than predictions based on equilibrium chemstries. A clear trend that the SiO fractional abundance decreases as the mass-loss rate of the star increases, as would be the case if SiO is accreted onto dust grains, is found. Further support for such a scenario comes from interferometric observations by Schöier *et al.* (2004) and Schöier *et al.* (2006b). In their analysis they found evidence of an inner compact component of high fractional abundance, consistent with predictions from stellar atmosphere chemistry in the case of the M-type objects but several orders of magnitude larger than expected for the carbon star IRC+10216, indicating also the importance of non-LTE chemical processes. These conclusions are further corroborated by recent chemical modeling by Cherchneff (2006). In contrast, for SiS the derived fractional abundances depend critically on the photospheric C/O-ratio and are more in line with expectations from LTE chemistries. There are also indications of adsorption of SiS onto dust grains in individual sources. In addition to constraining photospheric and circumstellar chemistry, it has become evident that the line emission also has the potential to probe the formation and evolution of dust grains in CSEs, as well as CSE dynamics.

3.2. HCN

In Figure 1 we show a best-fit to the available HCN single-dish data towards the M-type AGB star TX Cam. In order to obtain good constraints on the abundance distribution multi-transitional observations are required, By combining low-excitation ($\sim$5–20 K) rotational transitions at 3 mm with high-excitation transitions ($\sim$50–200 K) in the submillimeter regime a large radial range in the circumstellar envelope can be probed and existing chemical gradients can be detected. This is illustrated in Figure 2 (left panel) for the M-type AGB star TX Cam, where a good fit to the available observational data can be obtained using a Gaussian distribution ($f = f_0 \exp[-(r/r_\mathrm{e})^2]$) for the fractional abundance of HCN relative to H_2. In addition, available interferometric observations provide further constraints (Figure 2, right panel). Here, f_0 is sensitive to the chemistry occuring in and near the photosphere whereas r_e provide constraints on photodissociation models.

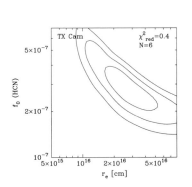 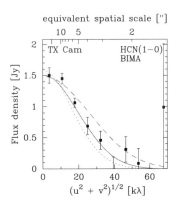

Figure 2. left – χ^2-map showing the quality of the fit to available HCN multi-transitional single-dish data for the M-type AGB star TX Cam when varying the adjustable parameters, f_0 and r_e, in the model. Contours are drawn at the 1, 2, and 3 σ levels. Indicated in the upper right corner are the reduced χ^2 of the best-fit model and the number of observational constrains used, N. **right** – Visibilities, averaged over 2 km s^{-1} around the systemic velocity, obtained for TX Cam using the BIMA interferometer (Marvel 2005). The observations are overlayed by model results using various abundance distributions: $f_0 = 1.1 \times 10^{-6}$ and $r_\mathrm{e} = 1.1 \times 10^{16}$ cm (dashed line); $f_0 = 5.0 \times 10^{-7}$ and $r_\mathrm{e} = 1.6 \times 10^{16}$ cm (solid line); $f_0 = 4.0 \times 10^{-7}$ and $r_\mathrm{e} = 2.1 \times 10^{16}$ cm (dotted line) (Schöier *et al.*, in prep.).

From left to right: Ewine van Dishoeck, Hans Olofsson, Åke Hjalmarson, and Bill Irvine at the reception.

Organic Matter in Space
Proceedings IAU Symposium No. 251, 2008
S. Kwok & S. Sandford, eds.

© 2008 International Astronomical Union
doi:10.1017/S1743921308021571

PAH anions as carriers of the mid-IR emission bands in planetary nebulae

Ryszard Szczerba[1], Christine Joblin[2], Olivier Berné[2], and Cezary Szyszka[3]

[1]N. Copernicus Astronomical Center,
Rabiańska 8, 87-100 Toruń, Poland
email: szczerba@ncac.torun.pl

[2]Centre d'Etude Spatiale des Rayonnements, Université Toulouse 3 et CNRS,
9 Av. du Colonel Roche, 31028 Toulouse cedex 04, France

[3]Nicolaus Copernicus University,
Gagarina 11, 87-100 Toruń, Poland

Abstract. We present results of the mid-IR spectra decomposition for planetary nebulae and compact H II regions in our Galaxy and Magellanic Clouds. The striking correlation between the required PAH component with "7.7" μm band shifted to about 7.8 μm and electron densities of the modeled sources allows us to argue that this PAH component may be in fact PAH anions (PAH^-).

Keywords. (ISM:) dust, extinction, (ISM:) planetary nebulae: general, infrared: ISM

1. Introduction

Among the different dust populations, the carriers of the so-called aromatic infrared bands (AIBs) are the subject of a long debate. The main AIBs are located at 3.3, 6.2, 7.7, 8.6, 11.3, 12.7 μm and are attributed to very small aromatic dust particles amongst which there are large polycyclic aromatic hydrocarbons (PAHs). However, these band positions are varying. In particular, the dominant band in the "7.7" μm complex arises at 7.8 μm in planetary nebulae (PNe) compared to 7.6 μm in compact-H II (c-H II) regions (Peeters *et al.* 2002) or in photo dissociation regions (PDRs). These shifts in the AIB positions may be related to chemical or physical changes of the carriers in different environments. To investigate possible evolution of the AIB carriers, Joblin *et al.* (2008) have analyzed spectra of several PNe and c-H II regions in the Milky Way (MW) and in the Large and Small Magellanic Clouds (LMC & SMC). Their analysis is based on the results of Berné *et al.* (2007) and Rapacioli *et al.* (2005) and uses a decomposition of the mid-IR spectra into a set of template spectra (see Sect.2).

In this paper we discuss results of the above analysis for selected PNe and c-H II regions in MW and LMC and show that the obtained decomposition may be understood by the presence of different amount of PAH anions in environments with different ionization parameter (U_{ion}).

2. Templates and the spectra decomposition

Rapacioli *et al.* (2005) and Berné *et al.* (2007) were able to extract the spectra of PAH neutrals (PAH^0), PAH cations (PAH^+) and very small grains (VSGs) from the mid-IR spectra of the analyzed PDRs. Joblin *et al.* (2008) have built three template spectra

Philippe Bréchignac and Rosario Brunettto.

Organic Matter in Space
Proceedings IAU Symposium No. 251, 2008
S. Kwok & S. Sandford, eds.

© 2008 International Astronomical Union
doi:10.1017/S1743921308021583

Aromatic, aliphatic, and the unidentified 21 micron emission features in proto-planetary nebulae

Bruce J. Hrivnak[1], Kevin Volk[2], T. R. Geballe[2], and Sun Kwok[3]

[1]Department of Physics & Astronomy, Valparaiso University,
Valparaiso IN, 46383, USA
email: bruce.hrivnak@valpo.edu

[2]Gemini Observatory, 670 North A'ohoku Place, Hilo, HI 96720, USA
email: kvolk@gemini.edu, tgeballe@gemini.edu

[3]Department of Physics, University of Hong Kong, Hong Kong, China
email: sunkwok@hku.hk

Abstract. Aromatic features at 3.3, 6.2, 7.7, 8.6, 11.3 μm are observed in proto-planetary nebulae (PPNe) as well as in PNe and H II regions. Aliphatic features at 3.4 and 6.9 μm are also observed; however, these features are often stronger in PPNe than in PNe. These observations suggest an evolution in the features from simple molecules (C_2H_2) in AGB stars to aliphatics in PPNe to aromatics in PNe. In the same carbon-rich PPNe, a strong, broad, unidentified 21 μm emission feature has been found. We will present recent observations of the aromatic, aliphatic, and 21 μm emission features, along with C_2H_2 (13.7 μm) and a new feature at 15.8 μm, and discuss correlations among them and other properties of these PPNe.

Keywords. Astrochemistry, circumstellar matter, ISM: lines and bands, infrared: ISM, stars: AGB and post-AGB, planetary nebulae: general

---

1. Background and New Observations

Aromatic hydrocarbon emission features at 3.3, 6.2, 7.7, 8.6, and 11.3 μm, often attributed to PAHs, are observed in the spectra of various objects with hot irradiating sources; planetary nebulae (PNe), H II regions, reflection nebulae. They have also been observed in proto-planetary nebulae (PPNe), objects in the short-lived ($\sim$1000 yr) transitional phase between AGB stars and PNe. In PPNe, the circumstellar envelope is detached but the central star is not hot enough to photo-ionize the nebula and is typically of spectral type F-G. Aliphatic emission features at 3.4 and 6.9 μm are also seen in PPNe and are often stronger than in PNe (Geballe 1997, Geballe *et al.* 1992). This suggests an evolution in the carbon chemistry of the circumsteller envelopes from C_2H_2 to aliphatics to aromatics as C-rich stars evolve rapidly from AGB to PPN to PN phases (Kwok 2004).

The unidentified 21 μm emission feature, first seen in *IRAS* spectra of four C-rich PPNe (Kwok *et al.* 1989), has subsequently been observed in additional C-rich PPNe with *ISO* (Volk *et al.* 1999) and recently with *Spitzer*. This 21 μm feature has been detected only in C-rich objects and essentially only in PPNe (and perhaps weakly in a few AGBs and young PNe). Suggested identifications include PAHs, TiC, SiC (see Speck & Hofmeister 2004 and references therein), and FeO (Guha Niyogi *et al.*, these proceedings).

New 3 μm spectra have been obtained of seven PPNe. All show the 3.3 μm and most show the 3.4 μm feature (Hrivnak *et al.* 2007). New mid-IR spectra have also been obtained of six carbon-rich PPNe using *Spitzer*. These reveal one new 21 μm source and give good observations of the others. Also seen are the 11.3 and 12.3 μm emission bands.

Table 1. Summary of the Spectral Features of Carbon-Rich PPNe and 21 μm Sources[a]

Object	SpT	C/O	C₂,C₃	3.3	3.4	6.2	6.9	7.7	8.6	8br	11.3	12.3	Class[b]	C₂H₂	15.8	21	30 μm
02229+6208	G8 Ia	...	Y,Y	Y	Y:*	Y:	Y	N	N	Y	Y	Y	A	...		Y	Y
20000+3239	G8 Ia	...	Y,...	Y	Y*	Y	Y	N	N	Y	Y	Y	A	...		Y	Y
05113+1347	G8 Ia	2.4	Y,Y	Y:	Y:	...	...	...	...	...	Y	Y	...	N:*	Y:*	Y	Y
22272+5435	G5 Ia	1.6	Y,Y	Y	Y	Y	Y	Y	N	Y	Y	Y	B		Y:	Y	Y
07430+1115	G5 Ia	...	Y,Y	Y	Y	...	...	...	...	Y:	Y:	...	A		...	Y*	Y
23304+6147	G2 Ia	2.8	Y,Y	Y*	Y:*	Y	Y	Y	Y	Y	Y	Y	A	Y:*	Y*	Y	Y
05341+0852	G2 Ia	1.6	Y,Y	Y	Y	Y	Y	N	N	Y	Y	Y*	B	Y*	Y*	Y	Y
22223+4327	G0 Ia	1.2	Y,Y	Y	N						Y		A			Y	Y
04296+3429	G0 Ia	...	Y,Y	Y	Y		Y			Y	Y		B			Y	Y
AFGL 2688	F5 Iae	1.0	Y,Y	Y	Y	Y	Y:	N	N	Y	Y	N:	A	Y		Y:	Y
06530−0230	F5 I	2.8	Y,Y	Y*	N	...	...	...	...	Y*	Y*	Y	A	Y*	Y*	Y*	Y*
07134+1005	F5 I	1.0	Y,N	Y	Y:	...	Y	Y	N	Y	Y	Y	A		Y:	Y	Y
19500−1709	F3 I	1.0	N,N	N	N					Y:	Y	Y:	...		Y:	Y	Y
16594−4656	B7	...	N,N	Y	N	Y	N	Y	Y	Y:	Y	Y:	A			Y	Y
01005+7910	B0 I	1.2	N,N	Y	Y*	Y	N	Y	Y	N	Y	N	A			N:	Y
22574+6609	...	...	...,...	...	...	Y	Y	Y	N	Y	Y*	...		Y*	N*	Y	Y
19477+2401	...	...	...,...	...	...	...	...	...	...	...	...	...	...	...	...	Y*	Y

Note 1: Colon indicates a marginal or uncertain detection, blank indicates lack of information, "..." indicates that the object has not been observed in this spectral region.

Note 2: Asterisk indicates a new detection from Hrivnak *et al.* (2007) or Hrivnak *et al.* (2008).

[a] Table does not include three newly discovered C-rich PPNe IRAS 08143−4406, 08281−4850, 14325−6428 (Reyniers *et al.* 2004, 2007) that have not been observed in the IR.

[b] Classification scheme of Geballe (1997) at 3.3, 3.4 μm.

Two other emission features are seen. At 15.8 μm is a new, relatively strong, unidentified feature seen in four sources; it is strongest in the two with the strongest 21 μm feature. At 13.7 μm is seen the C_2H_2 feature in four sources, including the first report of C_2H_2 in emission in a post-AGB object (Hrivnak *et al.* 2008). Results are listed in Table 1.

2. Summary

- 3.3, 3.4 μm: All C-rich PPNs have 3.3 μm and most have 3.4 μm emission features.
- 21 μm: (a) All have the same shape and central wavelength (20.1±0.1 μm) but differ in strength; (b) all are C-rich, (almost) all show C_2, C_3, 3.3, 11.3, 30 μm emission.
- C_2H_2: (a) Detected in four 21 μm sources; all show P-Cygni profiles; (b) first detection in emission in post-AGB stars.
- 15.8 μm: New feature seen in several of the PPNe including previous *ISO* spectra; unidentified; (b) correlated with 21 μm emission?
- Trends: (a) All 21 μm sources are C-rich, (almost) all show C_2, C_3, 3.3, 11.3, 30 μm emission; (b) no correlation found between 3.4/3.3 ratio and spectral type.

Acknowledgements

BJH acknowledges support from NASA (JPL/Caltech 1276197) and NSF (AST-0407087).

References

Geballe, T. R. 1997, in: Y. J. Pendleton & A. G. G. M. Tielens (eds.), *From Stardust to Planetesimals*, (ASP: San Francisco), p. 119

Geballe, T. R., Tielens, A. G. G. M., Kwok, S., & Hrivnak, B. J. 1992, *ApJ (Letters)*, 387, L89

Hrivnak, B. J., Geballe, T. R., & Kwok, S. 2007, *ApJ*, 662, 1059

Hrivnak, B. J., Volk, K., & Kwok, S. 2008, *ApJ*, submitted

Kwok, S. 2004, *Nature*, 430, 985

Kwok, S., Volk, K., & Hrivnak, B. J. 1989, *ApJ (Letters)*, 345, L51

Reyniers, M., van Winckel, H., Gallino, R., & Straniero, O. 2004, *A&A*, 417, 269

Reyniers, M., Van de Steene, G. C., van Hoof, P. A. M., & van Winckel, H. 2007, *A&A*, 471, 247

Speck, A. K. & Hofmeister, A. M. 2004, *ApJ*, 600, 986

Volk, K., Kwok, S., & Hrivnak, B. J. 1999, *ApJ (Letters)*, 516, L99

Organic Matter in Space
Proceedings IAU Symposium No. 251, 2008
S. Kwok & S. Sandford, eds.

© 2008 International Astronomical Union
doi:10.1017/S1743921308021595

On the inorganic carriers of the 21 micron emission feature in post-AGB stars

Ke Zhang[1], Biwei Jiang[1], and Aigen Li[2]

[1]Department of Astronomy, Beijing Normal University, Beijing 100875, China
emails: zhangke@mail.bnu.edu.cn, bjiang@bnu.edu.cn

[2]Department of Physics & Astronomy, University of Missouri, Columbia, MO 65211, USA
email: lia@missouri.edu

Abstract. The mysterious $21\,\mu m$ emission feature seen in only 12 C-rich proto-planetary nebulae (PPNe) remains unidentified since its discovery in 1989. Over a dozen materials have been suggested as the carrier candidates while none of them has received general acceptance. We investigate the inorganic carrier candidates by applying the observational constraints of the feature strength and associated features. It is found that: (1) three candidates, TiC clusters, fullerenes with Ti impurity atoms, and SiS_2, are not abundant enough to account for the emission power of the $21\,\mu m$ band, (2) five candidates, doped-SiC, SiO_2-mantled SiC dust, carbon and silicon mixtures, Fe_2O_3, and Fe_3O_4, all show associated features which are either not detected in the $21\,\mu m$ sources or detected but with a much lower strength, and (3) FeO, which satisfies the abundance constraints, does not display any associated features which are not seen in the $21\,\mu m$ sources. Moreover, FeO is more likely to survive in the C-rich environment than Fe_2O_3 and Fe_3O_4. Thus FeO seems to be the most plausible one among the inorganic carrier candidates.

Keywords. Infrared: stars, (stars:) circumstellar matter, stars: AGB and post-AGB

1. Introduction

The so-called "$21\,\mu m$ feature" has been well identified in 12 proto-planetary nebulae (PPNe; Kwok *et al.* 1999; also see Hrivnak *et al.* in this proceeding for two new $21\,\mu m$ sources). This feature, peaking at $\sim 20.1\,\mu m$ with a FWHM of ~ 2.2–$2.3\,\mu m$, has little shape variation among different sources (Volk *et al.* 1999). Most of these sources exhibit quite uniform characteristics: metal-poor, carbon-rich F and G supergiants with strong infrared (IR) excess and over abundant s-process elements (see Kwok *et al.* 1999 and Zhang *et al.* 2006).

Since its discovery by Kwok *et al.* (1989), over a dozen of both organic and inorganic carrier candidates have been proposed. The inorganic candidates are: (a) TiC nanoclusters (von Helden *et al.* 2000, but see Li 2003), (b) large-cage carbon particles (fullerenes) coordinated with Ti atoms (Kimura *et al.* 2005), (c) SiS_2 dust (Goebel 1993), (d) SiC dust with carbon impurities (Speck & Hofmeister 2004, but see Jiang *et al.* 2005), (e) carbon and silicon mixtures (Kimura *et al.* 2005), (f) SiC core-SiO_2 mantle grains (Posch *et al.* 2004), (g) FeO (Posch *et al.* 2004), (h) Fe_2O_3, and (i) Fe_3O_4 (Cox 1990). However, none of these carrier candidates have received general acceptance. The carrier of the $21\,\mu m$ feature remains unidentified. In this paper we report our recent efforts in testing the above-listed inorganic candidates in terms of the elemental abundance budget and spectral shape (with special emphasis on the possible accompanying features).

2. General Constraints: Band Strength and Associated Features

We assess the applicability of a proposed carrier candidate by examining (1) whether it is capable of emitting the observed large amount of energy in the $21\,\mu m$ band without requiring more dust material than available, and (2) whether the candidate carrier produces additional feature(s) with intensities inconsistent with those observed. HD 56126, a prototypical 21 μm feature source, is taken as the test case object.

The total energy emitted in the 21 μm band $E_{\rm tot}$ is related to the abundance (relative to H) of element X of the carrier by $[{\rm X/H}] = \frac{n_{\rm X}\, E_{\rm tot}}{\mu_{\rm d}\, M_{\rm H} \int_{21\,\mu {\rm m\ band}} \kappa_{\rm abs}(\lambda) \times 4\pi B_\lambda\,(T_{\rm d})\, d\lambda}$ (see Zhang *et al.* 2008). Knowing the dust temperature $T_{\rm d}$ and the mass absorption coefficient $\kappa_{\rm abs}$, one can calculate the element abundance required for a dust species (containing $n_{\rm X}$ X atoms in each molecule with a molecular weight $\mu_{\rm d}$) to account for the observed intensity of the 21 μm band. This constraint effectively excludes those candidates with scarce elements such as Ti. On the other hand, some of the proposed carrier candidates exhibit strong features in addition to the 21 μm band which are not seen in the 21 μm sources. Using lab optical constants and the stellar parameters, we calculate the intensity ratios of the associated features to the 21 μm band and compare with the observed spectra. In this way we exclude most of the candidates which display associated feature(s), such as SiC-bearing dust.

3. Results

Our results on individual carrier candidate are shown in Table 1. We investigate nine inorganic carrier candidates. Three of them are excluded due to abundance deficiency and another five suffer from producing strong associated features not seen in the 21 μm sources. At this moment, FeO nano particles seem to be the most promising candidate.

Table 1. Summary of the test on the carrier candidates ($\checkmark$: pass; $\times$: fail)

Candidate	Abundance (element)	Associated features
TiC nanoclusters	$\times$ (Ti)	$\checkmark$
fullerenes coordinated with Ti atoms	$\times$ (Ti)	$\checkmark$
SiS$_2$	$\times$ (S)	$\times$ (16.8 μm)
SiC dust with carbon impurities	$\checkmark$	$\times$ (11.3 μm)
carbon and silicon mixtures	$\checkmark$	$\times$ (9.5 μm)
SiC core-SiO$_2$ mantle grains	$\checkmark$	$\times$ (8.3 μm, 11.3 μm)
FeO	$\checkmark$	$\checkmark$
Fe$_2$O$_3$	$\checkmark$	$\times$ (9.2, 18, 27.5 μm)
Fe$_3$O$_4$	$\checkmark$	$\times$ (16.5, 24 μm)

References

Cox, P. 1990, *A&A* (Letters), 236, L29
Goebel, J. H. 1993, *A&A*, 278, 226
Jiang, B. W., Zhang, K., & Li, A. 2005, *ApJ* (Letters), 630, L77
Kimura, Y., Nuth, J. A., III, & Ferguson, F. T. 2005, *ApJ* (Letters), 632, L159
Kwok, S., Volk, K., & Hrivnak, B. J. 1999, in *Asymptotic Giant Branch Stars*, p. 297
Kwok, S., Volk, K. M., & Hrivnak, B. J. 1989, *ApJ*, 345, 51
Li, A. 2003, *ApJ* (Letters), 599, L45
Posch, T., Mutschke, H., & Andersen, A. 2004, *ApJ*, 616, 1167
Speck, A. K., & Hofmeister, A. M. 2004, *ApJ*, 600, 986
Volk, K., Kwok, S., & Hrivnak, B. J. 1999, *ApJ* (Letters), 516, L99
von Helden, G., Tielens, A. G. G. M., van Heijnsbergen, D. *et al.* 2000, *Science*, 288, 313
Zhang, K., Jiang, B. W., & Li, A. 2006, *Progress in Astronomy (Chinese)*, 24, 43
Zhang, K., Jiang, B. W., & Li, A. 2008, in preparation

Organic Matter in Space
Proceedings IAU Symposium No. 251, 2008
S. Kwok & S. Sandford, eds.

© 2008 International Astronomical Union
doi:10.1017/S1743921308021601

Dust properties in the circumstellar shells of evolved stars: Observational constraints from ISO and Spitzer infrared spectroscopy

P. García-Lario[1], J. V. Perea Calderón[2], D. A. García-Hernández[3], L. Stanghellini[4], D. Engels[5], A. Manchado[3], J. E. Davies[4], E. Villaver[6], R. A. Shaw[4], and M. Bobrowsky[7]

[1] Herschel Science Centre, European Space Astronomy Centre, ESA, Madrid, Spain
email: `Pedro.Garcia-Lario@sciops.esa.int`

[2] European Space Astronomy Centre, ESA, Madrid, Spain

[3] Instituto de Astrofísica de Canarias, Tenerife, Spain

[4] NOAO, Tucson, AZ, USA

[5] Hamburger Sternwarte, Germany

[6] Space Telescope Science Institute, Baltimore, MD, USA

[7] Computer Science Corporation / Space Telescope Science Institute, Baltimore, MD, USA

Abstract. We present the results of a systematic analysis of the solid state features identified in the circumstellar environments of a large sample of evolved stars with ISO/SWS and Spitzer/IRS spectroscopy. The sample includes several hundred stars with a wide variety of progenitor masses evolving from the early AGB phase to the PN stage. Our observations are used to propose an evolutionary scheme in which the results obtained can be interpreted as a consequence of the nucleosynthesis processes that take place in this short phase of the stellar evolution, in particular the third dredge-up and hot bottom burning, which in turn are also strongly modulated by the stellar metallicity.

Keywords. Circumstellar matter, stars: AGB and post-AGB, planetary nebulae

1. Selection of the sample

Three subsets of objects were analysed located in different metallicity environments: the Galactic Bulge, the Galactic Disk and the Magellanic Clouds. The overall efficiency of dust production, as well as the physical and chemical properties of the dust grains were analysed in each case.

2. Results

Galactic Bulge (high metallicity): 46 sources were observed with Spitzer/IRS. All of them showed a very strong dust continuum, indicative of efficient dust formation. Solid state features usually correspond to O-rich dust grains, in most cases crystalline silicates. The more luminous sources appear heavily obscured in the optical, indicative of a recent and strong mass loss. The higher frequency of O-rich shells in the Bulge is consistent with an inefficient third dredge-up operating during the AGB phase, predicted by nucleosynthesis models. The O-rich nature of the most luminous and obscured sources may correspond to a young population of high-mass bulge stars experiencing hot bottom burning. This is predicted to occur only for masses $\geqslant 3.5$–4.0 M$_\odot$ at these very high metallicities.

Galactic Disk (intermediate metallicity): 350 sources with available ISO/SWS spectra show a proportion of C-rich and O-rich shells close to unity. The chemical branching

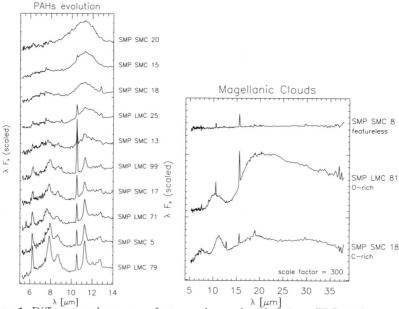

Figure 1. Different carbonaceous features observed with Spitzer/IRS in the sample of Magellanic Cloud PNe (left); at low metallicities these sources show unusual spectra (right).

observed is induced mainly by the third dredge-up and hot bottom burning, leading to a population of: *i*) low-mass O-rich PNe, the result of the evolution of O-rich bright Mira variables; *ii*) intermediate-mass C-rich PNe, the result of the evolution of carbon Miras; or *iii*) high-mass O-rich PNe, N-rich as well, also called *type I PNe*. The solid state features evolve from amorphous to crystalline as the AGB stars become PNe in the case of O-rich shells and from aliphatic to aromatic in C-rich shells.

Magellanic Clouds (low metallicity): 41 sources were observed with Spitzer/IRS show-ing very little dust continuum, indicative of an inefficient dust formation. Solid state features, when observed, correspond to C-rich dust grains in most cases, usually a com-bination of SiC and very small carbonaceous grains (VSGs), with only a few sources showing the characteristic PAH features observed in Galactic PNe. Among the O-rich sources (only 3), two of them display amorphous silicates, very rarely observed in their galactic analogues. Dust processing by the UV photon irradiation from the central star seems to be inhibited in LMC and SMC PNe, suggesting a more distant location for the dust in the shell. Little or no obscuration in the optical is observed even in the most luminous PNe studied, indicative of the relatively small mass loss experienced by these stars in the AGB phase. A much higher frequency of C-rich PNe versus O-rich sources (more extreme in the SMC, where metallicity is lower), is consistent with a very efficient 3rd dredge-up operating at low metallicities, as predicted by nucleosynthesis models. The O-rich nature of a few luminous PNe may be the consequence of these stars having experienced hot bottom burning in the AGB. At the low metallicity of the MCs this is predicted to occur for masses $\geqslant$2.5–3.0 M$_\odot$ by the theoretical models.

3. Conclusions

The chemical properties of the dust grains present in the circumstellar shells of evolved stars are confirmed to be strongly correlated with their progenitor masses, as predicted by nucleosynthesis models. In addition, our results show that the size and physical properties of the grains may also strongly depend on the metallicity.

Organic Matter in Space
Proceedings IAU Symposium No. 251, 2008
S. Kwok & S. Sandford, eds.

© 2008 International Astronomical Union
doi:10.1017/S1743921308021613

A survey of 3.3 micron PAH emission in planetary nebulae using FLITECAM

Erin C. Smith[1] and Ian S. McLean[2]

University of California, Los Angeles,
Division of Astronomy and Astrophysics, 430 Portola Plaza, Los Angeles CA 90095
email: [1]erincds@astro.ucla.edu, [2]mclean@astro.ucla.edu

Abstract. We have performed a study of 3.3 micron PAH emission in planetary nebulae using ground-based observations with FLITECAM, one of a suite of instruments designed for airborne astronomy aboard SOFIA, NASA's Stratospheric Observatory for Infrared Astronomy. The survey was performed on the Shane 3 meter telescope at Lick Observatory as part of the ground-based commissioning of the FLITECAM grism spectroscopy mode. Spectral resolution of $R \sim 1700$ was obtained with direct-ruled KRS-5 grisms. Targets included AGB stars and sources showing PAH emission in KAO, ISO or IRAS observations. Additionally, several oxygen-rich nebulae were observed in order to test methodology. Twenty objects were surveyed, of which 11 showed PAH emission. In objects exhibiting PAH emission, the relationship between the nebular C/O ratio and PAH equivalent width was found, showing a detectable PAH emission cutoff at a nebular C/O ratio of 0.65 ± 0.28. Selected objects with detected PAH emission were further investigated to trace PAH emission spectral variation within individual nebulae.

Keywords. Instrumentation: spectrographs, planetary nebulae: general, (ISM:) dust, extinction

We have carried out a preliminary survey of 3.3 μm PAH emission using FLITECAM on the Lick Observatory 3-m telescope. FLITECAM is a 1-5 μm camera developed by us at the UCLA Infrared Laboratory (P.I.: McLean) for NASA's SOFIA (Stratospheric Observatory for Infrared Astronomy) project (Mainzer & McLean 2003, McLean *et al.* 2006). SOFIA is a modified Boeing 747-SP airplane with a 2.5-meter f/19.6 bent-Cassegrain telescope operating at altitudes up to 45,000 ft and therefore above 99% of the atmosphere's water vapor content.

This study of 3.3 μm emission from PN was carried out as part of the commissioning and performance verification of the FLITECAM grism spectroscopy mode. This mode utilizes a suite of three KRS-5 grisms in conjunction with fixed slits of either 1 arcseconds or 2 arcseconds width and 60 arcseconds total length to achieve medium resolution ($R \sim 1700$ and 900) spectroscopy (Smith & McLean 2006). An airborne survey similar to this will be carried out using the same instrument on board the Stratospheric Observatory for Infrared Astronomy (SOFIA).

Observations were carried out over the course of five observing runs at Lick observatory. Targets were selected from ISO, KAO and IRAS observations of planetary nebulae, proto-planetary nebulae and post-AGB stars (Jourdain de Muizon *et al.* 1990, Rinehart *et al.* 2002). Out of 20 objects observed, 11 showed detectable 3.3 μm PAH emission. Telluric correction was achieved with the ATRAN atmosphere modeling package (Lord 1992). Each PAH detection was fitted with a gaussian to determine FWHM and central wavelength. We found the 3.3 μm feature to have an average FWHM of 0.042 μm ± 0.002 μm. and central wavelength of 3.288 μm ± 0.002 μm. These are consistent with values found by Tokunaga *et al.* (1991), Roche *et al.* (1996), and van Diedenhoven *et al.* (2004). All PAH emission was classified as type $A_{3.3}$ according to the system proposed by Peeters *et al.* (2002) and van Diedenhoven *et al.* (2004).

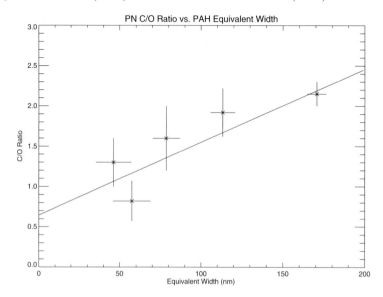

Figure 1. C/O Ratio versus PAH equivalent width.

The equivalent width of each PAH detection was also determined in order to investigate the correlation between the nebular C/O ratio and the PAH emission strength. Figure 1 shows the derived 3.3 μm emission equivalent width plotted against the nebular C/O ratio obtained from the literature (Kholtygin 1998, Liu *et al.* 2004) and found a correlation between C/O ratio and PAH emission strength. We calculate a threshold C/O ratio of 0.65 ± 0.28, which is consistent with that reported by Roche *et al.* (1996) and Cohen & Barlow (2005).

References

Cohen, M. & Barlow, M. J. 2005, *MNRAS*, 362, 1199

Jourdain de Muizon, M., Cox, P., & Lequeux, J. 1990, *A&AS*, 83, 337

Kholtygin, A. F. 1998, *A&A*, 329, 691

Liu, Y., Liu, X.-W., Barlow, M. J., & Luo, S.-G. 2004, *MNRAS*, 353, 1251

Lord, S. D. 1992, NASA TM103957

Mainzer, A. K. & McLean, I. S. 2003, *ApJ*, 597, 555

McLean, I. S., Smith, E. C., Aliado, T., Brims, G., Kress, E., Magnone, K., Milburn, J., Oldag, A., Silvers, T., & Skulason, G. 2006, in: I. S. McLean & M. Iye (eds.), *Proceedings of the SPIE, Volume 6269*

Peeters, E., Hony, S., Van Kerckhoven, C., Tielens, A. G. G. M., Allamandola, L. J., Hudgins, D. M., & Bauschlicher, C. W. 2002, *A&A*, 390, 1089

Rinehart, S. A., Houck, J. R., Smith, J. D., & Wilson, J. C. 2002, *MNRAS*, 336, 66

Roche, P. F., Lucas, P. W., Hoare, M. G., Aitken, D. K., & Smith, C. H. 1996, *MNRAS*, 280, 924

Smith, E. C. & McLean, I. S. 2006, in: I. S. McLean & M. Iye (eds.), *Proceedings of the SPIE, Volume 6269*, p. 50

Tokunaga, A. T., Sellgren, K., Smith, R. G., Nagata, T., Sakata, A., & Nakada, Y. 1991, *ApJ*, 380, 452

van Diedenhoven, B., Peeters, E., Van Kerckhoven, C., Hony, S., Hudgins, D. M., Allamandola, L. J., & Tielens, A. G. G. M. 2004, *ApJ*, 611, 928

Organic Matter in Space
Proceedings IAU Symposium No. 251, 2008
S. Kwok & S. Sandford, eds.

© 2008 International Astronomical Union
doi:10.1017/S1743921308021625

Molecules in nearby and primordial supernovae

Isabelle Cherchneff and Simon Lilly

Institut für Astronomie, ETH Hönggerberg
Wolfgang-Pauli-Strasse, 16, 8093, Zürich, Switzerland
email: isabelle.cherchneff@phys.ethz.ch

Abstract. We present new chemical models of supernova (SN) ejecta based on a chemical kinetic approach. We focus on the formation of inorganic and organic molecules including gas phase dust precursors, and consider zero-metallicity progenitor, massive supernovae and nearby core-collapse supernovae such as SN1987A. We find that both types are forming large amounts of molecules in their ejecta at times as early as 200 days after explosion. Upper limits on the dust formation budget are derived. Our results on dust precursors do not agree with existing studies on dust condensation in SN ejecta. We conclude that PMSNe could be the first non-primodial molecule providers in the early universe, ejecting up to 34% of their progenitor mass under molecular form to the pristine, local gas.

Keywords. Astrochemistry, molecular processes, supernova: general, individual (SN1987A), early universe

1. Introduction

Wether supernovae are major dust makers in the universe has been a long-standing debate. It was triggered by the explosion of SN1987A in the LMC two decades ago and the subsequent observation of dust and molecules like CO and SiO forming as early as 200 days after explosion. At high redshifts ($z \geqslant 6$), dust has been proposed to explain the reddening of background quasars and Lyman α systems. At these early times, only very massive, evolved stars can be dust makers, due to stellar evolution time constraints. Some theoretical studies have tackled the problem of molecule formation in nearby core-collapse SNe (hereafter CCSNe, see Lepp *et al.* 1990, Liu & Dalgarno 1994) or the condensation of dust in either CCSNe or primordial massive supernovae (PMSNe, see Kozasa *et al.* 1989, Clayton *et al.* 1999, Todini & Ferrara 2001, Nozawa *et al.* 2003, Schneider *et al.* 2004). These studies assume steady-state from time scale analysis and are often based on an incomplete treatment of the chemical processes. As for dust formation, the formalism used is that developed for the homogeneous nucleation of water droplets in the Earth's atmosphere. This approach has been seriously questioned by Donn & Nuth (1985) when applied to the formation of solids in circumstellar environments where equilibrium conditions do not apply and the condensation nuclei are often on molecular scale. We report in this proceeding preliminary results of a novel study on molecule and dust formation in supernova ejecta based on a chemical kinetic approach. We describe dust nucleation from the gas phase and the global ejecta chemistry using all chemical processes relevant to SNe ejecta. We then apply this chemical network to nearby CCSNe and PMSNe of zero-metallicity progenitor characteristic of Population III stars. Molecular contents and upper limits for dust yields are derived for these environments.

Table 1. Ejecta parameters versus time after explosion for both CCSN and PMSN models.

	CCSN	PMSN
Progenitor mass ($M_\odot$)	20	170
Helium core mass ($M_\odot$)	6	85
$M(^{56}Co)$ ($M_\odot$)	0.074	3.52
T_0 (100 days) (K)	5000	21000
T (500 days)	1490	1480
T (1000 days)	889	470
n_0 (100 days) (cm^{-3})	8.5×10^{10}	6.5×10^{11}
n (500 days)	5.6×10^8	5.2×10^9
n (1000 days)	7.0×10^7	6.5×10^8

2. Physical model of PMSN and CCSN ejecta

We consider two surrogates of SN explosions: a 20 $M_\odot$ CCSN representative of SN1987A with metallicity typical of the LMC, and a 170 $M_\odot$ PMSN of zero-metallicity progenitor. The gas temperature T is determined mainly by the explosion energy (1×10^{51} ergs and 2×10^{52} ergs for a typical CCSN and PMSN, respectively) and the CCSN surrogate T profile is that of Kozasa *et al.* (1991). For the PMSN, we use the T variation of Nozawa *et al.* (2003) for their 170 $M_\odot$ pair-instability SN. The ejecta expansion for both SN surrogates is homologous and the gas density varies with t according to $n(M_r, t) = n_0(M_r, t_0) \times (t/t_0)^{-3}$, where M_r is the mass coordinate and n_0 is the gas number density at t_0. Values for the various gas parameters are summarized in Table 1. The ejecta velocity for both surrogates is kept constant at 2000 km s^{-1} in good agreement with values derived from atomic emission lines in SN1987A and models from Schneider *et al.* (2004).

3. The ejecta chemistry

We are interested in studying the chemistry from time $t_0 = 100$ days to $t=1,000$ days, as molecules like CO and SiO and dust were observed over this time range in SN1987A (Spyromilio *et al.* 1988, Roche *et al.* 1991, Wooden *et al.* 1993). The light curve of SN1987A between 100 and 1000 days after explosion is dominated by the radioactive decay of ^{56}Co. This decay produces γ-rays that are Compton-scattered and creates a fast, energetic electron population in the gas. Collisions with such electrons are among the main destruction processes to molecules. We assume that similar radioactivity-induced processes take place in the ejecta of PMSNe. Compton electron collisions proceed with rates given by Liu & Dalgarno (1995) in the CCSN surrogate, whereas the rates are rescaled according to the ^{56}Co mass produced over the mass cut for the PMSN case (Umeda & Nomoto (2002). Evidence for strong mixing in SN ejecta is given by the early emergence of γ-rays from cobalt decay in SN1987A spectra (Pinto & Woosley 1988). Therefore, we consider two extreme cases for both SN surrogates: a microscopically fully-mixed ejecta, where we allow for hydrogen penetration from the progenitor envelope and consequent mixing, and a stratified ejecta that retains the onion-like structure resulting from the progenitor nucleosynthesis. Mixing in SN ejecta is expected to be macroscopic rather than microscopic, with the existence of homogeneous clumps and knots resulting from Rayleigh-Taylor instabilities after explosion and observed in SN remnants such as Cas A (Douvion *et al.* 2001). For the fully-mixed SN1987A surrogate, elemental abundances are those of Kozasa *et al.* (1989) whereas the elemental compositions of Nosawa *et al.* (2003) are used for both the PMSN fully-mixed and unmixed cases.

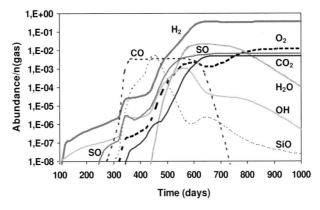

Figure 1. Molecular abundances versus time for the 170 $M_\odot$ PMSN fully-mixed case.

The chemical network includes tri-molecular reactions efficient in high density media, bi-molecular processes (neutral-neutral reactions with or without activation energy barriers, ion-molecule reactions) and radiative association reactions. In total, the system comprises 90 species and between 400 to 500 reactions, depending on the ejecta region under study. Among the chemical species, we consider astro-physically relevant molecules including CO, CO_2, O_2, SO, NO, C_2H_2, H_2O and H_2 and dust precursors (chains, rings and clusters) such as $(MgO)_n$, $(SiO_2)_n$, $(FeO)_n$, $(MgS)_n$, $(Si)_n$, and $(Fe)_n$ with n = 1-4. We also include carbon chains up to C_6 and benzene (C_6H_6).

4. Results and discussion

Molecular abundances for the fully-mixed case are illustrated in Figure 1 for the PMSN surrogate. Species form at early times and in large amounts by non-steady state processes. The main formation and destruction chemical pathways are neutral-neutral reactions up to $\sim$600 days whereas ion attack (He^+, O^+) and fast Compton electron destruction are effective at later times. Dominant molecules are O_2, CO_2, SO, and H_2 and H_2O if hydrogen penetration from the progenitor envelope is allowed. The total molecular budget is $\sim$57 $M_\odot$ while it is $\sim$42 $M_\odot$ in the unmixed PMSN case (Cherchneff & Lilly 2008). In any case, we find that 25 to 34% of the progenitor mass is ejected in molecular form in the local, pristine gas.

In CCSNe, molecular formation is also very efficient as illustrated in Figure 2. Again, main species include O_2, CO_2, SO, CO and NO, and H_2 and H_2O when H-mixing is considered. In this specific case, $\sim$5% of hydrogen penetrate the helium core. SiO also forms in large amount but is quickly removed from the gas at $t \geqslant 550$ days due to its inclusion into dust precursors. The total molecular budget is 7.3 $M_\odot$, equivalent to 36.5% of the progenitor mass. As for the most abundant dust precursors., AlO and $(SiO_2)_2$ start forming in large amounts at $t \geqslant 450$ days, a time in excellent agreement with the onset of dust formation in SN1987A (Lucy *et al.* 1989). This sequence indicates that corundum and quartz condense simultaneously at early times where the gas densities are high, leading to large dust amounts. Periclase, iron oxide and magnesium sulfite precursors form with large abundances but at much later times ($t \geqslant 750$ days), therefore leading to small amounts of their parent condensates. This fully-mixed case thus fosters the formation of corundum and quartz, but hampers silicate formation like forsterite (Mg_2SiO_4), for the abundance of periclase precursors is 4 orders of magnitude less than that of silica precursors at 650 days after explosion. These results disagree with condensation sequences

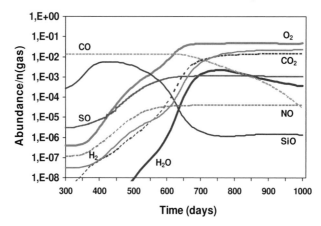

Figure 2. Molecular abundances versus time for the fully-mixed case of SN1987A.

derived by existing studies (Kozasa *et al.* 1989, 1991). These models predict the following condensate sequence for the fully-mixed case: corundum at 450 days, forsterite (or enstatite MgSiO$_3$) at 550 days, and magnetite (Fe$_3$O$_4$) at 630 days. This discrepancy points to the crucial role of chemical kinetics as the bottleneck to dust formation in circumstellar environments.

We conclude that SN ejecta are rich molecular environments. Their chemical composition depends on the degree of mixing of heavy elements in the helium core and of hydrogen from the progenitor envelope. Therefore molecules could be used as observational tracers of mixing in nearby SN ejecta. In the early universe, PMSN molecules may provide the cooling necessary to trigger Pop. II.5 star formation in the PMSN dense shell (Mackey *et al.* 2003, Salvaterra *et al.* 2004) if they survive the reverse shock some 10,000 years after the PMSN explosion.

References

Cherchneff, I. & Lilly, S. 2008, *ApJ* (Letters), submitted
Clayton, D. D., Liu, W., & Dalgarno, A. 1999, *Science*, 283, 1290
Donn, B. & Nuth, J. A. 1985, *ApJ*, 288, 187
Douvion, T, Lagage, P. O., & Pantin, E. 2001, *AA*, 369, 589
Kozasa, T., Hasegawa, H., & Nomoto, K. 1989, *ApJ*, 344, 325
Kozasa, T., Hasegawa, H. & Nomoto, K. 1991, *AA*, 249, 474
Lepp, S., Dalgarno, A., & McCray, R. 1990, *ApJ*, 358, 262
Liu, W. & Dalgarno, A. 1994, *ApJ*, 438, 789
Liu, W. & Dalgarno, A. 1995, *ApJ*, 454, 472
Lucy, L. B., Danziger, I. J., Gouiffes, C., & Bouchet, P. 1989, in: G. Tenorio-Tagle, M. Moles, & J. Melnick (eds.), *IAU Coll. 120, Structure and Dynamics of the Interstellar Medium* (Berlin: Springer-Verlag), p. 164
Mackey, J., Bromm, V., & Hernquist, L. 2003, *ApJ*, 586, 1
Nozawa, T, Kozasa, T., Umeda, H., Maeda, K., & Nomoto, K. 2003, *ApJ*, 598, 785
Pinto, P. A. & Woosley, S. E. 1988, *Nature*, 333, 534
Roche, P. F., Aitken, D. K., & Smith, C. H. 1991, *MNRAS*, 252, 39
Salvaterra, R., Ferrara, A., & Schneider, R. 2004, *Nature*, 10, 113
Schneider, R., Ferrara, A., Salvaterra, R. 2004, *MNRAS*, 351, 1379
Spyromilio, J., Meikle, W. P. S., Learner, R. C. M., & Allen, D. A. 1988, *Nature*, 334, 327
Todini, P. & Ferrara, A. 2001, *MNRAS*, 325, 726
Umeda, H. & Nomoto, K. 2002, *ApJ*, 565, 385
Wooden, D. H., Rank, D. M, Bregman, J. D., Witteborn, F. C., Tielens, A. G. G. M., Cohen, M., Pinto, P. A., & Axelrod, T. S. 1993, *ApJS*, 88, 477

Discussion

UNKNOWN: I would think the isotropic signature maybe particularly telling about what is going on in the supernovae that you considered the CNO isotopes at all?

CHERCHNEFF: Yes, I agree, but the chemistry so far does not include isotopes at all. I am just looking really at plain chemistry to try to identify which kind of molecules I can form.

UNKNOWN: Do these stars produce mostly 12-carbon and 16-oxygen?

CHERCHNEFF: Yes, they do.

MUMMA: I wonder if you consider the possibility of doing work on these molecules, water, CO and so forth in the infrared wavelengths before mm and submm facilities become available.

CHERCHNEFF: In terms of observing molecules at the infrared wavelengths, this field was very active 20 years ago when SN1987A exploded. If you look at the conditions in the ejecta of supernovae where dust forms, they are not so different from the conditions you find in AGB stars very close to the photosphere. We know that in AGB winds, dust forms along with large amounts of molecules. It is very much the same in supernovae. The problem with the infrared is that you really need to observe 'warm' molecules right after the explosion. On the other hand, if you look at objects that exploded two or three years ago at submm wavelengths, you should see molecules in the ejecta.

From let to right: Yi-Jehng Kuan, DeWayne Halfen, Lucy Ziurys, Erin Smith (photo by Dale Cruikshank).

Welcome reception. From left to right: Kasandra O'Malia, Oscar Martinez, Ted Snow, Ernst Zinner, Larry Nittler.

Organic Matter in Space
Proceedings IAU Symposium No. 251, 2008
S. Kwok & S. Sandford, eds.

© 2008 International Astronomical Union
doi:10.1017/S1743921308021637

Mapping the PAHs and H$_2$ in ρ Oph A

Kay Justtanont[1], René Liseau[1], and Bengt Larsson[2]

[1]Onsala Space Obseravtory, Chalmers University of Technology,
SE-439 92 Onsala, Sweden
email: `kay.justtanont@chalmers.se`, `rene.liseau@chalmers.se`

[2]Stockholm Observatory, AlbaNova University Center,
SE-106 91 Stockholm, Sweden
email: `bem@astro.su.se`

Abstract. We present an ISOCAM-CVF map of the ρ Oph A region, covering $3' \times 3'$. For each 6 arcsec^2 pixel, we extract the spectrum from 5–15 μm. We determine the fluxes of the main PAH features by fitting Lorentzian profiles to the spectrum. The peaks of the various PAH components correspond well with the known positions of the PDRs in this vicinity. The spectrum in several pixels exhibits strong rotational lines of molecular hydrogen which can be used to derive the physical properties of the cloud. The H$_2$ emission traces the hot gas of the bipolar CO outflow from VLA1623.

Keywords. Molecular processes, ISM: clouds, ISM: jets and outflows, Stars: formation

1. Introduction

At 120 pc (Lombardi *et al.* 2008), the ρ Oph A is one of the closest active star forming regions. Within this dense core, a couple of B stars ionize the surrounding gas, creating Photo-Dissociation Regions (PDRs). Using the data archive of the ISO mission (Kessler *et al.* 1996), we extracted ISOCAM (Cesarsky *et al.* 1996) CVF data which partly cover the SM1 cloud and the CO outflow from VLA 1623 (Dent *et al.* 1995), we extracted spectra of 32×32 pixels which extend from 5 to 15μm. Each pixel has a size of $6'' \times 6''$, i.e., 8.5×10^{-10} sr.

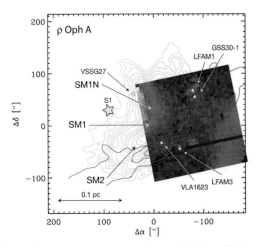

Figure 1. The ISOCAM PAH 6.2 μm map, together with the cold dust map (light contours) and the CO outflow (solid contours). The origin is at $16^h 23^m 27.5^s$, $-24° 23' 56''$, J2000.

2. PAH maps

The spectrum in most ISOCAM pixels shows that the emission is dominated by Poly-cyclic Aromatic Hydrocarbons (PAHs). We derive fluxes of each PAH band by fitting a Lorentzian profile to the continuum subtracted spetrum. Figure 1 shows that the peak emission of the $6.2 \,\mu$m band arises from the region heated by the star S1, and lies close to the peak of the cold dust emission (Motte *et al.* 1998). Another emission peak comes from the PDR region excited by the star HD 147889 (Abergel *et al.* 1996). Different emission bands of these PAHs have a similar distribution (Figure 2), having two peaks on the north-east and south-west of the map. The region of the CO outflow shows minimum emission from PAHs in all the bands.

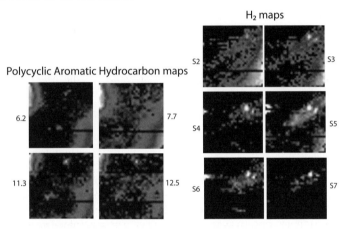

Figure 2. Left : The map of main PAH bands. One peak emission is a region close to the exciting star S1. The emission from the lower right arises from the bight western filament, excited by the star HD 147889. Right : The ISOCAM-CVF H_2 maps showing that the higher transition lines exclusively trace the CO outflow from VLA 1623 (see Figure 1).

3. H₂ maps

Several pixels also show strong emission due to rotational transitions of H_2. We fitted the H_2 lines using Gaussian profiles with the width coresponding to the resolution at that wavelength. The maps of the pure H_2 lines show that they trace the CO outflow. The lower transition lines S(2) and S(3) also show strong emission due to the PDR region on the western part of the cloud (see Figure 2), The line S(3) line is serverely affected by deep silicate dust absorption which is present in all pixels. Also the lines S(6) and S(4) are blended with the strong PAHs at 6.2 and $7.7 \,\mu$m.

From these observations, we derive iteratively the dust extinction, the gas temperature and the column density of the ortho- and para-states of warm H_2, i.e., for the brightest H_2 pixel, $A_v = 10^{\mathrm{mag}}$, $T = 1380 \pm 120$ K, o/p = 0.728 ± 0.002. In that pixel, the total H_2 luminosity is 4×10^{-11} L$_\odot$ and the H_2 mass is 3×10^{-3} M$_\odot$ ($\Omega = 8.5 \times 10^{-10}$ sr).

References

Abergel, A., Bernard, J. P., Boulanger F., *et al.* 1996, *A&A* (Letters), 315, L329
Cesarsky, C. J., Abergel, A., Agnèse, P., *et al.* 1996, *A&A* (Letters), 315, L32
Dent, W. R. F., Matthews, H. E., & Walther, D. M. 1995, *MNRAS*, 277, 193
Kessler, M. F., Steinz, J. A., & Anderegg, M. E. 1996, *A&A* (Letters), 315, L27
Lombardi, M., Lada, C. J., & Alves, J. 2008, *A&A*, 480, 785
Motte, F., André, P., & Neri, R. 1998, *A&A*, 336, 150

Organic Matter in Space
Proceedings IAU Symposium No. 251, 2008
S. Kwok & S. Sandford, eds.

© 2008 International Astronomical Union
doi:10.1017/S1743921308021649

Organic compounds in galaxies

Takashi Onaka[1], Hiroko Matsumoto[1], Itsuki Sakon[1], and Hidehiro Kaneda[2]

[1]Department of Astronomy, Graduate School of Science, The University of Tokyo, Bunkyo-ku,
Tokyo 113-0033, Japan
email: onaka@astron.s.u-tokyo.ac.jp, matsumoto@astron.s.u-tokyo.ac.jp,
isakon@astron.s.u-tokyo.ac.jp

[2]Institute of Space and Astronautical Science, Japan Aerospace Exploration Agency,
Sagamihara, Kanagawa 229-8510, Japan
email: kaneda@ir.isas.jaxa.jp

Abstract. The unidentified infrared (UIR) emission bands in the near- to mid-infrared are thought to originate from organic compounds in the interstellar medium. Recent space observations with *Spitzer* and *AKARI* have clearly revealed that the UIR bands are commonly seen in external galaxies, including elliptical galaxies, except for very metal-poor dwarf galaxies. They are also detected in extended structures of galaxies, such as extra-planar components and filaments produced by outflows, suggesting that the band carriers are ubiquitous organic compounds in galaxies. Since the UIR bands are prominent features in the infrared spectrum of galaxies and are linked to the star-formation activity, it is highly important to understand the nature, formation, processing, and destruction of the UIR band carriers in galaxies. While there is no systematic variation detected in the UIR spectrum in normal galaxies, significantly low values are derived for the ratio of the 7.7 μm to 11.2 μm bands in elliptical galaxies as well as in galaxies with low-luminosity AGNs compared to normal star-forming galaxies. Relatively low band ratios are also seen in the UIR band spectrum of extended structures in galaxies. If the same mechanism leads to the low band ratio, it would provide important information on the band carrier properties. It should also be noted that the band carriers are believed to be destroyed in a short time scale in environments where low band ratios are detected. The survival and supply processes in these environments are a key to understand the nature of the band carriers.

Keywords. Galaxies: ISM, infrared: galaxies, infrared: ISM, dust, extinction

1. Introduction

A set of emission bands at 3.3, 6.2, 7.6–7.8, 8.6, 11.2, 12.7, and 16–18 μm, have been observed in various celestial objects and are called the unidentified infrared (UIR) bands. The exact nature of the carriers has not yet been understood completely, but it is generally believed that the emitters or emitting atomic groups containing polycyclic aromatic hydrocarbons (PAHs) or PAH-like atomic groups of carbonaceous materials are responsible for the UIR bands. The UIR bands have been observed in a wide range of objects and they are commonly seen in diffuse Galactic emission (Onaka 2004) as well as in external galaxies (Helou *et al.* 2000, Smith *et al.* 2007), suggesting that the band carriers are a major component of the interstellar matter. Since their intensities are well correlated to far-infrared emission or to star-formation activity (Onaka 2000, Peeters *et al.* 2004) and since the UIR bands are distinct features in the infrared spectrum of galaxies, the understanding of their origin and nature is of great importance for the study of star-formation activities in remote galaxies as well.

Where the band carriers are formed and how they are processed and destroyed in the interstellar medium (ISM) are important issues for the study of the UIR bands in addition to the exact nature of the carriers. The band carriers may be formed in circumstellar envelopes of carbon stars (Frenklach & Feigelson 1989, Latter 1991, Cherchneff *et al.* 1992, Galliano *et al.* 2008a). The UIR bands have, however, been detected only in a handful of carbon stars (Boersma *et al.* 2006, Speck *et al.* 2008). It is not yet clearly understood how efficiently the carriers are formed in stellar sources. They could also be produced by fragmentation of large carbonaceous grains in the interstellar medium (Jones *et al.* 1996, Greenberg *et al.* 2000) or formed in situ in dense clouds (Herbst 1991).

Small band carriers are thought to be destroyed rapidly in hot plasma by thermal sputtering (Dwek & Arendt 1992, Tielens *et al.* 1994). They will also be destroyed in strong radiation fields (Allain *et al.* 1996), although the radiation destruction is only effective in the vicinity of exciting sources. Peeters *et al.* (2002) indicate that there are at least three classes of the 6–9 μm UIR band spectrum in Galactic sources on the basis of the peak wavelengths of the bands. Processing from 'fresh' band carriers to 'matured' ones is not detected in a planetary nebula, suggesting that the change must occur in the ISM (Matsumoto *et al.* 2008a). On the other hand, the relative band strengths and profiles of the UIR bands do not show systematic variations in the mid-infrared (MIR) spectrum of diffuse Galactic radiation (Chan *et al.* 2001, Kahanpää *et al.* 2003) except that small changes are seen between those in the inner and outer Galaxy (Sakon *et al.* 2004). External galaxies provide a much wider range of physical conditions and thus variations with environments should appear more clearly, if any. In this study, the variations of the UIR band spectrum among external galaxies that include elliptical and dwarf galaxies together with their spatial variations in galaxies are discussed based on latest observations with *Spitzer* (Werner *et al.* 2004) and *AKARI* (Murakami *et al.* 2007).

2. UIR Bands in Galaxies

Elliptical galaxies are deficient in interstellar matter and they are believed to provide hostile environments against dust grains: they are associated with hot plasma, which destroys dust grains in a short time scale by thermal sputtering. Recent *Spitzer* and *AKARI* observations, however, indicate that the presence of the UIR bands is not uncommon in elliptical galaxies although the appearance of the UIR bands is quite different from those familiar to spiral galaxies (Kaneda *et al.* 2005, 2007). Figure 1 shows some examples of spectra taken with the Infrared Spectrograph (IRS) on board *Spitzer* (Houck *et al.* 2004) and the Infrared Camera (IRC) on board *AKARI* (Onaka *et al.* 2007, Ohyama *et al.* 2007). The 11.2 μm band is clearly seen together with the 17 μm complex, whereas the 6.2, 7.7, and 8.6 μm bands are faint or almost absent (see also Kaneda *et al.* 2008a,b).

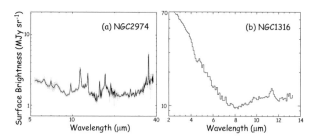

Figure 1. Examples of elliptical galaxy spectra. (a) NGC2974 with IRS on board *Spitzer* (Kaneda *et al.* 2005) and (b) NGC1316 with IRC on board *AKARI* (Kaneda *et al.* 2007).

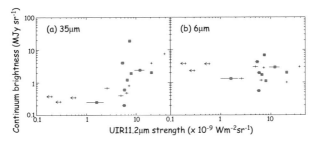

Figure 2. Correlation of the continuum emission with the UIR 11.2 μm strength for elliptical galaxies (Kaneda *et al.* 2008b) (a) 35 μm continuum emission vs. the UIR band and (b) 6 μm continuum emission vs. the UIR band. The boxes indicate galaxies with low-luminosity AGNs.

The 3.3 μm band is shown to be absent in NGC1316 based on *AKARI* IRC observations (Figure 1b), suggesting the dominance of large neutral PAHs (Kaneda *et al.* 2007).

While the presence of the UIR bands in elliptical galaxies is well established, the formation and survival of the band carriers in them are not yet clearly understood. There seem to be three potential origins for the band carriers in elliptical galaxies; stellar mass loss, recent merger events, and cooling flows. Kaneda *et al.* (2008b) indicate that the band intensity is well correlated with the dust continuum emission at 35 μm, but the correlation is poor with the continuum emission at 6 μm for 18 elliptical galaxies (Figure 2). Since the 6 μm continuum is supposed to come from stellar photospheric emission in these objects, it is suggested that the stellar mass loss is not a dominant source for the UIR band carriers in elliptical galaxies. Correlation with the dust continuum strongly suggests the interstellar origin of the band carriers.

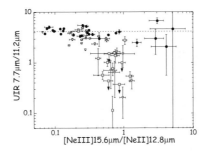

Figure 3. Band intensity ratio of the UIR 7.7 to 11.2 μm against the line ratio of [NeIII]15.6 μm/[NeII]12.8 μm. The SINGS galaxy data are taken from Smith *et al.* (2007). They are indicated either by the filled circles for those with HII-dominated nuclei or by the open triangles for those with low-luminosity AGNs. Elliptical galaxies from Kaneda *et al.* (2008) are indicated by the open squares. The dotted line shows the median of the HII-dominated sources.

Smith *et al.* (2007) studied the UIR band spectra of the Spitzer Infrared Nearby Galaxies Survey (SINGS) sample galaxies (Kennicutt *et al.* 2003). Figure 3 shows the band ratio for the SINGS galaxies together with the elliptical galaxies (open squares) in Kaneda *et al.* (2008b). Smith *et al.* (2007) found the intensity ratio of the 7.7 to 11.2 μm UIR bands is lower for galaxies harboring low-luminosity AGNs (open triangles) than galaxies with HII-dominated nuclei (filled circles). Since most elliptical galaxies harbor AGNs, this trend is compatible with the results of elliptical galaxies as indicated in Figure 3. The important question is whether or not AGNs play a key role in making the band ratio low. Kaneda *et al.* (2007) point out that the AGN activity of NGC1316, which shows a low band ratio in its MIR spectrum (Figure 1b), ceased 100 Myr ago and thus its AGN should not make an appreciable effect on the present UIR band spectrum of this galaxy.

Recently Shi *et al.* (2007) show that the UIR band ratios of AGN host galaxies are not different from those of star-forming galaxies. Taking account of these facts, it seems to be unlikely that AGNs play a major role for the low band ratio. There must be other factors that produce the peculiar UIR band spectrum.

Figure 3 also shows that the band ratio, if the UIR bands are present, does not change even for dwarf galaxies with hard radiation fields characterized by high [NeIII]/[NeII] ratios. On the other hand, observations clearly indicate that the UIR bands are absent in metal-poor dwarf galaxies. There seems to be a threshold, such as log(O/H)+12 ~ 8, in the metallicity, below which the UIR bands are not present (Smith *et al.* 2007, Engelbracht *et al.* 2008). Since those low-metallicity galaxies also have very strong radiation fields, the absence of the UIR bands could be related either to the metallicity, the radiation field strength, or to a combination of them (Wu *et al.* 2006). On the other hand, O'Halloran *et al.* (2006) attributed the absence to efficient destruction of the band carriers in shocks based on the correlation with the [FeII]/[NeII] line ratio. Recently Galliano *et al.* (2008a) indicated that if carbonaceous dust is mainly supplied from carbon stars, young galaxies may not have an appreciable amount of carbonaceous dust since they do not have a sufficient time for less massive stars to evolve to carbon stars. The absence of the UIR bands can thus be accounted for by the delayed formation of carbonaceous dust in low-metallicity environments. *Spitzer* and *AKARI* observations have revealed that the UIR band carriers are commonly present in most galaxies including ellipticals except for metal-poor dwarf galaxies and that they are major organic compounds in the ISM.

3. Spatial Variation of the UIR bands in Galaxies

Sakon *et al.* (2004) reported the first detection of a systematic variation in the UIR bands in diffuse Galactic emission that the UIR 7.7 to 11.2 μm band ratio is slightly lower in the outer Galactic plane than in the inner region based on observations of diffuse Galactic emission with *IRTS* (Onaka *et al.* 1996). Sakon *et al.* (2007) also indicated that the band ratio is lower in the interarm region than in the arm region in NGC 6946 based on observations with *AKARI* IRC (see Figure 4). Galliano *et al.* (2008b) made detailed analysis on ISOCAM observations of several galaxies, suggesting that the UIR 7.7 to 11.2 μm band ratio is lower in the outer region than in the disk region of galaxies. Irwin & Madden (2006) reported the detection of the UIR bands in the halo of NGC 5907 and suggested that the 7.7 to 11.2 μm band ratio becomes lower in the halo region than in the disk region. *ISO* and *Spitzer* observations further suggest that the UIR bands are present in regions even several kpc away from the galactic disk (Engelbracht *et al.* 2006, Irwin *et al.* 2007). The environmental conditions at such distant regions are not favorable for the band carriers and the survival of the carriers is an interesting problem to be investigated.

Recently Matsumoto *et al.* (2008b) reported the detection of the UIR bands in the MIR spectrum of one of the Hα filaments in the dwarf galaxy NGC 1569 (Hunter *et al.* 1993)

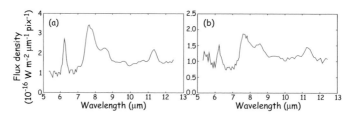

Figure 4. *AKARI* IRC spectra of (a) arm and (b) interarm region of NGC 6946 (Sakon *et al.* 2007).

based on *AKARI* IRC observations. NGC 1569 is known to harbor several super star clusters (e.g. Tokura *et al.* 2006) and the latest star-formation event is estimated to have occurred about 10 Myr ago (Greggio *et al.* 1998). The UIR band emission is detected over the entire galactic disk (Madden *et al.* 2006). HI gas inflow from an outer cloud is also suggested, which may have triggered the star-formation and the creation of the Hα filaments (Stil & Israel 1998, Mühle *et al.* 2005). The X-ray distribution shows a good correspondence to the Hα filaments, suggesting that the filaments are produced by the outflow from the galaxy disk triggered by the recent star-formation activity (Martin 1998, Martin *et al.* 2002).

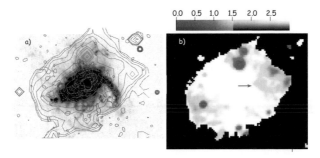

Figure 5. (a) Images of NGC 1569. The blue contours indicate the 7 μm band image of the *AKARI* IRC observations and the red contours show the X-ray intensity taken by *Chandra* (Martin *et al.* 2002). The gray scale shows the Hα image taken from Hunter *et al.* (1993). (b) The 15 to 7 μm band ratio image of NGC 1569. The arrow indicates the position of the filamentary structure corresponding to the Hα filament.

Figure 5a shows the 7 μm band intensity of the *AKARI* IRC in blue contours together with the Hα in gray scale and the X-ray emission in red contours (see the color figure in the electronic version). A prominent Hα filament seen in the west part of the galaxy encloses the X-ray emission, just outside of which a filamentary structure is seen in the 7 μm contours. Figure 5b shows an image of the 15 to 7 μm color. The arrow indicates the position of the filamentary structure corresponding to the Hα filament. The filament appears bright at 7 μm relative to 15 μm. The IRC 7 μm band effectively probes the UIR 6.2 and 7.7 μm bands and the 15 μm band intensity comes mostly from the MIR continuum emission (Sakon *et al.* 2007). Figure 5b suggests the presence of the UIR bands in the filament.

Subsequent spectroscopic observations with the IRC confirmed the presence of the UIR bands in the MIR spectrum (Matsumoto *et al.* 2008b). Figure 6 shows the comparison of the filament spectrum with that in the galaxy disk taken by *Spitzer* IRS. After fitting with Lorentzians (e.g. Smith *et al.* 2007), the intensity ratio of the 7.7 and 11.2 μm bands is derived to be 3.5 ± 0.4 and 1.4 ± 0.6 for the disk and the filament, respectively. The band ratio of the galaxy disk is compatible with those of star-forming galaxies (Figure 3), whereas that of the filament shows as lower a value as seen in extended structures of other galaxies. The UIR 7.7 μm band consists of at least two components, 7.6 μm and 7.8 μm. A closer look of the spectrum further indicates that the 7.8 μm component seems to be absent in the filament spectrum. A similar "narrow" and weak 7.7 μm band has been reported for the low-metallicity dwarf galaxy IC2574 (Walter *et al.* 2007). Further investigations are needed to understand the link between the two objects.

The age of the filamentary structure is estimated to be several Myr based on the expansion velocity of the Hα filament ($\sim$90 km s^{-1}, Westmoquette *et al.* 2008) and the distance to the filament from the youngest super star cluster ($\sim$320 pc), which is consistent

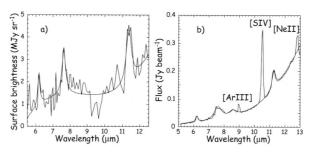

Figure 6. (a) *AKARI* IRC spectrum of a Hα filament of NGC 1569 (Matsumoto *et al.* 2008b) and (b) *Spitzer* IRS spectrum of the NGC 1569 disk (Tajiri *et al.* 2008). The thin lines indicate the observed spectra, while the thick lines indicate the fitted spectra. The ionic forbidden lines are labeled.

with the age of the recent star-formation event. The destruction time scale by thermal sputtering in hot plasma can be approximated by $\sim 1 \times 10^3 \mathrm{yr}\, (a/1\,\mathrm{nm})\, (1\,\mathrm{cm}^{-3}/n_\mathrm{H})$ for $10^6 < T < 3 \times 10^8$ K, where a is the particle radius and n_H is the proton density (Tielens *et al.* 1994). The density of the X-ray emitting gas is estimated to be 0.035 cm^{-3} (Ott *et al.* 2005a, b). Thus for $a = 1$ nm, the destruction time scale becomes about 3×10^4 yr, too short compared to the age of the filament. There must be an efficient supply source for the band carriers in the filament region. No significant star-formation is seen in the region and thus neither formation in dense clouds nor production in the stellar outflow seems to be a dominant supply source. It is most likely that the carriers are produced by fragmentation from large carbonaceous grains in shocks, which produce the Hα emission.

4. Summary

Recent *Spitzer* and *AKARI* observations provide new and interesting data for the study of the UIR bands in external galaxies. They have clearly revealed that the UIR bands are commonly seen in galaxies, including elliptical galaxies, except for very metal-poor dwarf galaxies. They also detect the UIR bands in extra-planar structures and filamentary structures that are formed by outflows from the galactic disk. The UIR bands have been detected even several kpc away from the galactic disk, indicating the UIR band carriers are major and ubiquitous organic compounds in the universe. The formation, processing, and destruction of the band carriers in various environments are key issues for the understanding of the UIR bands.

The ratio of the UIR 7.7 to 11.2 μm band intensity is quite constant for star-forming galaxies, including dwarf galaxies with strong radiation fields. However the ratio is systematically lower in extended structures of galaxies. It is also indicated to be low in the interarm region. The extreme case is seen in elliptical galaxies, where some of their spectra do not show the 6.2 and 7.7 μm emission at all. The low value is usually attributed to the low ionization degree of the band carriers in these environments since the 6.2 and 7.7 μm bands are not strong in neutral PAHs and are enhanced when they become ionized. The objects that show low band ratios have a common characteristic that they are mostly low density and free from significant stellar components. Some of them are associated with Hα emission and/or X-ray emission. If these low ratios have a common origin, it would be very important to understand the physical conditions that make the ratio low for the study of the UIR band carriers. There is also an indication that the 7.7 μm band lacks its longer wavelength component in some objects when they are weak. High S/N ratio spectra are required for further study.

The band carriers are easily destroyed in hot plasma and thus the presence of the UIR bands in extended structures of galaxies suggests revisit of the survival/destruction process of the band carriers. Recent observations also indicate likely supply sources for the UIR band carriers in galaxies. Correlation study of the elliptical galaxy sample indicates that mass loss from evolved stars is not a dominant source. It is suggested that fragmentation from large carbonaceous grains play a role in a filament of NGC1569.

Acknowledgements

This work is based on observations with *Spitzer*, which is operated by Jet Propulsion Laboratory, California Institute of Technology, under NASA contract 1407, and on observations with *AKARI*, a JAXA project with the participation of ESA. The authors thank all the members of the *AKARI* project and the members of the Interstellar Medium and Nearby Galaxy team for their continuous encouragement. This work was supported by a Grant-in-Aid for Scientific Research on Priority Areas from the Ministry of Education, Culture, Sports, Science and Technology, Japan and Technology of Japan and a Grant-in-Aid for Scientific Research from the Japan Society of Promotion of Science.

References

Allain, T., Leach, S., & Sedlmayr, E. 1996, *A&A*, 305, 602
Boersma, C., Hony, S., & Tielens, A. G. G. M. 2006, *A&A*, 447, 213
Chan, K.-W. *et al.* 2001, *ApJ*, 546, 273
Cherchneff, I., Barker, J. R., & Tielens, A. G. G. M. 1992, *ApJ*, 401, 269
Dwek, E. & Arendt, R. G. 1992, *ARAA*, 30, 11
Engelbracht, C. W. *et al.* 2006, *ApJ (Letters)*, 642, L127
Engelbracht, C. W. *et al.* 2008, *ApJ*, 678, 804
Frenklach, M. & Feigelson, E. D. 1989, *ApJ*, 341, 372
Galliano, F., Dwek, E., & Chanial, P. 2008a, *ApJ*, 672, 214
Galliano, F., Madden, S. C., Tielens, A. G. G. M., Peeters, E., & Jones, A. P. 2008b, *ApJ*, 679, 310
Greenberg, J. M. *et al.* 2000, *ApJ (Letters)*, 531, L71
Greggio, L. *et al.* 1998, *ApJ*, 504, 725
Helou, G. *et al.* 2000, *ApJ (Letters)*, 532, L21
Herbst, E. 1991, *ApJ*, 366, 133
Hunter, D. A., Hawley, W. N., & Gallagher, J. S. 1993, *AJ*, 106, 1797
Houck, J. R. *et al.* 2004, *ApJS*, 154, 18
Irwin, J. A., Kennedy, H., Parkin, T., & Madden, S. 2007, *A&A*, 474, 461
Irwin, J. A. & Madden, S. 2006, *A&A*, 445, 123
Jones, A. P., Tielens, A. G. G. M., & Hollenbach, D. J. 1996, *ApJ*, 469, 740
Kahanpää, J., Mattila, K., Lehtinen, K., Leinert, C., & Lemke, D. 2003, *A&A*, 405, 999
Kaneda, H., Onaka, T., & Sakon, I. 2005, *ApJ (Letters)*, 632, L83
Kaneda, H., Onaka, T., & Sakon, I. 2007, *ApJ (Letters)*, 666, L21
Kaneda, H., Onaka, T., Sakon, I., Matsumoto, H., & Suzuki, I. 2008a, this volume
Kaneda, H., Onaka, T., Sakon, I., Kitayama, T., Okada, Y., & Suzuki, T. 2008b, *ApJ*, in press
Kennicutt, R. C. Jr. *et al.* 2003, *PASP*, 115, 928
Madden, S., Galliano, F., Jones, A. P., & Sauvage, M. 2006, *A&A* 446, 877
Martin, C. L. 1998, *ApJ*, 491, 561
Martin, C. L., Kobulnicky, H. A., & Heckman, T. M. 2002, *ApJ*, 574, 663
Matsumoto, H. *et al.* 2008a, *ApJ*, 677, 1120
Matsumoto, H., Onaka, T., Sakon, I., & Kaneda, H. 2008b, this volume
Mühle, S., Klein, U., Wilcots, E. M., & Hüttenmeister, S. 2005, *AJ*, 130, 524
Murakami, H. *et al.* 2007, *PASJ*, 59, S369
O'Halloran, B., Satyapal, S., & Dudik, R. P. 2006, *ApJ*, 641, 795

Ohyama, Y. *et al.* 2007, *PASJ*, 59, S411

Onaka, T. 2000, *Adv. Sp. Res.*, 25, 2167

Onaka, T. 2004, in: A. N. Witt, G. C. Clayton, & B. T. Draine (eds.), *Astrophysics of Dust*, ASP Conf. ser. 309, p. 163

Onaka, T., Yamamura, I., Tanabé, T., Roellig, T. L., & Yuen, L. 1996, *PASJ* (Letters), 48, L59

Onaka, T. *et al.* 2007, *PASJ*, 59, S401

Ott, J., Walter, F., & Brinks, E. 2005a, *MNRAS*, 358, 1423

Ott, J., Walter, F., & Brinks, E. 2005b, *MNRAS*, 358, 1453

Peeters, E., Spoon, H. W. W., & Tielens, A. G. G. M. 2004, *ApJ*, 613, 986

Peeters, E. *et al.* 2002, *A&A*, 390, 1089

Sakon, I. *et al.* 2004, *ApJ*, 609, 203 (Erratum: *ApJ*, 625, 1062)

Sakon, I. *et al.* 2007, *PASJ*, 59, S483

Shi, Y. *et al.* 2007, *ApJ*, 669, 841

Smith, J. D. *et al.* 2007, *ApJ*, 656, 770

Speck, A., Barlow, M., Wesson, R., Glayton, G., & Volk, K. 2008, this volume

Stil, J. M. & Israel, F. P. 1998, *A&A*, 337, 64

Tajiri, Y. Y. *et al.* 2008, in: R. Chary, H. I. Teplitz, & K. Sheth (eds.), *Infrared Diagnostics of Galaxy Evolution*, ASP Conf. ser. 381, p. 50

Tielens, A. G. G. M., McKee, C. F., Seab, C. G., & Hollenbach, D. J. 1994, *ApJ*, 431, 321

Tokura, D. *et al.* 2006, *ApJ*, 648, 355

Walter, F. *et al.* 2007, *ApJ*, 661, 102

Werner, M. W. *et al.* 2004, *ApJS*, 154, 309

Westmoquette, M. S., Smith, L. J., & Gallagher, J. S. 2008, *MNRAS*, 383, 864

Wu, Y. *et al.* 2006, *ApJ*, 639, 157

Discussion

PAPOULAR: To my knowledge, the effect of UV or visible radiation, if any, is to aromatize the particles. So what is the experimental laboratory evidence for a possible destruction of these particles to explain the disappearance of some of the features?

ONAKA: I do not know very well about the laboratory data, but observationally there are data indicating the UIR bands are not present in the H II regions. But there are some exceptions also, as there are data indicating the presence of the UIR bands together with the [Ne II] line emission. But there are also data indicating that they are quite absent inside diffuse H II regions.

PAPOULAR: Yes, this is an observational effect, but that does not mean that the radiation can destroy these particles.

CECCARELLI: I am curious you didn't show the sizes. How far away from the galactic plane you see these PAHs?

ONAKA: They are somewhere around several kilo parsecs from the galactic plane of M82.

Organic Matter in Space
Proceedings IAU Symposium No. 251, 2008
S. Kwok & S. Sandford, eds.

© 2008 International Astronomical Union
doi:10.1017/S1743921308021650

Diffuse interstellar bands in the Local Group: From the Milky Way, the Magellanic Clouds to the Andromeda galaxy

N. L. J. Cox[1] and M. A. Cordiner[2]

[1]Herschel Science Centre, European Space Astronomy Centre, European Space Agency, Villanueva de Cañada, Madrid, Spain; email: nick.cox@sciops.esa.int

[2]Astrophysics Research Centre, School of Mathematics and Physics, Queen's University, Belfast, U.K

Abstract. We report on recent developments in the study of diffuse interstellar bands in the Local Group galaxies. We present preliminary results on the detection of the 5780 Å DIB toward 17 targets in the vicinity of NGC 206 in M31.

Keywords. Local Group, ISM: lines and bands, ISM: molecules

1. Diffuse interstellar bands

The diffuse interstellar bands (DIBs) constitute a group of over 200 optical interstellar (IS) absorption features observed toward reddened stars (see review by Herbig 1995). For Galactic lines-of-sight their equivalent widths correlate with the amount of interstellar reddening, E_{B-V}, as well as with the Na I column density (e.g., Herbig 1993). To this day it is still debated whether they arise from the dust or the gas component of the ISM. However, although not one of the DIBs has been identified, there is strong evidence that the carriers are composed of organic material (Sarre 2008). The substructure present in many of the DIB profiles indicate that the carriers are large gas-phase molecules (Sarre 1995, Ehrenfreund & Foing 1996). Likely candidates include UV-resistant organics such as PAHs, fullerenes and carbon chains (see, e.g., Salama 1996). Unfortunately it is not yet possible to compare directly the observational spectra with laboratory spectra of large ionised PAHs in the gas-phase (Salama 2008).

In the last decade, observations of diffuse interstellar bands (DIBs) have provided important insights into the chemical characteristics of possible carrier compounds (Sarre 2006). Recent observations show a strong dependence of DIB strength on the local environment in terms of cloud density and exposure to the interstellar radiation field (Cox & Spaans 2006). In particular, extragalactic environments have contributed recently to our knowledge of DIB behavior (Cox *et al.* 2006, 2007).

Our goal is to examine the link between the physical and chemical conditions in the ISM and shed light on the nature of the large molecules and very small grains believed to give rise to the DIBs.

2. The Local Group diffuse ISM

The Local Group is an ideal laboratory for detailed studies of the ISM over a broad range of physical and chemical conditions. Previous research of DIBs in our nearest neighbouring galaxies has focused on the Large and Small Magellanic Clouds (e.g., Ehrenfreund *et al.* 2002, Cox *et al.* 2006, 2007, Welty *et al.* 2006) where the effects on DIB strengths of the higher gas-to-dust ratios, lower metallicities, lower R_V and higher interstellar radiation fields were examined. Beyond the Magellanic Clouds, studies are sparse, confined to sightlines probed by sufficiently bright supernovae (e.g., Sollerman *et al.* 2005, Cox & Patat 2008) or background quasars (e.g., York *et al.* 2006, Ellison *et al.* 2008). Cordiner *et al.* (2008) detected for the first time DIBs in the spectra of two bright stars in M31, the Andromeda galaxy. In the next section we present new preliminary results on the 5780 DIB detection in 15 additional M31 sightlines.

The LMC, SMC and M31 have different global properties of the ISM. The Large Magellanic Cloud (LMC) and the Small Magellanic Cloud (SMC) have metallicities that are smaller by factor 4 and 10, respectively, with respect to the Milky Way. M31 has a metallicity close to solar (e.g., Urbaneja *et al.* 2005). A similar trend can be seen for the dust-to-gas ratio. Also their dust extinction curves are very different. The dust in the Magellanic Clouds contains more smaller dust particles (stronger FUV extinction) than the MW. The strength of the UV radiation field is also stronger in the LMC and SMC. The M31 galaxy, on the other hand shows a low interstellar UV flux (Pagani 1999) – accordingly, the M31 ISM has been subject to relatively little UV-processing which may also explain the anomalous mid-IR spectrum. Xu & Helou (1994) find a lack of small graphitic dust grains in the ISM of M31.

3. New developments: low-resolution spectra of M31 (super)giants

In the study by Cordiner *et al.* (2008), low-resolution ($R \approx 3300$), optical spectra of 72 early-to-mid-type giants and supergiants in the vicinity of the M31 OB78 association (NGC206, Van den Bergh 1964) were obtained with the Keck DEIMOS spectrograph. Five of the strongest narrow DIBs, $\lambda\lambda$ 5780, 5797, 6203, 6283 and 6613 Å, were detected in the ISM of M31. The exposure of the red spectra was 1.5 hours during excellent conditions, with seeing $0.5 - 0.8''$. 17 targets were selected from this set of spectra. These spectra, with highest S/N and earliest spectral-type (less stellar lines), showed evidence of absorption consistent with the 5780 DIB. The radial velocities, between -530 and -590 km s^{-1}, of the interstellar gas towards these targets was derived from the the Na I doublet, and subsequently used to fit synthetic DIB profiles to the observed spectra. The spectra of the selected sightlines and the profile fits are shown in figure 1. Note that these spectra also include MAG 63885 and MAG 70817 from the study by Cordiner *et al.* (2008).

In figure 2 the observed DIB strengths are compared with the interstellar sodium column density and the interstellar reddening, respectively. For comparison the Galactic relationships are also shown. Due to the low spectral resolution there is likely to be contamination of the Na I column densities due to stellar sodium, particularly for the later spectral types. The uncertainties in the stellar photometry and spectral types make an accurate determination of the reddening also difficult.

With these possible sources of error in mind, it is still clear from Figure 2 that the DIBs so far detected in M31 are, on average, stronger than Galactic DIBs relative to E_{B-V}. The ISM in the vicinity of OB78 evidently provides environments which are highly favourable for the production of DIB carriers but may have join up atypical properties compared to the rest of M31 due to the abundance of early-type stars compared with the rest of M31. The spectrum of DIBs observed in the vicinity of OB78 in M31, where the metallicity is approximately equal to solar, is found to be comparable to (or stronger than) that observed locally in the ISM of the Milky Way.

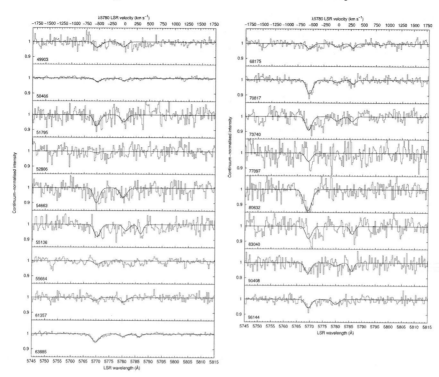

Figure 1. The λ5780 (left) and λ6283 (right) DIB spectral ranges of 17 targets in the OB78 (NGC206) association in the southern outskirts of M31. Target numbers are from Magnier 1992. The 5780 DIB is detected in 17 sightlines and the 6283 DIB in 7 sightlines.

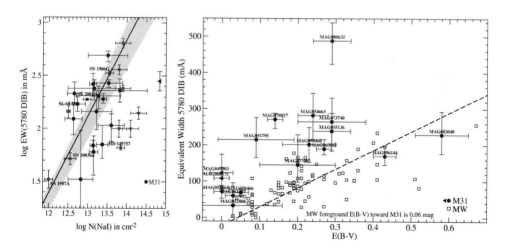

Figure 2. Correlation of λ5780 DIB equivalent widths with neutral sodium column density (left) and reddening (right). The linear fits represent the Galactic behaviour (see, e.g., Herbig 1993). The small dots in the right panel are from a Galactic DIB study (Vos *et al.* 2008).

4. Progress and future work

The low resolution of the spectra presented precludes a detailed analysis of the structure and composition of the M31 ISM. Nevertheless, the results here (preliminary) and in Cordiner *et al.* (2008) verify the feasibility of studying in more detail the atoms, molecules and diffuse bands in the ISM of the Local Group members M31 and M33. The next step will be to obtain high-resolution, high signal-to-noise spectra in order to consolidate these results and to identify whether the unusually strong DIBs observed are a common feature of M31 or are unique to the OB78 region. The chemical properties of the sightlines studied in this article may be further elucidated by examining the nature of the dust, particularly the UV extinction, which will be possible with the new 'COS' and the refurbished 'STIS' instruments on the HST.

References

Cordiner, M. A., Cox, N. L. J., Trundle, C., Evans, C. J., Hunter, I., Przybilla, N., Bresolin, F., & Salama, F. 2008, *A&A*, 480, L13

Cox, N. L. J., Cordiner, M. A., Cami, J., Foing, B. H., Sarre, P. J., Kaper, L., & Ehrenfreund, P. 2006, *A&A*, 447, 991

Cox, N. L. J., Cordiner, M. A., Ehrenfreund, P., Kaper, L., Sarre, P. J., J., Foing, B. H., Spaans, M., Cami, J., Sofia, U. J., Clayton, G. C., Gordon, K. D., & Salama, F. 2007, *A&A*, 470, 941

Cox, N. L. J. & Patat, F. 2008, *A&A*, accepted

Cox, N. L. J. & Spaans, M. 2006, *A&A*, 451, 973

Ehrenfreund, P. & Foing, B. H. 1996, *A&A*, 307, L25.

Ehrenfreund, P., Cami, J., Giminez-Vicente, J., *et al.* 2002, *ApJ*, 576, L117

Ellison, S. L., York, B. A., Murphy, M. T., Zych, B. J., Smith, A. M., & Sarre, P. J. 2008, *MNRAS* (Letters), 383, L30

Herbig, G. H. 1993, *ApJ*, 407, 142

Herbig, G. H. 1995, *ARA&A*, 33, 19

Magnier, E. A., Lewin, W. H. G., van Paradijs, J., Hasinger, G., Jain, A., Pietsch, W., & Truemper, J. 1992, *A&AS*, 96, 379

Pagani, L., Lequeux, J., Cesarsky, D., Donas, J., Milliard, B., Loinard, L., & Sauvage, M. 1999, *A&A*, 351, 447

Salama, F. 2008, *this volume*

Salama, F., Bakes, E. L. O., Allamandola, L. J., & Tielens, A. G. G. M. 1996, *ApJ*, 458, 621

Sarre, P. J. 2006, *Journal of Molecular Spectroscopy*, 238, 1

Sarre, P. J. 2008, *this volume*

Sarre, P. J., Miles, J. R., Kerr, T. H., Hibbins, R. E., Fossey, S. J., & Somerville, W. B. 1995, *MNRAS*, 277, L41

Sollerman, J., Cox, N. Mattila, S. Ehrenfreund, P., Kaper, L., Leibundgut, B., & Lundqvist, P. 2005, A&A, 429, 559.

Urbaneja, M. A., Herrero, A., Kudritzki, R. P., Najarro, F., Smartt, S. J., Puls, J., Lennon, D. J., & Corral, L. J. 2005, *ApJ*, 635, 311

Van den Bergh, S. 1964, *ApJS*, 9, 65

Vos, D. A. I., Cox, N. L. J., Kaper, L., & Ehrenfreund, P. 2008, *A&A*, submitted

Welty, D. E., Federman, S. R., Gredel, R., Thorburn, J. A., & Lambert, D. L. 2006, *ApJS*, 165, 138

Xu, C. & Helou, G., 1994, *ApJ*, 426, 109

York, B. A., Ellison, S. L., Lawton, B., Churchill, C. W., Snow, T. P., Johnson, R. A., & Ryan, S. G. 2006, *ApJ*, 647, L29

Discussion

IRVINE: A this point, what is the most distant Galaxy in which you have seen DIBs?

COX: We have detected narrow DIBs in NGC1448 toward a supernova, and also recently in M100 (also toward a supernova), which is a little bit closer. Also two narrow DIBs have been detected in a DLA system at $z \sim 0.5$ (work by York, Ellison, Lawton, Snow and collaborators).

Organic Matter in Space
Proceedings IAU Symposium No. 251, 2008
S. Kwok & S. Sandford, eds.

Properties of polycyclic aromatic hydrocarbons in the star forming environment in nearby galaxies

Itsuki Sakon[1], Takashi Onaka[1], Daisuke Kato[1], Hidehiro Kaneda[2], Hirokazu Kataza[2], Yoko Okada[2], Akiko Kawamura[3], Toshikazu Onishi[3], and Yasuo Fukui[3]

[1] Department of Astronomy, Graduate School of Science, University of Tokyo, 7-3-1 Hongo, Bunkyo-ku, Tokyo 113-0033, Japan
email: isakon@astron.s.u-tokyo.ac.jp

[2] Institute of Space and Astronautical Science, Japan Aerospace Exploration Agency, 3-1-1 Yoshinodai, Sagamihara, Kanagawa 229-8510, Japan

[3] Department of Astrophysics, Nagoya University, Chikusa-ku, Nagoya 464-8602, Japan

Abstract. We have carried out the mid-infrared slit spectroscopic observations of sources in the LMC and in NGC 6946 with AKARI/IRC. We investigate the properties of the UIR bands in terms of the star forming activities. We find systematically larger ratios of UIR bands in 6–9 μm to 11.2 μm band in active star forming regions than in the quiet regions. This behavior is consistent with the photo-ionization model of PAHs. Our results suggest that the ratios of UIR bands in 6–9 μm to 11.2 μm band can be used as more efficient and vigorous tools to measure the extent of on-going star formation in remote galaxies rather than just the presence or absence of the features themselves.

Keywords. Dust, extinction, ISM: lines and bands, galaxies: ISM, infrared: ISM

1. Introduction

Ionization of interstellar polycyclic aromatic hydrocarbons (PAHs) is one of the most notable processes to cause the variations in the UIR band spectra. Many studies based on quantum chemical calculations as well as laboratory experiments have shown that the strengths of the PAH features in 6–9 μm relative to 11.2 μm feature increase when PAHs are ionized (Allamandola *et al.* 1985, Hudgins & Allamandola 1999, DeFrees *et al.* 1993). Actually several recent observations have reported the decreasing band ratios of 7.7 μm/11.2 μm and 8.6 μm/11.2 μm with the increasing distance from the heating source within the reflection nebulae NGC 1333 (Bregman & Temi 2005, Joblin *et al.* 1996) and also the higher 6–9 μm bands to 11.2 μm band ratios for the Herbig Ae/Be stars with earlier stellar spectral types (Sloan *et al.* 2005), which are reasonably explained by the ionization model of PAHs.

In this proceedings, we present some of our recent observational results on mid-infrared spectroscopy of interstellar PAHs in Large Magellanic Cloud (LMC) and in the nearby starforming galaxy NGC 6946 using Infrared Camera (IRC, Onaka *et al.* 2007) onboard AKARI.

Catherine Cesarsky (left) and Yvonne Pendleton (right) by the Jumbo floating restaurant in Aberdeen (photo by Dale Cruikshank).

Organic Matter in Space
Proceedings IAU Symposium No. 251, 2008
S. Kwok & S. Sandford, eds.

Polycyclic aromatic hydrocarbons in elliptical galaxies

Hidehiro Kaneda[1], Takashi Onaka[2], Itsuki Sakon[2], Hiroko Matsumoto[2], and Toyoaki Suzuki[1]

[1]Institute of Space and Astronautical Science (ISAS), JAXA,
Sagamihara, Kanagawa 229-8510, Japan
email: kaneda@ir.isas.jaxa.jp

[2]Dept. of Astronomy, Graduate School of Science, The University of Tokyo,
Bunkyo-ku, Tokyo 113-0003, Japan

Abstract. We have observed 18 nearby dusty elliptical galaxies in the near- to far-infrared with Spitzer and AKARI. We have found that polycyclic aromatic hydrocarbons are present in 14 out of the 18 elliptical galaxies.

Keywords. Infrared: ISM, ISM: lines and bands, galaxies: elliptical and lenticular, cD

1. Introduction

Elliptical galaxies provide a unique interstellar environment, old stellar radiation fields with little UV light and interstellar space dominated by hot plasma. Far-IR emission has been detected from many elliptical galaxies (e.g., Goudfrooij & de Jong 1995, Temi *et al.* 2004). The dust masses seem to be larger than those determined by the balance between replenishment by stars and sputtering destruction by plasma (Goudfrooij & de Jong 1995), calling for additional dust-supplying sources. Recent studies show even the presence of polycyclic aromatic hydrocarbon (PAH) emission features in elliptical galaxies (e.g., Kaneda *et al.* 2005, 2007).

2. Observations

We observed 18 IRAS-detected nearby elliptical galaxies with AKARI and Spitzer; we have derived near- to far-IR 10-band images and near-IR spectra from the AKARI nearby galaxy project and mid-IR spectra from the Spitzer GO1&3 programs.

3. Results

We have detected PAH emission from 14 elliptical galaxies; Figure 1 shows the Spitzer/IRS spectra of several elliptical galaxies. Most of them show unusual PAH interband strength ratios; usually strongest 7.7 μm emission (e.g., Allamandola *et al.* 1985) is weak in contrast to prominent features at 11.3 and 17 μm. This may reflect the peculiar physical conditions of the ISM, i. e. dominance of large PAHs due to destruction in hot plasma, and/or dominance of neutral PAHs due to soft radiation fields from evolved stars.

The left panel in Figure 2 shows the spectral energy distribution of NGC 4589 overlaid with the Spitzer spectrum. In particular, the PAHs exhibit a prominent 17 μm feature relative to the continuum emission, which makes the AKARI/IRC 16-μm-band image an excellent tracer of large PAHs. The right panel in Figure 2 displays the AKARI 16-μm-band image shown after subtraction of the 7-μm image with its intensity adjusted on the basis of the spectrum, which is thus likely to be the spatial distribution of PAHs. The PAH distribution appears to be different from the stellar distribution, while it has some resemblance to that of cool dust in the far-IR (Kaneda *et al.*, in prep.). NGC 4589 is known to have a minor-axis dust lane as well as a complex stellar rotation field with the P.A. difference of 45° between the kinematic and morphological minor axes, which suggests that the galaxy is a merger remnant (Mollenhoff & Bender 1989). Since the distribution of the PAHs extends along the direction of the major stellar rotation, the PAHs are likely to be brought in by a merger event rather than internally produced.

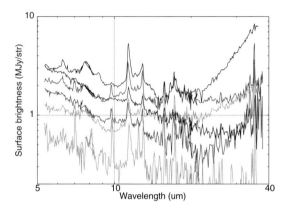

Figure 1. Spitzer/IRS spectra of elliptical galaxies obtained from our GO1&GO3 programs. Background spectra estimated from nearby blank skies have been subtracted.

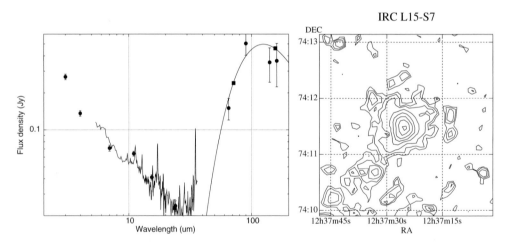

Figure 2. (Left) Spectral energy distribution of NGC4589, constructed from the AKARI (circles) and Spitzer/MIPS (boxes) photometric data, overlaid with the Spitzer/IRS spectrum. (Right) The AKARI/IRC 16-μm-band image shown after subtraction of the 7-μm-band image.

References

Allamandola, L. J., Tielens, A. G. G. M., & Barker, J. R. 1985, *ApJ* (Letters), 290, L25
Goudfrooij, P. & de Jong, T. 1995, *A&A*, 298, 784
Kaneda, H., *et al.* 2007, *PASJ*, 59, 107
Kaneda, H., Onaka, T., & Sakon, I. 2005, *ApJ* (Letters), 632, L83
Mollenhoff, C. & Bender, R. 1989, *A&A*, 214, 61
Temi, P., Brighenti, F., Mathews, W. G., & Bregman, J. D. 2004, *ApJS*, 151, 237

Organic Matter in Space
Proceedings IAU Symposium No. 251, 2008
S. Kwok & S. Sandford, eds.

© 2008 International Astronomical Union
doi:10.1017/S1743921308021686

Detection of the unidentified infrared bands in a filament of the dwarf galaxy NGC1569

H. Matsumoto[1], T. Onaka[1], I. Sakon[1], and H. Kaneda[2]

[1]Department of Astronomy, Graduate School of Science, The University of Tokyo,
Tokyo 113-0033, Japan

[2]Institute of Space and Astronautical Science, Japan Aerospace Exploration Agency,
Kanagawa 229-8510, Japan

Abstract. We made near- to mid-infrared imaging and spectroscopic observations of the dwarf galaxy NGC 1569 with the Infrared Camera (IRC) on board *AKARI*. The unidentified infrared (UIR) band features at 6.2, 7.7, and 11.2 μm, which are generally attributed to polycyclic aromatic hydrocarbons (PAHs), are clearly detected in a structure associated with an Hα filament. The filament is filled with X-ray emission and is thought to be formed by outflow from the galaxy. Since PAHs are destroyed rapidly in hot plasma, it is most likely that PAHs in the filament are produced from fragmentation of large carbonaceous grains in the shock. We also detect excess emission in 2–5 μm in the filament, which may come from very small grains.

Keywords. Infrared: galaxies, dust, extinction, galaxies: individual (NGC1569)

1. Introduction

The unidentified infrared (UIR) bands, whose major features appear at 3.3, 6.2, 7.7, 8.6, and 11.2 μm, are seen in the diffuse Galactic emission as well as in external galaxies. Polycyclic aromatic hydrocarbons (PAHs) or PAH-like aromatic groups of carbonaceous materials are thought to be the carriers of the UIR bands. There remains an interesting question where and how the band carriers are produced in galaxies, in particular in dwarf galaxies, which may represent the characteristics of the early universe because of their low-metallicity. We report here the results of near- to mid-infrared imaging and spectroscopic observations of the starburst dwarf galaxy NGC 1569 with the Infrared Camera (IRC) on board *AKARI* (Murakami *et al.* 2007). NGC 1569 is a well-studied nearby galaxy with the metallicity of about a quarter of solar, which harbors several super star clusters (SSCs) in it (e.g., Tokura *et al.* 2006). A number of filamentary structures are detected in Hα (Hunter *et al.* 1993), which delineate the X-ray emission. It is suggested that the interaction between the galactic wind and the surrounding gas results in superbubbles and Hα emitting shocked gas (Martin 1998, Martin *et al.* 2002). There is a HI cloud about 5 kpc from the galaxy, which may fuel HI gas into NGC 1569 (Stil & Israel 1998, Mühle *et al.* 2005). We report here the detection of the UIR bands in one of the Hα filaments and discuss the possible supply source of the band carriers.

2. Observations

NGC 1569 was observed in the imaging (at 3, 4, 7, 11, 15, and 24 μm) and spectroscopic (2–13 μm) modes of the IRC (Onaka *et al.* 2007, Ohyama *et al.* 2007) on 2006 September 9 and on 2007 March 8, respectively. A western filament appears prominently in the 7 μm image (Figure 1a), which correlates well with the Hα emission (Hunter *et al.* 1993), while structures corresponding to the filament are not clearly seen in the 15 and 24 μm images. Since the 7 μm band probes the UIR 6.2 and 7.7 μm bands efficiently (Sakon *et al.* 2007), this result suggests the presence of the UIR bands in the filament and subsequent spectroscopic observations of the filament were carried out. The spectrum of the western filament is shown in Figure 1b after subtracting the sky background. It confirms the presence of the UIR bands at 6.2, 7.7, and 11.2 μm in the filament. There may be weak emission at 3.3 and 8.6 μm, but further observations are needed to confirm their presence.

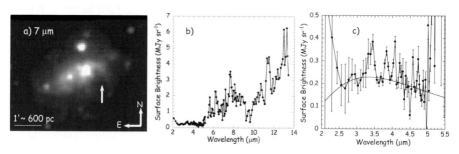

Figure 1. (a) *AKARI*/IRC 7 μm image of NGC 1569. The white arrow indicates the western filament. (b) IRC spectrum of the filament. (c) An enlargement of the near-infrared spectrum of the filament. The solid line indicates a fit with a blackbody (868 K) with a λ^{-2} emissivity.

3. Discussion

The UIR bands have been detected in halos of galaxies (Irwin & Madden 2006, Irwin *et al.* 2007) and in filaments produced by outflows (Tacconi-Garman *et al.* 2005) based on imaging observations. The present observations detect the UIR 6.2, 7.7, and 11.2 μm bands in a filamentary structure of a galaxy by spectroscopy for the first time. Since PAHs are destroyed rapidly in hot plasmas, it is most likely that PAHs in the filament are produced by fragmentation of large carbonaceous grains in the shock. Figure 1c also suggests the presence of excess continuum emission at 2–5 μm in the filament. Similar excess emission, which is possibly due to very small grains, has been observed in external galaxies (Lu *et al.* 2003) as well as in the diffuse Galactic emission (Flagey *et al.* 2007).

The work is based on observations with *AKARI*, a JAXA project with the participation of ESA. It is supported by Grants-in-Aid for Scientific Research from MEXT and JSPS.

References

Flagey, N. *et al.* 2006, *A&A*, 453, 969
Hunter, D. A., Hawley, W. N., & Gallagher, J. S. 1993, *AJ*, 106, 1797
Irwin, J. A., Kennedy, H., Parkin, T., & Madden, S. 2007, *A&A*, 474, 461
Irwin, J. A. & Madden, S. 2006, *A&A*, 445, 123
Lu, N. *et al.* 2003, *ApJ*, 588, 199
Martin, C. L. 1998, *ApJ*, 491, 561
Martin, C. L., Kobulnicky, H. A., & Heckman, T. M. 2002, *ApJ*, 574, 663
Murakami, H. *et al.* 2007, *PASJ*, 59, S369
Mühle, S., Klein, U., Wilcots, E. M., & Hüttenmeister, S. 2005, *AJ*, 130, 524
Ohyama, Y. *et al.* 2007, *PASJ*, 59, S411
Onaka, T. *et al.* 2007, *PASJ*, 59, S401
Sakon, I. *et al.* 2007, *PASJ*, 59, S483
Stil, J. M. & Israel, F. P. 1998, *A&A*, 337, 64
Tacconi-Garman, L. E. *et al.* 2005, *A&A*, 432, 91
Tokura, D. *et al.* 2006, *ApJ*, 648, 355

Organic Matter in Space
Proceedings IAU Symposium No. 251, 2008
S. Kwok & S. Sandford, eds.

Spectra of nearby galaxies measured with a new very broadband receiver

Gopal Narayanan[1,2], Ronald L. Snell[1], Neal R. Erickson[1], Aeree Chung[1], Mark H. Heyer[1], Min Yun[1], and William M. Irvine[1,3]

[1] Astronomy Department, University of Massachusetts, Amherst, MA 01003 USA

[2] email: gopal@astro.umass.edu [3] The Goddard Center for Astrobiology

Abstract. Three-millimeter-wavelength spectra of a number of nearby galaxies have been obtained at the Five College Radio Astronomy Observatory (FCRAO) using a new, very broadband receiver. This instrument, which we call the Redshift Search Receiver, has an instantaneous bandwidth of 36 GHz and operates from 74 to 110.5 GHz. The receiver has been built at UMass/FCRAO to be part of the initial instrumentation for the Large Millimeter Telescope (LMT) and is intended primarily for determination of the redshift of distant, dust-obscured galaxies. It is being tested on the FCRAO 14 m by measuring the 3 mm spectra of a number of nearby galaxies. There are interesting differences in the chemistry of these galaxies.

Keywords. Instrumentation: spectrographs, techniques: spectroscopic, galaxies: ISM

1. Introduction

The Large Millimeter Telescope (LMT) is a 50-meter diameter millimeter-wavelength single-dish telescope being built jointly by the University of Massachusetts, Amherst in the USA and the Instituto Nacional de Astrofísica, Óptica y Electrónica in Mexico (Perez-Grovas *et al.* 2006). The telescope uses recent advances in structural design and active control of surface elements, and aims to reach an overall effective surface accuracy of ~ 70 μm and an ultimate pointing accuracy of better than $1''$. The LMT is sited at 4600 m elevation at a latitude of $19°$ N in the Mexican state of Puebla and offers good sky coverage of both hemispheres. The normally low humidity will allow operation at frequencies as high as 345 GHz. Telescope construction is well advanced. Three of the planned five rings of surface panels are in place. The initial complement of instruments will include SEQUOIA, a 32 element heterodyne focal plane array for 3 mm that is currently in use at FCRAO; AzTEC, a large format, focal plane bolometer array that has had successful runs on the JCMT and ASTE; a dual-polarization receiver for the 1 mm band; and a unique wide band receiver and spectrometer, the Redshift Search Receiver (RSR), the instrument utilized in the present paper.

The RSR (Erickson *et al.* 2007) utilizes very wideband indium phosphide MMIC amplifiers operated at 20 K and has two dual polarization beams (thus a total of 4 receivers). Heterodyne receivers require a fast beam switch to produce flat spectral baselines over such a wide bandwidth, and the RSR uses a novel polarization switch operating at 1 kHz (Erickson & Grosslein 2007). This is the first wide band, low loss electrical switch operating at a wavelength as short as 3 mm. The fast polarization switch is followed by a broadband orthomode transducer (OMT) that splits the polarizations into two independent receivers. One beam of each polarization of the RSR is always on source.

The principal motivation for the construction of the RSR is to measure the spectra, and particularly the redshift **z**, of very distant galaxies. Galaxies are believed to form in the very early universe with the first episode of star formation $\sim 10^9$ years after the Big Bang, corresponding to ($z \gtrsim 6$). Understanding the formation process requires a catalog of a significant population out to $z \sim 10$ (5×10^8 yrs). Although these objects are relatively easy to detect in continuum, their distance and age are not easy to determine.

A few of these galaxies have redshifts measured in the visible, but most have no visible counterpart. The strongest emission lines from galaxies at roughly mm-wavelengths are the rotational transitions of CO, and the RSR bandwidth is large enough that there is a very high probability that one line of CO will fall in the observing band for redshifts $z > 1$, and that two CO rotational lines will be in the band for $z > 3.2$. When two lines of the CO rotational ladder are detected within the RSR band, the redshift is uniquely determined.

Since the LMT is not yet complete (we are hoping for initial 3 mm commissioning during 2008), the RSR was installed on the FCRAO 14 m telescope during spring 2007. The receiver frontend worked very well with the spectrometer to give very flat baselines. CO emission from 22 ULIRGs at moderate redshift was detected (Chung *et al.* 2008), and broadband 3 mm spectra of many nearby galaxies were obtained. The latter are presented here.

2. Observations

Each pixel of the RSR is sent to 6 different backend cards, each of which can process ~ 6.5 GHz of bandwidth. Twenty-four backend cards are required to process all four pixels over the 75–111 GHz bandwidth. While all four frontend pixels were available for the Spring 2007 run, only 4 out of the required 24 sections of the RSR backend spectrometer had been fabricated at that time, and were ready to use. For the nearby galaxy work, these four spectrometer cards were hooked up to the best pixel of the frontend to cover first a bandwidth range of 85 – 111 GHz, and then in separate observations a frequency range of 75 – 92 GHz. The two sets of observations are combined and averaged together (using special-purpose software written specifically for the RSR) to produce the spectra presented here.

3. Results and Discussion

In Figures 1 and 2, we show the full 3 mm band spectra towards two galaxies, M82 and IC342. The emission lines from a number of molecular species are detected in the observed galaxies (see Table 1). Note that the $J = 1 - 0$ transition of ^{12}CO, normally the strongest 3 mm emission line from galaxies, is not included in the observed bandwidth, but would move into the band for moderate values of the redshift z. As has been found by previous observers (e.g., Aalto 2006, Aalto 2007, Martin 2008), the ratios of the integrated intensities of such important chemical tracers as HCN, HNC, HCO$^+$, N$_2$H$^+$, CS and ^{13}CO vary among the galaxies. It has been proposed that such variations may be due to differences in the relative importance of photon-dominated regions (PDRs); X-ray dominated regions (XDRs), perhaps in the vicinity of AGNs; shocked regions; warm, dense molecular clouds; and infrared pumping for HCN and HNC (e.g., Aalto 2008, Imanishi *et al.* 2007, Perez-Beaupuits *et al.* 2007, Turner & Meier 2008). Because of the rather large beam-size of the FCRAO 14 m telescope (from 45 – 65 arcsec), it is difficult to definitely separate these effects for our galaxies; this will be much more clear cut with the LMT, whose corresponding beam size will be in the range 12 – 18 arcsec. Nonetheless, we note that our results for the ratios of the integrated intensities

Ratio	NGC 253	M82	IC342	Arp 220
HNC/HCN	0.51 (0.03)	0.43 (0.02)	0.37 (0.06)	0.81 (0.11)
HCO$^+$/HCN	0.84 (0.03)	1.47 (0.04)	0.64 (0.07)	0.50 (0.11)
HCN/^{13}CO	0.92 (0.03)	0.87 (0.03)	0.55 (0.03)	1.8 (0.6)
CS/HCN	0.43 (0.02)	0.58 (0.03)	0.33 (0.05)	0.55 (0.14)
N$_2$H$^+$/HCN	0.06 (0.02)	< 0.02	0.24 (0.07)	0.45 (0.08)

Line	Frequency (GHz)	Transition
C^{18}O	109.8	1-0
HC$_3$N	109.1	12-11
CH$_3$C$_2$H	102.5	6(n)-5(n)
HC$_3$N	100.1	11-10
SO	99.3	3(2)-2(1)
CH$_3$OH	96.7	2(0,2)-1(0,1) A+, & blend
C$_2$H	87.3	1-0, 3/2-1/2
CH$_3$C$_2$H	85.5	5(n)-4(n)
C$_3$H$_2$	85.3	2(1,2)-1(0,1)
CH$_3$OH	84.5	5(-1,5)-4(0,4) E
HC$_3$N	81.9	9-8
CH$_3$OH	81.0	7(2,6)-8(1,7) A-

Table 1. (a) Molecular Line Ratios in 4 Galaxies ($J = 1 - 0$ except CS, $J = 2 - 1$). (b) Other detected lines.

HCO$^+$/HCN, HNC/HCO$^+$, and HNC/HCN (see Table 1a) agree well for M82, NGC 253 and Arp 220 with those reported by Baan *et al.* (2008) (observations with the IRAM 30 m and the SEST 15m), although those latter authors give somewhat higher values for the HNC/HCN ratio in NGC 253.

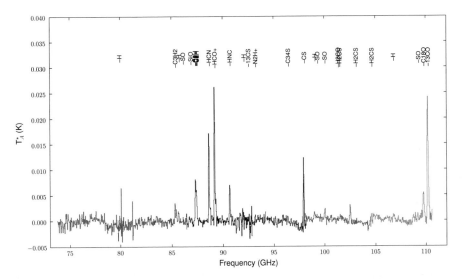

Figure 1. Full 3 mm band spectrum of M82 obtained with the redshift search receiver. The different colors represent distinct spectrometer boards that were used in the measurment. The names of important molecular lines typically seen in the ISM are denoted at the appropriate frequencies in the plot. M82 has strong continuum emission, and that accounts for the non-flat baselines (compare with Figure 2).

4. Summary

Spectra in the 3-mm wavelength band covering 75 to 111 GHz have been observed for about 10 galaxies using a new, very broadband receiver (RSR) with the FCRAO 14 m radio telescope. Interesting differences in line ratios are found, consistent with previous observations. When the RSR is mounted on the LMT in Mexico, surveys of the chemistry

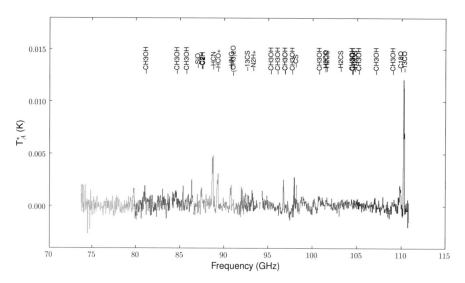

Figure 2. Full 3 mm band spectrum of IC342 (see Figure 1).

of external galaxies will be possible. We note in this connection the suggestions that the variety of environments near the center of the Milky Way may serve as templates for unraveling the processes in external galaxies (Martin *et al.* 2008, Jones & Burton 2008).

Acknowledgements

We are grateful for partial support of this research by the NSF (AST-0096854 and AST-0704966) and by NASA GSFC Cooperative Agreement NNG04G155A; and to Mike Brewer, Ron Grosslein, Kamal Souccar, Don Lydon and Gary Wallace for assistance during the engineering commissioning of the RSR at the FCRAO 14m telescope.

References

Aalto, S. 2006, *Proc. IAU Symposium 231*, ed. Lis, D. C., Blake, G. A. and Herbst, E., Cam-
    bridge Univ. Press.
Aalto, S. 2007, *A&A*, 475, 479
Aalto, S. 2008, *Ap&SS*, 313, 273
Baan, W. A., Henkel, C., Loenen, A. F., Baudry, A., & Wiklind, T. 2008, *A&A*, 477, 747.
Chung, A., Narayanan, G., Yun, M., Heyer, M. H., & Erickson, N. R. 2008, *in prep. ApJ*.
Erickson, N. R., Narayanan, G., Goeller, R., & Grosslein, R. 2007, *From Z-Machines to ALMA:
    (Sub)Millimeter Spectroscopy of Galaxies*, 375, 71
Erickson, N. R. & Grosslein, R. 2007, *IEEE Trans. Microwave Theory Tech*, 55, 2495.
Imanishi, M., Nakanishi, K., Tamura, Y., Oi, N., & Kohno, K. 2007, *AJ*, 134, 2366.
Jones, P. A., & Burton, M.G. 2008, *these Proceedings*, in press.
Martin, S., Requena-Torres, M. A., Martin-Pintado, J., & Mauersberger, R. 2008, *Ap&SS*, 313,
    303.
Martin, S. 2008, in: Meech, K., Mumma,M., Siefert, J., and Werthimer, D. (eds.), *Bioastronomy
    2007: Molecules, Microbes and Extraterrestrial Life*, (Astronomical Society of the Pacific)
Perez-Beaupuits, J. P., Aalto, S., & Gerebro, H. 2007, *A&A*, 476, 177.
Perez-Grovas, A. S., Schloerb, P. F.,Hughes, D., & Yun, M. 2006, *SPIE*, 6267, 1.
Turner, J. L. & Meier, D. S. 2008, *Ap&SS*, 313, 267.

Discussion

OLOFSSON: How is the site compared to, for instance, Mauna Kea?

IRVINE: It is certainly seasonal. There is a rainy season in the summer, when we would confine ourselves to the 3 mm region. It is a good site, probably not as good as Mauna Kea. We won't be capable of observing at wavelengths shorter than, say, 0.8 mm.

CECCARELLI: When will we start to operate the telescope?

IRVINE: We are in the process of re-aligning the surface for the inner 3 rings of panels. The telescope had a good appearance in some of those pictures, taken when we had a dedication for President Vicente Fox of Mexico, before he finished his term about a year ago. We hope to begin some 3 mm tests this year with the inner 3 rings of panels.

CECCARELLI: A scientific question. I saw that the ratio of N_2H^+/HCN is the largest difference between the galaxies that you showed. Can you comment on this?

IRVINE: No, I really can't. I should point out that the beam size of the 14 m telescope is pretty large on these galaxies, so we are probably looking at a range of different regions in these galaxies. That makes it a little difficult to interpret.

Brad Dalton showing his poster.

Peter Strittmatter and Lucy Ziurys having fun on the dance floor of the Bauhinia.

Organic Matter in Space
Proceedings IAU Symposium No. 251, 2008
S. Kwok & S. Sandford, eds.

© 2008 International Astronomical Union
doi:10.1017/S1743921308021704

A 3-mm molecular line study of the Central Molecular Zone of the Galaxy

Paul A. Jones, Michael G. Burton, and Vicki Lowe

School of Physics, University of New South Wales, NSW 2052, Australia
email: pjones@phys.unsw.edu.au

Abstract. We are studying the Central Molecular Zone (CMZ) in the inner few degrees around the Galactic Centre, by mapping multiple 3-mm molecular lines, with the 22-m Mopra telescope. During 2006, we covered a 5×5 arcmin^2 area of the Sagittarius B2 molecular cloud complex (Jones *et al.* 2008). We find substantial differences in chemical and physical conditions within the complex. We show some results here of Principal Component Analysis (PCA) of line features in this Sgr B2 area. During 2007 we covered the larger region of longitude -0.2 to 0.9 deg. and latitude -0.20 to 0.12 deg., including Sgr A and Sgr B2, in the frequency range 85.3 to 91.3 GHz. This includes lines of C_3H_2, CH_3CCH, $HOCO^+$, SO, $H^{13}CN$, $H^{13}CO^+$, SO, $H^{13}NC$, C_2H, HNCO, HCN, HCO^+, HNC, HC_3N, ^{13}CS and N_2H^+.

Keywords. ISM: molecules, radio lines: ISM, astrochemistry

1. Introduction

The Central Molecular Zone (CMZ) is the region within a few hundred parsec (a few degrees) of the Galactic Centre, which has strong molecular emission (Morris & Serabyn 1996), around 10% the molecular content of the whole Galaxy. The CMZ has extensive star formation, as shown by the mid-IR (Price *et al.* 2001) to sub-millimetre (Pierce-Price *et al.* 2000) dust emission. The conditions in the CMZ are characterised by high densities, large velocity dispersions, and high temperatures, compared to molecular clouds elsewhere in the Galaxy. The study of the CMZ provides a local analogue for the active nuclear regions of other galaxies.

The chemistry in the CMZ is complex, comparable to that of compact (~ 0.1 parcsec) "hot molecular cores", but extended over a scale several orders of magnitude larger, as shown, for example, in CH_3OH (Gottlieb *et al.* 1979) and HNCO (Dahmen *et al.* 1997)

The CMZ has been well studied in multiple CO lines (e.g. Nagai *et al.* 2007, Martin *et al.* 2004) giving a good estimate of the physical conditions. Particular, rich molecular regions of the CMZ, notably the hot cores Sgr B2(N) and Sgr B2(M), have had line surveys (e.g. Turner 1989, Belloche *et al.* 2007). However, due to observational limitations, there have not yet been studies which combine *large area coverage* of the CMZ and *large numbers of lines*, which is really needed to study the chemistry throughout the CMZ.

We are making a multi-line mapping survey of the CMZ, in the 3-mm band, with the Mopra 22-m telescope of the Australia Telescope National Facility. This project uses the new MMIC receiver (allowing easy tuning) and UNSW MOPS digital filterbank, which can cover up to 8 GHz bandwidth simultaneously. The broad band mode of the MOPS has up to 8192 channels of 0.27 MHz (~ 0.8 km s^{-1}), for each of four 2.2 GHz sub-bands. The zoom mode has up to four 137 MHz wide spectra of 4096 channels (~ 0.1 km s^{-1}) in each 2.2 GHz sub-band, making a maximum of 16 spectra over the 8 GHz. We use

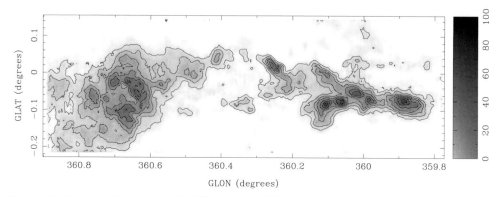

Figure 3. Integrated emission of N_2H^+ at 93.17 GHz, in the region of Sgr B ($l = 0.7$ deg.) and Sgr A ($l = 0.0$ deg.), from 2007 Mopra observations.

3. The Central Molecular Zone

During 2007 we started the large-scale imaging of the CMZ, with a single wide-band tuning covering the frequency range 85.3 to 91.3 GHz, and the area longitude -0.2 to 0.9 deg. and latitude -0.20 to 0.12 deg. This area includes Sgr A and Sgr B2, and we aim in 2008 to extend the longitude coverage to the area -0.7 to 1.7 deg., to include Sgr C and G1.6-0.025. These observations are with the full MOPS capability giving ~ 0.8 km s^{-1} channels, which is quite sufficient given the large line-widths in the CMZ area. The frequency tuning was chosen to include the strongest lines in the 3-mm band, other than the CO lines which have been well covered by other projects (e.g Oka *et al.* 1998). We have imaged 18 lines: c-C$_3$H$_2$ (85.34 GHz), CH$_3$CCH (85.46 GHz), HOCO$^+$ (85.53 GHz), SO (89.06 GHz), H^{13}CN (86.34 GHz), H^{13}CO$^+$ (86.75 GHz), SiO (86.85 GHz), HN^{13}C (87.09 GHz), C$_2$H (87.32, 87.40 GHz), HNCO (87.93 GHz), HCN (88.63 GHz), HCO$^+$ (89.19 GHz), HNC (90.66 GHz), HC$_3$N (90.98 GHz), CH$_3$CN (91.99 GHz), ^{13}CS (92.49 GHz) and N$_2$H$^+$ (93.17 GHz).

Figure 3 shows the integrated N$_2$H$^+$ (93.17 GHz) emission, as an example of the Mopra CMZ survey data.

We are currently extending the Principal Component Analysis to the larger CMZ area, and studying images of line ratios, for example HCO$^+$/HCN and HNC/HCN, which are diagnostics of high-density conditions (cf. Baan *et al.* 2008).

References

Baan, W. A., Henkel, C., Loenen, A. F., Baudry, A., & Wiklind, T. 2008, *A&A*, 477, 747

Bains, I. *et al.* 2006, *MNRAS*, 367, 1609

Belloche, A., Comito, C., Hieret, C., Menten, K.M., Müller, H. S.P., & Schilke, P. 2007, in: J. L. Lemaire & F. Combes (eds), *Molecules in Space & Laboratory*, p. 10

Dahmen, G., *et al.* 1997, *A&AS*, 126, 197

Gottlieb C.A., Ball J.A., Gottlieb E.W. & Dickinson D. F. 1979, *ApJ*, 227, 422

Hunt, M. R., Whiteoak, J. B., Cragg, D. M., White, G. L., & Jones, P. A. 1999, *MNRAS*, 302, 1

Jones, P. A., Burton, M. G., Cunningham, M. R., Menten, K. M., Schilke, P., Belloche, A., Leurini, S., Ott, J., & Walsh, A. J. 2008, *MNRAS*, in press, (arXiv:0712.0218)

Ladd, N., Purcell, C., Wong, T., & Robertson, S. 2005, *PASA*, 22, 62

Martin, C. L., Walsh, W. M., Xiao, K., Lane, A. P., Walker, C. K. & Stark, A. A. 2004, *ApJS*, 150, 239

Meier, D. S., & Turner, J. L. 2005, *ApJ*, 618, 259

Morris, M., & Serabyn, E. 1996, *ARA&A*, 34, 645

Nagai, M., Tanaka, K., Kamegai, K. & Oka, T. 2007, *PASJ*, 59, 25

Oka, T., Hasegawa, T., Sato, F., Tsuboi, M., & Miyazaki, A. 1998, *ApJS*, 118, 455

Pierce-Price, D., *et al.* 2000, *ApJ*, 545, L121

Price, S. D., Egan, M. P., Carey, S. J., Mizuno, D. R., & Kuchar, T. A. 2001, *AJ*, 121, 2819

Turner, B. E. 1989, *ApJS*, 70, 539

Ungerechts, H., Bergin, E. A., Goldsmith, P. F., Irvine, W. M., Schloerb, F. P., & Snell, R. L. 1997, *ApJ*, 482, 245

Discussion

CECCARELLI: I am curious what is the sensitivity that you reach in this survey?

JONES: It is a trade off between area and sensitivity. We're doing in this project on a large area, recognizing we are not probing the sensitivity as in a spectral line survey. For example, Sgr B2 has been observed with the IRAM 30 m where the 3mm band has much greater sensitivity. This means that we're getting things like methyl cyanide and methanol but not the more complex molecules. We see about hundred and fifty lines, but most of them are located in one point in Sgr B2 (N). We establish complex molecules there.

ZIURYS: Did you ever compare your maps with interferometer maps of the Galactic center such as those done by BIMA?

JONES: With 30 arc sec resolution, the BIMA interferometer maps correspond to about one or two beam areas, but they do match up.

ZIURYS: I'm just curious how much flux was resolved out in their data versus what you have?

JONES: One of the advantages are having single dish observations is that we can put the single dish flux back into the interferometric data to get good interpretation of the data.

OLOFSSON: 8 GHz is fairly broad bandwidth. Do you have reliable baseline so that you can observe also very weak lines?

JONES: When we look at the strong continuum source such as Sgr B2 you do get some ripples in the baseline due to standing waves, so that can become a problem.

OLOFSSON: How about continuum free sources such as circumstellar envelopes?

JONES: The baseline is pretty flat over the full 8 GHz.

Dale Cruikshank (left) and Cliff Matthews (right).

Organic Matter in Space
Proceedings IAU Symposium No. 251, 2008
S. Kwok & S. Sandford, eds.

© 2008 International Astronomical Union
doi:10.1017/S1743921308021716

ISM spectrum by cosmic dust?

Thomas V. Prevenslik

Discovery Bay, Hong Kong
email: thomas.prevenslik@gmail.com

Abstract. The interstellar medium (ISM) spectrum is usually explained by the response of dust particles (DPs) to the absorption of ultraviolet (UV) and visible (VIS) photons from nearby stars. With regard to the unidentified infrared (UIR) bands, the DPs are thought heated by UV and VIS photons to about 100 K thereby exciting the polycyclic aromatic hydrocarbons (PAHs). However, the UIR bands may be explained with the DPs at 2.7 K. To wit, the UIR bands form by the direct excitation of PAHs by infrared (IR) radiation induced from the absorption of cosmic microwave background (CMB) radiation in DPs by quantum electrodynamics (QED).

Keywords. (ISM:) dust, ISM: lines and bands, infrared: ISM, cosmology: theory

1. Introduction

In the ISM, the UIR bands have been linked (Li 2004) to the heating of DPs by UV photons (Li & Draine 2001). The UIR bands observed in UV-poor regions (Uchida, *et al.* 1998) was explained by (Li & Draine 2002) with VIS photons alone assuming thermal emission from DPs heated to the same 100 K temperature assumed for UV photons in (Li & Draine 2001). However, the assumed 100 K temperature of DPs may be fortuitous given not only are UV and VIS intensities not uniform in the ISM, but the only electromagnetic (EM) radiation ubiquitous to the ISM is the cosmic microwave background (CMB) at 2.7 K.

2. Theory and Analysis

CMB radiation at 2.7 K may explain the ISM spectrum given a frequency up-conversion process in DPs. One such process is QED induced EM radiation (Prevenslik 2008). Finding analogy with creating photons of wavelength λ by supplying EM energy to a quantum mechanical box having walls separated by $\lambda/2$, the DPs having diameter D produce QED photons of wavelength $2D$. In the ISM, EM energy is supplied to DPs upon the absorption of CMB radiation (Mie 1908).

But Mie absorption efficiency is low in single interactions of CMB radiation with a DP (Bohren & Huffman 1983). The ISM spectrum by CMB radiation therefore relies on multiple Mie absorptions and QED emissions in a continuous distribution of DPs of diameter D, each interaction successively increasing the frequency of QED emissions to produce the ISM spectrum at near unity Mie absorption efficiency (Prevenslik 2008).

Moreover, the Einstein specific heat of DPs depends on the vibration frequency of the absorbed CMB photon under total internal reflection (TIR) as it adjusts to the EM confinement imposed by the DP geometry. Since EM confinement occurs at TIR frequencies beyond the UV, the specific heats already very low at 2.7 K vanish during CMB photon absorption. Absent specific heat, the absorbed CMB radiation cannot be conserved by an increase in temperature (Prevenslik 2008). Instead, conservation of absorbed CMB radiation proceeds by producing a continuum of QED induced EM radiation within the DP, the IR content of which exciting the PAHs to produce the UIR bands.

Beyond excluding Mie absorption efficiency, the classical Stefan-Boltzmann (SB) equation assumes the DP has specific heat c_p to conserve the absorbed blackbody (BB) radiation by an increase in temperature T. Since the Einstein specific heat c_p vanishes, the modified SB equation for DPs including Mie absorption efficiency Q_{abs} is,

$$\sigma \pi D^2 Q_{abs} T_{BB}^4 - E_p \frac{dN_p}{dt} = Mc_p \frac{dT}{dt} = 0 \qquad (2.1)$$

where, σ is the SB constant, T_{BB} is 2.7 K, Q_{abs} is the Mie absorption efficiency taken to be unity in multiple interactions, M is the DP mass, dT/dt is rate of DP temperature change, dN_p/dt is the rate of QED induced photons produced having Planck energy $E_p = hc/2D$. Here, h is Planck's constant, and c is the speed of light.

3. Discussion

Historically, DPs having an entire heat capacity comparable to the Planck energy of a single UV or VIS photon were thought (Purcell 1976) to cause temperature fluctuations, and even more recently, the UIR bands have been explained (Li & Draine 2002) by temperatures of about 100 K upon heating by absorbed VIS photons. In contrast, the QED induced IR continuum directly excites the PAHs to produce the UIR bands with the DPs remaining at 2.7 K. Moreover, DP temperatures of 100 K from absorbed UV and VIS photons are unlikely because the DP specific heats already low at 2.7 K vanish as the CMB photons upon absorption vibrate at optical TIR frequencies.

4. Conclusions

The ISM spectrum is produced by QED induced EM radiation from the multiple interactions of a continuum of DP diameters. DPs always remain at 2.7 K. Neither UV nor VIS photons from nearby stars are necessary to produce the UIR bands by heating DPs to temperatures of 100 K. In fact, CMB radiation of DPs may produce in part the UV and VIS now thought to originate in nearby stars.

The ISM spectrum produced by QED induced EM radiation finds origin in the blueshift of CMB radiation in submicron DPs.

The converse of blueshift of CMB radiation in DPs is also true in the redshift of VIS light from distant galaxies claimed in Hubbles Law. The galaxy light observed on Earth and thought to be redshift by the Doppler effect in an expanding universe may be nothing more than numerous QED induced blue and redshift of galaxy light in DPs that is finally observed on Earth as redshift. Hence, the Hubble theory as experimental evidence for an expanding universe is therefore held in question by QED induced EM radiation in DPs.

References

Bohren, C. F. & Huffman, D. R. 1983, *Absorption and Scattering of Light by Small Particles*, J. Wiley & Sons, New York

Li, A. 2004, *Astrophysics of Dust*, ASP Conference Series, 309, 417

Li, A. & Draine, B. T. 2001, *ApJ*, 554, 778

Li, A. & Draine, B. T. 2002, *ApJ*, 572, 232

Mie, G. 1908, *Ann. Phys.*, 330, 337

Prevenslik, T. V., 2008, www.nanoqed.net

Purcell, E. M. 1976, *ApJ*, 206, 685

Uchida, K. I., Selgren, K., & Werner, M. W. 1998, *ApJ (Letters)*, 493, L109

Session II

Organic compounds within
the Solar System

Organic Matter in Space
Proceedings IAU Symposium No. 251, 2008
S. Kwok & S. Sandford, eds.

Organic matter in interplanetary dust particles

G. J. Flynn[1], L. P. Keller[2], S. Wirick[3], and C. Jacobsen[3]

[1]Dept. of Physics, SUNY-Plattsburgh, 101 Broad St, Plattsburgh NY 12901 USA
email: flynngj@plattsburgh.edu

[2]NASA Johnson Space Center, NASA Rt. 1, Houston TX, 77058 USA

[3]Dept. of Physics, SUNY- Stony Brook, Stony Brook NY 11794

Abstract. Anhydrous interplanetary dust particles (IDPs), which are the most mineralogically primitive extraterrestrial materials available for laboratory analysis, contain several percent organic matter. The high O:C and N:C ratios suggest the organic matter in the anhydrous IDPs is significantly less altered by thermal processing than the organic matter in meteorites. X-ray Absorption Near-Edge Structure (XANES) spectroscopy and infrared spectroscopy demonstrate the presence of C=C, most likely as C-rings, C=O, and aliphatic C-H$_2$ and C-H$_3$ in all the IDPs examined. A D-rich spot, containing material that is believed to have formed in a cold molecular cloud, has C-XANES and infrared spectra very similar to the organic matter in the anhydrous IDPs, possibly indicating a common formation mechanism. However the primitive organic matter in the IDPs differs from the interstellar/circumstellar organic matter characterized by astronomical infrared spectroscopy in the relative strengths of the asymmetric aliphatic C-H$_2$ and C-H$_3$ absorptions, with the IDP organic having a longer mean chain length. If both types of organic matter originated by the same process, this may indicate the interstellar organic matter has experienced more severe radiation processing than the organic matter in the primitive IDPs.

Keywords. Astrochemistry, astrobiology, infrared: general

1. Introduction

Interplanetary dust particles (IDPs) are small fragments from asteroids and comets that enter the Earth's atmosphere, decelerating slowly over distances of tens of kilometers, and settling towards the Earth's surface. Since the 1970's, NASA has collected IDPs from the Earth's stratosphere (Brownlee 1985), where the concentration of terrestrial dust is relatively low. These IDPs span the size range from ~ 5 μm to > 50 μm, though most of the particles that are collected are ~ 10 μm in diameter.

The IDPs have a similar elemental composition to that of the CI carbonaceous chondrite meteorites, which are similar to the composition of the Solar photosphere for all but the most volatile elements (Anders & Grevesse 1989) and are believed to represent the elemental composition of the non-volatile material of the Solar Nebula. However, the IDPs contain, on average, about 12 wt-% C (Thomas *et al.* 1994), which is more than three times the C-abundance of the most C-rich meteorite, and are similarly enriched relative to CI in several moderately-volatile minor elements (Flynn *et al.* 1996).

Anders (1989) suggested that, since many of the IDPs decelerate so gently that they are not significantly heated, the IDPs might have contributed a significant amount of pre-biotic organic matter to the surface of the early Earth, providing important material for the origin of life. At the time of Anders paper little was known about the amount or the type(s) of organic matter in the IDPs. Because of their small size, the organic

Figure 1. Scanning electron microscope image of an interplanetary dust particle (IDPs), measuring $\sim 10\ \mu m$ in its largest dimension, collected by NASA from the Earth's stratosphere. The texture shows the individual subunits, which are glass or discrete mineral grains, typically sub-micron in size, held together by an organic coating. (NASA photo)

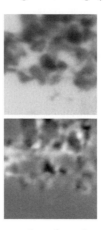

Figure 2. X-ray absorption image of a portion of an ultramicrotome section of L2008R15 (top), and (bottom) a carbon map derived by comparing the absorption just below the C K-edge, which C absorbs only weakly, with the absorption just above the C K-edge, which C absorbs strongly. Bright areas have high C. (Image width $\sim 1\ \mu m$.)

matter in the IDPs cannot be studied using the conventional laboratory techniques that are applied to much larger samples, such as the meteorites.

We characterize the carbonaceous material in the IDPs using two synchrotron-based instruments at the National Synchrotron Light Source (NSLS) at Brookhaven National Laboratory (Upton, New York USA). A Scanning Transmission X-ray Microscope (STXM), having a spatial resolution of ~ 35 nm, is employed to determine the carbon abundance and map the distribution of carbon in ~ 100 nm thick ultramicrotome sections of the IDPs (see Figure 2). At carbon-rich spots, X-ray Absorption Near-Edge Structure (XANES) spectroscopy is used to determine the C, N, and O functional groups, as described in Flynn *et al.* (2003). A microscope-based Fourier Transform Infrared Spectrometer (μ-FTIR) is used to identify functional groups with absorptions in the mid-IR. Combining data from the two instruments allows us to detect a variety of organic functional groups at the micron or sub-micron scale.

The majority of the IDPs are anhydrous, showing no evidence of hydrous silicates, but some contain hydrous minerals, suggesting these IDPs experienced aqueous alteration after incorporation into a parent body such as an asteroid. The anhydrous IDPs are generally fine-grained, porous aggregates, called chondritic porous (or CP) IDPs. A typical $\sim 10\ \mu m$ CP IDP contains $\sim 10^4$ or more individual subunits (see Figure 1). The mean bulk density of the CP IDPs is ~ 0.6 gm/cc (Flynn & Sutton 1991), although most of the anhydrous minerals they are made of have densities > 3 gm/cc. This indicates a typical CP IDP has a very high porosity. Carbon maps of the CP IDPs show that most of the individual grains are coated with carbonaceous material, which appears to be the "glue" that holds these porous structures together (see Figure 2). The hydrous IDPs are significantly more compact, with a mean density of ~ 1.9 gm/cc (Flynn & Sutton 1991).

Comparison of the mineralogy of IDPs with that of other types of extraterrestrial materials, including various types of meteorites and the samples of Comet 81P/Wild 2 collected by the Stardust spacecraft, indicates that the "CP IDPs remain the most cosmochemically primitive astromaterials" (Ishii *et al.* 2008). Thus, the fine-grained, porous, anhydrous IDPs are the best samples to examine in order to characterize the primitive organic matter of our Solar System.

2. Organic Matter in Primitive, Anhydrous IDPs

The organic matter in 8 anhydrous IDPs was characterized by C-XANES spectroscopy and μ-FTIR spectroscopy. The C-XANES analyses are performed on ultramicrotome sections, typically 70 to 100 nm thick. The sections were prepared by embedding each particle in liquid elemental S, allowing the S to cool, then sectioning the S-bead containing the sample. The sections are deposited on TEM grids having a silicon monoxide substrate, thus avoiding exposure of the samples to carbon-bearing materials, such as embedding epoxy and substrates, during the preparation.

In general, C is inhomogeneously distributed, as shown in Figure 2. Different C-rich spots in the same ultramicrotome section frequently exhibit different C-XANES spectra, as shown in Figure 3. Most C-rich regions show both the C=C absorption, at ~ 285 eV, and the C=O absorption feature, at ~ 288.5 eV. The C-rich regions also show a weaker absorption at ~ 286.5 eV, attributed to O on a C-ring, C-C, or C-N bonding. Some spectra show other features, such as the aliphatic C-H* Rybergh resonances at ~ 289 eV (Stohr 1992). Variations in the center position and the width of a specific peak indicate differences in the bonding environment of the particular functional group. Differences in the relative areas of two peaks indicate a variation in the ratio of the two functional groups. Thus, the organic matter in these IDPs exhibits differences in the ratios of the various organic functional groups at the micron-scale, suggesting it is a mixture of several organic molecules.

The mean ratio of the area of the C=C to C=O absorption in the 8 anhydrous IDPs we measured is 0.92 (Flynn *et al.* 2003). This is quite different from the ratio of 1.29 that we find for organic matter in an ultramicrotome section of the Murchison meteorite, and suggests that the organic matter in the anhydrous IDPs contains more O than the organic matter in Murchison. As a direct test of this result, we can determine the C:N:O ratio in the IDP organic matter by measuring the increase in x-ray absorption at the K-edge energies of C, N, and O. However, determining the O:C ratio is complicated by O being present in both organic matter and minerals.

The N-XANES and O-XANES spectra were measured in a few of the anhydrous IDPs. Figure 4 shows the combined C-, N-. and O-XANES spectra of an anhydrous IDP, L2009D11. Because of the low N-concentration, the N-XANES spectra are generally noisy. The N absorption features in two of the three anhydrous IDPs measured, L2005*A3 and L2011R11, are very broad and difficult to assign to a specific functional group. The third anhydrous IDP, L2008H9, has a much stronger N-XANES spectrum. No carbon is found at the N-rich spots in L2008H9, and the N-XANES spectrum is consistent with a nitride.

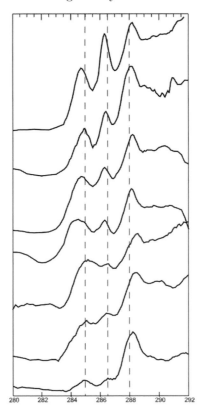

Figure 3. C-XANES spectra of seven different areas on an ultramicrotime section of L2009D11. an anhydrous IDP. Differences in the center positions and widths of the C=C absorption ($\sim$ 285 eV) and the C=O absorption ($\sim$ 288.5 eV) indicate differences in the nearest neighbor atoms of these functional groups at different locations while differences in the areas of the two peaks indicate different ratios of the C=C and C=O functional groups.

The dominant O-host in most of the anhydrous IDPs is a silicate, either olivine or pyroxene. However, O is also found co-located with C. The pre-edge absorption in the O-XANES spectra of these C-rich areas is consistent with C=O, confirming our attribution of the 288.5 eV feature in the C-XANES spectrum to C=O.

In one anhydrous IDP, L2011*B2, the organic coating was pushed away from the edge of the silicate grain during microtoming, providing an isolated area of IDP organic matter for analysis. We determined the C:N:O ratio to be $C_{100}N_{10}O_{50}$ for the organic matter in this C-rich area by measuring the increase in absorption at the K-edge of each of these elements and using the known x-ray absorption cross-sections. The inferred composition of $C_{100}N_{10}O_{50}$ has much higher O:C and N:C ratios than are reported in the acid insoluble organic matter extracted from the Murchison meteorite, which has $C_{100}N_{1.8}O_{12}$ (Zinner 1988).

Based on measurements by Cody, Sandford *et al.* (2006) have shown that the O:C and N:C ratios of the organic matter extracted from meteorites increase with increasing primitiveness (i.e., a lesser degree of thermal metamorphism). The organic matter in L2011*B2 exhibits even higher O:C and N:C ratios than the organic matter in the least thermally metamorphosed meteorite. This suggests the IDP organic matter has experienced even less thermal metamorphism since its formation, consistent with the

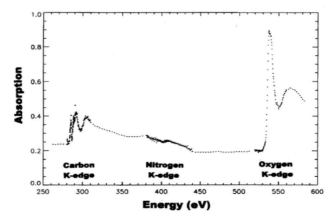

Figure 4. Combined C-, N-, and O-XANES spectra of L2009D11, an anhydrous IDP with a large D hot spot.

mineralogical observation that the anhydrous, porous IDPs are more primitive than any type of meteorite.

The high intensity of the synchrotron light allows us to perform μ-FTIR analyses with a small beamspot. However, the size-scale for the infrared analysis is still limited by diffraction. We can obtain good quality spectra of the C-H stretching region, at a wavelength of ~ 3.4 μm, with 3 to 5 μm analysis spots. However, good quality spectra of the C=O and Si-O regions of the spectrum require analysis spots of 7 to 10 μm, comparable to the size of the whole ~ 10 μm IDP. Even for the C-rich IDPs, the infrared spectrum is typically dominated by the Si-O absorption, which is near 10 μm (see Figure 5). The aliphatic C-H$_2$ and C-H$_3$ stretching absorptions, near 3.4 μm, have been detected in all IDPs examined thus far. However, the aromatic C-H absorption is below the detection limit in most IDPs. The C-C and C=O absorption features are only detected by infrared spectroscopy in IDPs having a relatively high C-content (see Figure 5).

3. Organic Matter in Hydrous IDPs

The hydrous IDPs contain, on average, about the same amount of C as the anhydrous IDPs (Thomas *et al.* 1994). We characterized the organic matter in 5 hydrous IDPs. In general, the C-XANES spectra of the C-rich spots in the hydrous IDPs are quite similar to those in the anhydrous IDPs, exhibiting strong C=C and C=O absorptions and a weaker absorption feature at ~ 296.5 eV. The mean C=C to C=O peak area ratio for the 5 hydrous IDPs is 0.96, very similar to the 0.92 value for the 8 anhydrous IDPs. The C-H stretching region of the infrared spectra of the hydrous IDPs is also quite similar to that of the anhydrous IDPs, These similarities in the types and abundance of organic matter in the anhydrous and the hydrous IDPs suggest that the bulk of the organic matter in the IDPs was formed prior to the aqueous alteration event, which most likely took place after incorporation of the IDPs into their parent bodies. Thus, the organic matter in the IDPs appears to have formed very early in Solar System history, or may even precede the formation of the Solar System.

4. Non-Solar Organic Matter in the IDPs

The isotopic ratios of most elements vary over only a narrow range for Solar System materials. Material that preserved the record of its pre-Solar history is usually identified

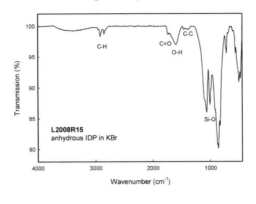

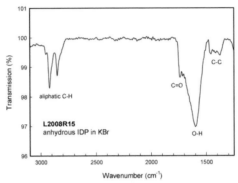

Figure 5. (top) The mid IR-spectrum of L2008R15, an anhydrous IDP, showing the organic and silicate features. (bottom) The region of this mid-IR spectrum from 1300 to 3100 cm^{-1}, showing the organic signatures of C-C, C=O, and C-H in the spectrum of L2008R15.

by isotopic ratios outside the Solar range. In the case of organic matter, isotopic ratios of H, C, N, or O that are well outside the range seen in the Solar System point to a non-Solar origin for the material. Some IDPs have localized hot-spots showing extremely high D:H ratios, suggesting the D-carrier was formed in a cold molecular cloud (Messenger 2000). We analyzed the D-hot spot in a particularly D-rich IDP, L2009D11, using C-XANES. We also analyzed L2009D11 using μ-FTIR (Keller *et al.* 2004). The D hot-spot in L2009D11 is coincident with a C-rich region of the IDP, indicating the D-host is organic, rather than water, which is the other likely H-rich phase. The C-XANES spectrum of the D hot-spot shows the same three absorption features, at $\sim$ 285 eV, 286.5 eV and 288.5 eV, that we detect in the other IDPs. The C=C and C=O peak area ratio is 0.76 in the D hot-spot, lower than the mean of 0.92 for the anhydrous IDPs, but within the range of values for the 8 anhydrous IDPs. The infrared spectrum of L2009D11 has a C-H stretching region consistent with the other anhydrous IDPs, showing a strong aliphatic C-H$_2$ absorption and a weaker C-H$_3$ absorption feature. The aromatic C-H feature, at $\sim$ 3.3 μm, is below our detection limit, suggesting that the D-carrier is an aliphatic hydrocarbon (Keller *et al.* 2004).

We can directly compare the C-H stretching region of the infrared spectrum of the primitive organic matter in the anhydrous IDPs, including L2009D11, with astronomical spectra of interstellar/circumstellar dust. The astronomical spectra generally show roughly equal depths for the asymmetric aliphatic C-H$_2$ and C- H$_3$ absorptions (Sandford *et al.* 1991), while the primitive organic matter in the IDPs has a mean ratio of $\sim$ 2.5 for the asymmetric aliphatic C-H$_2$ and C-H$_3$ absorptions. This suggests the mean length for the aliphatic chains in the anhydrous IDPs is significantly longer than has been observed for the organic matter in interstellar/circumstellar sites. If both types of organic matter were produced in similar processes, the difference in aliphatic chain length might have resulted from radiation processing of the interstellar/circumstellar organic matter.

5. Delivery of Organic Matter to the Earth by the IDPs

The orbits of particles in the size range of the IDPs, generally 5 to 50 μm in diameter, are significantly perturbed by solar radiation. The Poynting-Robertson drag force causes an ~ 10 μm IDP to spiral inwards, towards the Sun. A particle starting in the asteroid belt reaches 1 AU in 20,000 to 100,000 years (Flynn 1989). Particles released by comets have a wider range of space exposure times. However, even IDPs produced by collisions in the Kuiper Belt spiral into the Sun on a time-scale short compared to the age of the Solar System. Thus, IDPs spend most of the life incorporated in larger bodies, most likely asteroids and comets, both of which are associated with bands or trails of dust detected in the infrared. During this time they are shielded from ionizing radiation.

However, once they are released from the parent body, the IDPs are exposed to ionizing radiation. The most damaging radiation is likely to be UV or x-rays, which can cause photo-ionization, but this radiation penetrates only a few hundred nanometers into the sample. Thus, even in an ~ 10 μm IDP, we expect the interior to be shielded from the most damaging radiation.

However, modeling by Zagorski (2007) suggests that after 100 years in space in our Solar System the radiation dose experienced by a 10 μm radius silicate particle is sufficient to sterilize the interior, and that much longer exposures will alter the organic matter, eventually resulting in complete degradation. Our results demonstrate that a significant amount of organic matter survives exposure to the radiation experienced in the inner Solar System for times of $\sim$20,000 to $\sim$100,000 years in particles as small as $\sim$ 10 μm. Thus, the organic matter delivered to the primitive Earth by the IDPs may have been important to the origin of life on Earth.

6. Conclusions

Organic matter is abundant in both the anhydrous and hydrous IDPs. Both the mineralogy and the high O and N content of the organic matter in anhydrous IDPs indicates these particles contain the least processed samples of the early organic matter of our Solar System. This organic matter survived the heating experienced by IDPs during atmospheric deceleration, demonstrating that IDPs contribute a significant amount of pre-biotic organic matter to the surface of the Earth, which (Anders 1989) suggested may be important for the origin of life. Silicates and the aliphatic C-H, carbonyl (C=O), and C-ring functional groups dominate the infrared and C-XANES spectra of both the anhydrous and the hydrous IDPs, but the aromatic C-H functional group is usually below the detection limit. The infrared and C-XANES spectra of the organic matter in both the anhydrous and the hydrous IDPs are very similar to each other, indicating that most of the pre-biotic organic matter of our Solar System formed by a process that preceded aqueous processing. The aliphatic C-H_2 to C-H_3 ratio in the IDPs is much higher than in astronomical spectra of interstellar/circumstellar dust regions, indicating the aliphatic organic matter in the IDPs occurs in longer chains and suggesting the interstellar/circumstellar organic matter may have experienced significant radiation processing.

George Cody arguing a point of view (photo by Sze-Leung Cheung).

Organic Matter in Space
Proceedings IAU Symposium No. 251, 2008
S. Kwok & S. Sandford, eds.

© 2008 International Astronomical Union
doi:10.1017/S1743921308021741

Unraveling the chemical history of the Solar System as recorded in extraterrestrial organic matter

George D. Cody[1], Conel M. O'D. Alexander[2], A. L. David Kilcoyne[3], and Hikaru Yabuta[1]

[1]GL, Carnegie Institution of Washington,
5251 Broad Branch, Washington DC 20015, USA
email: g.cody@gl.ciw.edu, hyabuta@gl.ciw.edu

[2]DTM, Carnegie Institution of Washington,
5241 Broad Branch, Washington DC 20015, USA
email: alexande@dtm.ciw.edu

[3]Advanced Light Sorce,
Lawrence Berkeley Laboratory, Berkeley CA, USA
email: alkilkoyne@lbl.gov

Abstract. We have initiated an extensive program of molecular analysis of extraterrestrial organic matter isolated from a broad range of meteorites (spanning multiple classes, groups, and petrologic types), including recent molecular spectroscopic analyses of the organic matter in the Comet 81P/Wild 2 samples. The results of these analyses clearly reveal the signature of multiple reaction pathways that transformed extraterrestrial organic matter away from its primitive roots. The most significant molecular transformation occurred in the post-accretionary phase of the parent body. However, each of the various chemical transformation trajectories point unambiguously back to a common primitive origin. Applying a wide range of spectroscopic techniques we find that the primitive organic precursor is striking in its chemical complexity exhibiting a broad array of oxygen- and nitrogen-bearing functional groups. The π-bonded carbon exists as predominately highly substituted single ring aromatics, there exists no evidence for abundant, large, polycyclic aromatic hydrocarbons (PAHs). We find that the molecular structure of primitive extraterrestrial organics is consistent with synthesis from small reactive molecules, e. g. formaldehyde, whose random condensation and subsequent rearrangement chemistry at low temperatures leads to a highly cross-linked macromolecule.

Keywords. Comets, asteroids, Solar System, molecular data, ISM: dust

1. Introduction

The organic matter in carbonaceous chondrites potentially records a succession of chemical histories that started with reactions in the interstellar medium, followed by reactions that accompanied the formation and evolution of the early solar nebula, and, ultimately, ended with reactions driven by hydrothermal alteration in the meteorite parent bodies. One of the challenges in meteorite research is establishing whether one can identify the chemical signatures of these reactions in the organic fractions and whether any relationship exists between the molecular structure(s) of organic matter within a given meteorite group and the degree of parent body alteration.

There currently exists considerable information regarding the inventory of soluble organic matter derived from numerous studies across a number of carbonaceous chondritic groups (e.g., reviews by Hayes 1967, Cronin *et al.* 1988, Botta & Bada 2002, Sephton 2002). It is well known that the vast proportion ($> 70\%$) of organic matter in carbonaceous chondrites exists as an insoluble, perhaps macromolecular, solid; defined here as Insoluble Organic Matter (IOM) (Hayes 1967, Cronin *et al.* 1988). Over the past three decades, substantial progress has been made at characterizing IOM using various analytical techniques. For example, pyrolysis gas chromatography (pyr-GCMS) has been used extensively as a means of characterizing the IOM fraction (Hayatsu & Anders 1981, Sephton & Gilmour 2002, Kitajima *et al.* 2002).

The analysis of various carbonaceous chondrites revealed a predominance of single ring aromatic compounds with a range of alkyl substituents, alkylated phenols, and naphthalenes and alkyl naphthalenes (Hayatsu & Anders 1981, Sephton & Gilmour 2002). More recently, pyr-GCMS data were used to characterize the effects of thermal metamorphism on a suite of CM chondrites (Kitajima *et al.* 2002). Pyrolysis GCMS is an excellent tool for this purpose as these analyses are reproducible and reliably reflect differences in chemical structure among different macromolecules. This stated, it is not always a simple task to reconstruct an accurate structure of a given macromolecule from the molecular constituents in the pyrolysate. This is because considerable chemical modification can occur during the pyrolytic liberation of molecules. There is also the issue of disproportionation, wherein the pyrolysate maybe chemically very different from the non-pyrolyzable residue.

Solid-state nuclear magnetic resonance (NMR) spectroscopy has been shown to be a highly effective analytical tool for establishing the types and distributions of organic functional groups in amorphous organic solids. While lacking the molecular detail of pyr-GCMS, solid-state ^{13}C NMR can provide a quantitative, albeit averaged, structural picture of organic macromolecules. Solid State ^{13}C NMR was first applied to meteoritic IOM by Cronin *et al.* (1987) who analyzed partially demineralized IOM fractions of Orgueil (CI1), Murchison (CM2), and Allende (CV3). Later, Gardinier *et al.* (2000) revisited the analysis of Murchison (CM2) and Orgueil (CI1) IOM with Solid State ^{13}C NMR. In 2002, Cody *et al.* provided a comprehensive analysis of Murchison employing both ^{1}H and ^{13}C solid state NMR experiments. In particular, they performed several independent, yet complementary, experiments to better refine the structural picture of Murchison's IOM fraction.

An example of the solid-state ^{13}C NMR spectrum of Murchison (CM2) IOM is presented in Figure 1. The likely presence of various organic functional groups is clearly evident, where the most abundant functional group is likely aromatic carbon (at ~ 129 ppm). It is important to note that the source of the line broadening is not due to any resolution limit of solid state NMR. Rather, the broad lines reflect considerable chemical complexity, i. e., any given functional group exists in a multiplicity of local electron environments. Cody *et al.* (2002) developed a self-consistent, albeit statistical, picture of the organic structure of Murchison's IOM. What is concluded is that the IOM structure is dominated by small highly substituted aromatics, highly branched and oxygen substituted aliphatics, and abundant carbonyl and carboxyl moieties. How such a structure could have evolved and where the most likely environment was for its synthesis is not possible to determine from the study of a single meteorite.

2. Low temperature alteration of IOM

Cody & Alexander (2005) published an extensive NMR study of IOM from four type 1 and 2 chondrites spanning multiple groups (CR, CI, CM, and Tagish Lake an ungrouped C2). What was observed was a spectacular range in organic structure where the apparent

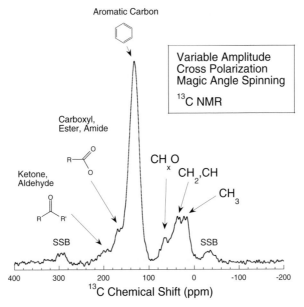

Figure 1. The Solid-state ^{13}C NMR spectrum of Murchison IOM. The various spectral features are labeled with the most likely organic functional groups. Note that the small peaks labeled SSB, correspond to spinning side-bands derived from the most intense peak at a frequency of ~ 129 ppm. The chemical shift is recorded as the parts per million shift relative to the ^{13}C frequency (~ 75 MHz at 7.05 T). The shift is normalized such that the frequency of methyl groups in tetramethyl silane are defined as having a shift of 0 ppm.

aromatic carbon abundance rose from a low of $\sim 48\%$ (for the CR) up to $\sim 80\%$ in the case of Tagish Lake (Figure 2). Remarkably none of the type 1 or 2 chondrites have ever experienced heating in excess of ~ 100 °C during parent body alteration. This raises the question as to what type of chemistry could lead to such a remarkable difference in molecular structure, assuming that all IOM has a common precursor.

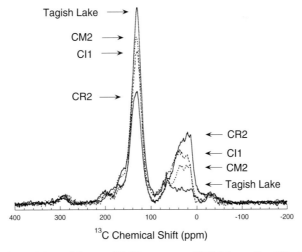

Figure 2. An over-lay of the solid state NMR spectra of IOM from the CR, CI, CM, and Tagish Lake groups of carbonaceous chondrites. Note the tremendous range in apparent aromatic carbon content (at ~ 129 ppm).

Through a series of independent NMR experiments Cody & Alexander (2005) proved that the significant differences in the chemistry of IOM across the four meteorite groups (Figure 2) reflect primarily differences in the abundance and types of sp^3 bonded carbon. The most likely interpretation for this chemical variation is that the differences in type 1 and 2 IOM record a progressive trend in chemical oxidation that would require the presence of a soluble oxidant in the aqueous phase during parent body alteration. One possibility is hydrogen peroxide, a likely constituent in the icy mantles of dust grains subjected to UV photolysis. In this study of meteoritic IOM, therefore, the CR2 is apparently the least and Tagish Lake is apparently the most chemically degraded. The selective enrichment in aromatic moieties and nano-diamond suggested that these constituents of IOM are relative inert to this oxidation and that their relative abundance may provide a measure of the extent of oxidation. The organic matter contained within the matrices of type 1 and 2 carbonaceous chondrites, therefore, may provide a unique record of the oxidative nature of aqueous fluids that altered each meteorite's parent body.

3. The effect of thermal metamorphism on IOM

In order to study the organic structure of the thermally metamorphosed type 3 chondrites, one is faced with the problem that there is generally much less organic matter and we can rarely obtain sufficient amounts of IOM to acquire solid state NMR spectra. To circumvent this difficulty we apply micro Carbon X-ray Absorption Near Edge Structure (C-XANES) spectroscopy. C-XANES provides similar information to that of solid state NMR. While C-XANES lacks the ease of quantification that solid state NMR provides, samples as small as a few microns in diameter can easily be analyzed.

Through the analysis of ~ 25 different IOM samples obtained from type 3 CV, CO, and Ordinary chondrites, a clear picture of the effect of thermal metamorphism on IOM has been obtained (Cody *et al.*, submitted). Prolonged exposure to elevated temperatures ($T > 200$ °C) in the chondritic parent body progressively transforms primitive IOM into a highly aromatic material that exhibits semi-conductor-like electronic properties. In addition to showing a clear correlation between IOM structure and mineralogic indicators of metamorphic grade, Cody *et al.* (submitted) also showed that laboratory heating of primitive IOM (e.g., from Murchison) leads to the formation of organic structures that are identical to that in heated chondrites. These laboratory derived transformation kinetics can be used as a cosmothermometer revealing that the organic matter spanning the type 3 to type 4 chondrites records a temperature range up to 950 °C.

Finally, a significant result of the study of the type 3 chondrites is the unambiguous demonstration that the constituent primitive organic matter in type 1 and 2 chondrites may be thermally transformed into a structural state that is nearly identical to that observed in thermally metamorphosed type 3 chondrites. As it is generally agreed that type 1 and 2 chondrites never experienced temperatures much in excess of 100 °C; these results require that, in general, meteoritic organic matter was synthesized cold and, at least in the case of the type 1 and 2 chondrite parent bodies, accreted cold.

4. Comparison with comet 81P/Wild 2 organics

The Stardust comet sample return mission from Comet 81P/Wild 2 provides an unprecedented opportunity to assess the organic chemistry of what may be the most primitive Solar System material, providing a link to the molecular cloud material that was the ultimate source of the matter from which our Solar System originated. It is immediately compelling to see whether any chemical connection exists between the cometary organic matter and meteoritic IOM. Only very small particles were collected by necessity and, therefore, C-XANES provides a robust method for organic structure analysis. Combining C-, N-, and O-XANES also allows for the robust determination of elemental ratios, e. g. O/C and N/C.

To date, eight organic cometary particles analyzed have been using C-XANES (Sandford *et al.* 2006, Cody *et al.* 2008). In Figure 3b, the elemental chemistry of these particles are shown along with a single measure of an interplanetary dust particle, as well as a representative group of chondritic IOM (Alexander *et al.* 2007). In general, the comet organics are enriched in both oxygen and nitrogen relative to meteoritic IOM. The C-XANES spectra reveal what appears to be a chemical evolutionary trend moving from the most oxygen-rich cometary organics (Comet A) through less oxygen-rich organics (Comet B) to Murchison IOM (Figure 3a).

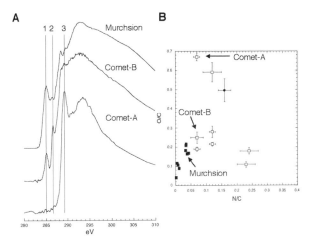

Figure 3. A) Carbon X-ray Absorption Near Edge Structure Spectra of two comet particles compared with IOM isolated from the Murchison (CM2) chondrite. Peak 1 corresponds to sp^2 bonded carbon, i. e., either aromatic or olefinic; Peak 2, corresponds to a keto-substituted olefin and is characteristic of the degradation of polyol (e.g. sugars). Peak 3 highlights the presence of sp^3 carbon bonded to oxygen, e. g. alcohol or ether. B) particle "Comet-A" is very oxygen-rich with chemistry dominated by alcohol and/or ether; particle "Comet-B" has an elemental composition not far from chondritic IOM; the elemental composition of Murchison IOM still contains appreciable oxygen and nitrogen. The loss of oxygen moving from particle "Comet-A" to Murchison is consistent with the progressive elimination of OH groups from sugar-like molecules yielding keto, carboxyl, and olefinic functional groups.

Any chemical pathway connecting the most oxygen-rich cometary organic matter to the molecular matter that is chondritic IOM requires the formation of considerable sp^2 bonded carbon (e.g., peak 1, Figure 3a) at the expense of oxygen-bearing sp^3 carbon (e.g., Peak 3, Figure 3a). Such a transformation would have to occur at very low temperatures, even type 2 chondritic IOM never experienced temperatures in excess of 100 °C. The apparent trend in chemical evolution observed in Figure 3a is potentially informative. In particular, the growth of the peak at 286.5 eV (Peak 2, Figure 3a) is consistent with the transformation of complex sugar-like molecules into unsaturated keto-bearing structures through the elimination of OH. Support for this pathway is provided by the observation of the progressive growth of a peak at 286.5 eV resulting from secondary electron damage of polysaccharides (Cody 2000).

that you are always observing an ensemble of compounds. If the actual chemistry were as simple as the examples I showed you, you would observe a nest of really sharp resonance lines.

SLOAN: My concern about is how we're going to drive the mixture of aromatics and aliphatics. Whatever mechanism does this, it has to happen pretty quick, because we see these unusual spectra in post-AGB objects. You don't have to go to molecular clouds to get this mixed chemistry.

CODY: I have been listening with great intrigue to the PAH story. I am outside of this field, which is why this meeting has been superb for me. My astronomy colleagues at the Carnegie often ask me why I don't observe a lot of PAH's in IOM and I cannot provide them an answer yet. I also cannot answer your question, one which I agree is profound. What I can say is that if we want to connect IR spectra observed in diffuse molecular clouds and that observed in Murchison IOM, it is not beyond the realm of possibility that there is similar chemistry going on out there that yields products that mimic what we observe in meteoritic IOM.

MUMMA: Just a comment on formaldehyde in comets. The situation is that the monomer of formaldehyde is typically on the order of half percent relative to water, so there's lot of that. But in Halley, for example, the polymer of formaldehyde would be about 7%, and in Hale Bopp it is about 12% relative to water. So you can imagine these icy mantle grains with that composition falling in the nebula and some vaporizing near the asteroid region and forming later into the IOM. You might have some really profound source of material you are talking about here. You could be dead on.

CODY: I agree. Thinking about polymerization theory, we can begin to ask how much of the carbon that we observe as IOM is related to the abundance of primitive chemistry out there. In other words how efficient could the IOM forming chemistry be. Are we looking at sub percent, 5, or 10% polymerization? I should note that one thing I did not emphasize in this talk is that IOM is very tough in regards to temperature. If IOM resides in hot regions of the solar nebula, high oxygen or hydrogen abundance will chemically erode it away. If the oxygen and hydrogen content are low, IIOM will hang around even to very high temperatures. I never would have expected this.

Organic Matter in Space
Proceedings IAU Symposium No. 251, 2008
S. Kwok & S. Sandford, eds.

© 2008 International Astronomical Union
doi:10.1017/S1743921308021753

Organic matter in the Solar System: From colors to spectral bands

Dale P. Cruikshank

NASA Ames Research Center
MS 245-6
Moffett Field, CA 94035 USA
email: Dale.P.Cruikshank@nasa.gov

Abstract. The reflected spectral energy distribution of low-albedo, red-colored, airless bodies in the outer Solar System (planetary satellites, Centaur objects, Kuiper Belt objects, bare comet nuclei) can be modeled with spectral models that incorporate the optical properties of refractory complex organic materials synthesized in the laboratory and called tholins. These materials are strongly colored and impart their color properties to the models. The colors of the bodies cannot be matched with plausible minerals, ices, or metals. Iapetus, a satellite of Saturn, is one such red-colored body that is well matched with tholin-rich models. Detection of aromatic and aliphatic hydrocarbons on Iapetus by the Cassini spacecraft, and the presence of these hydrocarbons in the tholins, is taken as evidence for the widespread presence of solid organic complexes aromatic and aliphatic units on many bodies in the outer Solar System. These organic complexes may be compositionally similar to the insoluble organic matter in some classes of the carbonaceous meteorites, and thus may ultimately derive from the organic matter in the interstellar medium.

Keywords. Astrochemistry, Planets and satellites, Kuiper Belt, Molecules

1. Introduction

Small bodies in the outer Solar System (planetary satellites, comet nuclei, Kuiper Belt objects, and Centaurs) are highly diverse in terms of orbital characteristics, dimensions, surface structures, and spectral characteristics. Many small bodies are thought to be remnant planetesimals or collisional fragments of planetesimals, preserved largely intact since their condensation in the solar nebula. As such, to the degree that they are chemically unaltered, they represent primordial material that can offer clues to the composition of the Solar System's nascent molecular cloud and the processes that occurred within the solar nebula during the epoch of planet formation. In addition to rocky material, metals, and ices, it is clear that organic molecular material is also often present. This organic component must be identified and understood as we progress toward a complete view of the original chemistry and processes in the early Solar System that led to the compositions we now see, and to the origin of life, at least on one planet.

This paper discusses new observations and insight into the presence of complex organic macromolecular material as a component of the low albedo surfaces that are typically found on small planetary satellites, the nuclei of comets, and on members of the newly identified populations of bodies beyond the orbit of Neptune.

These objects are typically too distant and faint for the application of infrared spectroscopy in the spectral regions where organic molecules have characteristic spectral bands ($\lambda > 3\ \mu$m). They can, however, be observed photometrically over a sufficient wavelength interval to determine their colors, typically in a portion of the spectral region in which diffusely reflected sunlight dominates the spectral energy distribution ($\lambda < 5\ \mu$m).

For very faint objects the observable spectral interval includes $\sim$ 0.3-1 μm, while for brighter objects the range can be extended to 2.5 μm and beyond. Photometry of large samples of the asteroids shows a wide variation in color, from neutral (or slightly blue) with respect to the Sun, to very red (Chapman *et al.* 1975, Zellner *et al.* 1985). These observational studies extended to small planetary satellites (Degewij *et al.* 1980), comets (Hartmann *et al.* 1982), and Kuiper Belt objects (Barucci *et al.* 2001, 2004) revealed similar trends, with a preponderance of red objects in these populations. At about the same time, parallel observations of the thermal emission of many of these objects at longer infrared wavelengths (typically 10 and 20 μm; e.g., Morrison 1973, Cruikshank & Morrison 1974) revealed the overall very low albedos ($<$ 0.08 at the V-band wavelength) in the majority of cases.

2. Organic Solids and the Surface Colors of Solar System Bodies

In an effort to understand the compositions of the reddest asteroids (the D class), Gradie and Veverka (1980) introduced the concept of complex organic solid materials on their surfaces, suggesting the presence of '... very red, very opaque, polymer-type organic compounds, which are structurally similar to aromatic-type kerogen...' Laboratory work by Sagan and Khare (1979) in progress at Cornell University, Gradie and Veverka's home institution, was at that time showing the importance and possible relevance of such organic compounds in explaining certain features of interstellar dust and the aerosol component of Titan's atmosphere. This synthetic organic solid, showing strong colors ranging from yellow through red to black, was termed 'tholin' by Sagan.

The optical properties (complex refractive indices) measured for several tholins (e.g., Khare *et al.* 1984, MacDonald *et al.* 1994) have enabled the use of these materials in calculations of synthetic spectra of diffusely scattered sunlight from a particulate solid surface for comparison with the observational data (e.g., Cruikshank & Dalle Ore 2003, Barucci *et al.* 2005). In the case of the extremely red Centaur object, 5145 Pholus, and by extension to other red bodies, it was demonstrated that the complex organic solid material (tholin) is an essential ingredient of the models to produce the observed color (Cruikshank *et al.* 1998). Recent work on the synthesis and analysis of tholins, primarily in the context of Titan's aerosol, is reported by Imanaka *et al.* (2004) and Bernard *et al.* (2006). The importance of tholins as coloring agents on bodies in the outer Solar System is discussed in detail by Cruikshank *et al.* (2005).

While minerals and ices do not give satisfactory spectral fits to the strongly red-colored surfaces of Solar System bodies, whereas models containing tholins do, it has been assumed for many years that organic solids are present on those surfaces, even though specific and diagnostic spectral absorption bands could not be detected. With the in situ study of the satellites of Saturn with VIMS (Visible-Infrared Mapping Spectrometer) aboard the Cassini spacecraft, that situation has changed with the detection of aromatic and aliphatic hydrocarbon bands in the red-colored, low-albedo material on Iapetus, Phoebe, and Hyperion (Clark *et al.* 2005, Cruikshank *et al.* 2008).

3. Results from the Satellites of Saturn

Iapetus. In this paper we focus on the surface compositions of the reddest Solar System bodies, using the measurements of three of Saturn's satellites, most specifically Iapetus, obtained by the VIMS instrument on the Cassini spacecraft, and then generalize to other surfaces whose spectra can be satisfactorily modeled by the use of tholins as the principal coloring agent.

Iapetus has the unique property among planetary satellites that the hemisphere centered on the apex of its locked synchronous orbital motion around Saturn has a geometric albedo of 0.02-0.06, while most of the remainder of the surface is some ten times more reflective and shows characteristic spectral absorption bands of H_2O ice. The mean density of Iapetus is 1.1 g cm^{-3}, suggestive of a bulk composition dominated by H_2O ice, but note that organic macromolecular solids also have a similar density. The pattern of distribution of the ice and the low-albedo material is seen to be more complex in the detailed images obtained by Cassini in December 2004 and in September 2007 than that deduced from ground-based data and images from Voyagers 1 and 2, but the dominance of an almost uninterrupted low albedo surface centered on the leading hemisphere

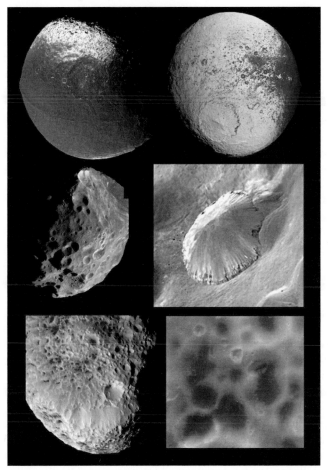

Figure 1. Three satellites of Saturn for which near-infrared reflectance spectra reveal aromatic and aliphatic hydrocarbon bands. Top two images show two different hemispheres of Iapetus (mean diameter 1470 km); the low-albedo material on the left image is distributed around the apex of the satellites locked-synchronous revolution around Saturn. The hydrocarbon bands occur mainly in the low-albedo material. The bright material is water ice. The central two images show Phoebe (long dimension 230 km). The hydrocarbon bands are distributed non-uniformly across the surface, which also shows water ice absorption bands. Center right image shows an apparent shallow subsurface layer of especially low-albedo material in a crater rim (crater named Euphemus, diameter about 40 km). Bottom left is a portion of Hyperion (dimensions 360 x 280 x 225 km) showing a largely water-ice surface with low-albedo material concentrated in some crater interiors (enlarged on bottom right; crater diameters about 6 km). NASA images from the Cassini spacecraft.

remains a unique and intriguing feature of this satellite. Because no other known object shows such a global albedo contrast, an explanation of its cause on Iapetus must also address the absence of such a brightness distribution on any of the other satellites of the major planets (Buratti *et al.* 2005) or transneptunian bodies known to date.

The overall spectral reflectance of the low-albedo material on Iapetus shows a very red color, a weak absorption band at 0.67 μm, and a strong absorption band at 3 μm [see Buratti *et al.* (2005) and Owen *et al.* (2001) for details and references]. Owen *et al.* (2001) modeled the low-albedo material with an intimate mixture of (nitrogen-rich) tholin, amorphous carbon, and H_2O ice. In their models the tholin provided the red color and a strong absorption band (N-H stretching mode) at 3 μm, matching the albedo, overall spectral shape, and the profile of the 3-μm band throughout the spectral region 0.3-3.8 μm.

The first complete coverage of the spectrum of Iapetus from 0.4-5.1 μm, without the interruptions induced by the Earth's atmosphere, was afforded by the VIMS instrument (Brown *et al.* 2004) on the Cassini spacecraft in December, 2004. Weak residual H_2O ice absorption bands and the 4.27 μm CO_2 band were clearly seen (Buratti *et al.* 2005, Brown *et al.* 2006). Initial models of the Cassini VIMS spectrum presented in Buratti *et al.* (2005) use tholins as the principal coloring agent, and also include HCN polymer, which acts to lower the albedo in the model to match that of Iapetus closely. Further detailed analysis of the 3.0-3.7 μm spectral region yielded the detection of the 3.3-μm C-H stretching mode band in aromatic hydrocarbons and components of the 3.4-μm aliphatic C-H stretching mode band complex (Cruikshank *et al.* 2008). This detection is corroborated and strengthened in additional spectra of the low-albedo surface units obtained at another close flyby of Iapetus in September, 2007.

Phoebe and Hyperion. The VIMS data for two other satellites of Saturn also give indications of the aromatic and aliphatic C-H bands (Clark *et al.* 2005, Cruikshank *et al.* 2007, Coradini *et al.* 2008), also distributed nonuniformly, but the signatures are weaker than on Iapetus, and a detailed analysis has not yet been completed. Other common characteristics of the low-albedo material on Iapetus, Phoebe, and Hyperion include the presence of CO_2 absorption band at 4.27 μm, a prominent but unidentified absorption band at 2.42 μm, and residual H_2O ice absorption. The band attributed to CO_2 has not been fully elucidated because of shifts in the band's central wavelength outside the range expected for pure CO_2 ice (Cruikshank *et al.* 2007, Clark *et al.* 2008). This matter is currently under study.

4. Discussion

The aromatic and aliphatic hydrocarbon signatures clearly seen on Iapetus are consistent the presence of aromatic units of small size (1-10 rings) linked with short aliphatic bridging units similar to the kind envisioned for meteoritic organic complexes (e.g., Kerridge *et al.* 1987). The aromatic units are thought to be small because the strength of the band indicates the presence of a relatively large number of H-bonding sites per unit C atom. Because meteoritic insoluble organic matter (IOM) commonly shows a strong aliphatic absorption and little or no aromatic band (e.g., Pendleton *et al.* 1994), the aromatic units in the Iapetus material maybe significantly smaller than those in the IOM. Note that the absolute strength of the aliphatic band complex (a mixture of $-CH_3$ and $-CH_2$ units) is $\sim$ 10 times that of the aromatic band, so the strength of the 3.3-μm band on Iapetus indicates that aromatic material is abundant.

The low-albedo material on Iapetus and the other Saturnian satellites has the characteristics of hydrogenated amorphous carbon (HAC) that is similar in composition and structure to the carbon-rich organic dust particles in the diffuse interstellar medium (Pendleton & Allamandola 2002). Interstellar HAC is composed of a mixture of aromatic regions (sp^2 bonded systems with a polycyclic aromatic hydrocarbon (PAH) structure) within a matrix of diamond-like (sp^3 bonded) or polymeric (sp^3, sp^2, and sp^1 bonded) material (Jones 1990). The conditions of formation and subsequent processing by UV, GCR, and charged particles from the magnetosphere of Saturn influence the instantaneous relative abundances of the various structural components as well as the evolution of this material over time.

Mennella *et al.* (2002) have shown in the laboratory that hydrogenation of carbon particles by H atoms is the principal mechanism that determines the presence of the 3.4-μm aliphatic C-H band in the diffuse interstellar dust. This hydrogenation can occur when small carbon grains are exposed to a flux of H atoms produced by a microwave discharge. In this process, only the 3.4-μm aliphatic C-H structure is produced, but when these same particles are heated, the strength of the aliphatic band decreases and the 3.3-μm aromatic band becomes prominent. The aromatic band is most prominent at $T \sim 500$ C; at higher temperatures dehydrogenation begins to occur and the C-H bands rapidly disappear (V. Mennella private communication).

If the low-albedo material on the Saturn satellites originated as interstellar HAC with an intrinsic 3.4-μm aliphatic hydrocarbon band of sufficient strength to allow spectroscopic detection, and if the mechanism of hydration was the same as Mennella *et al.* (2002) reproduced in the laboratory, then the appearance of the 3.3-μm aromatic hydrocarbon band suggests a degree of thermal processing. At the same time, the absence of detectable bands in large regions of the surfaces of the satellites that are otherwise low in albedo suggests the loss of C-H bonds through the effects of processing by solar UV, GCR, and charged particles from Saturn's magnetosphere. The dehydrogenation of HAC leaves abundant elemental carbon of graphitic structure and with low albedo and more neutral color than its hydrogenated counterpart. The darkening and dehydrogenation of HAC films by UV irradiation in the laboratory (see Jones 1990 for discussion and references) may represent the same basic mechanism that occurs on planetary surfaces upon prolonged exposure to the space environment, and may be a major factor in accounting for the wide variation in the degree of reddening of solid body surfaces among low-albedo planetary satellites, KBOs, and Centaurs.

The detection of aromatic and aliphatic hydrocarbon bands on Iapetus has implications for the presence of hydrocarbons in the low-albedo material that covers or constitutes the surfaces of a great number of other small bodies in the outer Solar System. The Iapetus material is closely modeled with tholins as the principal refractory component and as the component giving the overall reflected spectral energy distribution (color) over the wavelength range 0.3-4 μm (Owen *et al.* 2001). Similarly, red-colored KBOs and Centaurs are well modeled with tholins serving the same roles (Barucci *et al.* 2001, 2005, Cruikshank *et al.* 1998). At the same time, tholins are known to contain, among other things, both aliphatic and aromatic hydrocarbons (Imanaka *et al.* 2004, Bernard *et al.* 2006). It is therefore reasonable assert that the very red-colored surface materials of the wider range of bodies also consist of hydrocarbon-bearing organic complexes, even though the exact structure, composition, and nature are most probably different from the synthetic tholins made in the laboratory.

Naturally occurring organic complexes in meteorites contain aliphatic and aromatic units, as noted above (see also Yabuta *et al.* 2007), and while the aliphatic component is most prominent in the near-infrared spectra, the insoluble organic matter from some meteorites (C. Alexander, private communication) also shows a prominent aromatic band at ~ 3.3 μm. Future investigations will include the modeling of the Iapetus spectrum and the spectra of other red-colored bodies in the Solar System using the optical constants derived for meteoritic organic matter. Successful modeling of the spectral shape and albedo levels with meteoritic organics will strengthen the hypothesis that organic solid material is a basic and wide-spread component of the surfaces of many outer Solar System bodies.

Acknowledgements

I thank Drs. Max Bernstein, Conel Alexander, George Cody, Yvonne Pendleton, Vito Mennella, and Louis Allamandola for many stimulating discussions. The Iapetus spectral information from the VIMS instrument is a part of the NASA-ESA Cassini mission to the Saturn system. I thank my many colleagues on the Cassini mission for their efforts over many years to obtain these and other data.

References

Barucci, M. A., Cruikshank, D. P., Dotto, E., Ferlin, F., Poulet, F., Dalle Ore, C., Fornasier, S., & de Bergh, C. 2005, *A&A* (Letter), 439, L1

Barucci, M. A., Doressoundiram, A., & Cruikshank, D. P. 2004, in: M. Festou, H. Keller, and H. Weaver (eds.), *Comets II*, (Tucson: Univ. Arizona Press), p. 647

Barucci, M. A., Fulchignoni, M., Birlan, M., Doressoundiram, A., Romon, J., & Boehnhardt, H. 2001, *A&A*, 371, 1150

Bernard, J-M, Quirico, E., Brissaud, O., Montagnac, G., Reynard, B., McMillan, P., Coll, P., Nguyen, M-J., Raulin, F., & Schmitt, B. 2006, *Icarus*, 185, 301

Brown, R. H. *et al.* 2004, *Space Sci. Rev.*, 115, 111

Brown, R. H. *et al.* 2006, *A&A*, 446, 707

Buratti, B. J., Cruikshank, D. P., & the VIMS Team 2005, *ApJ* (Letters), 622, L149

Chapman, C. R., Morrison, D., & Zellner, B. 1975, *Icarus*, 25, 104

Clark, R. N. *et al.* 2005, *Nature*, 435, 66

Clark, R. N. *et al.* 2008, *Icarus*, 193, 372

Coradini, A. *et al.* 2008, *Icarus*, 193, 233

Cruikshank, D. P. & Dalle Ore, C. M. 2003, *Earth, Moon, Planets*, 92, 315

Cruikshank, D. P., Imanaka, H., & Dalle Ore, C. M. 2005, *Adv. Space Res.*, 36, 178

Cruikshank, D. P. & Morrison, D. 1974, *Icarus*, 20, 477

Cruikshank, D. P., *et al.* 1998, *Icarus*, 135, 389

Cruikshank, D. P., *et al.* 2007, *Nature*, 448, 54

Cruikshank, D. P., *et al.* 2008, *Icarus*, 193, 344

Degewij, J., Cruikshank, D. P., & Hartmann, W. K. 1980, *Icarus*, 44, 541

Gradie, J. & Veverka, J. 1980, *Nature*, 283, 840

Hartmann, W. K., Cruikshank, D. P., & Degewij, J. 1982, *Icarus*, 52, 377

Imanaka, H. *et al.* 2004, *Icarus*, 168, 344

Jones, A. P. 1990, *MNRAS*, 247, 305

Kerridge, J., Chang, S., & Shipp, R. 1987, *Geochim. Cosmochim. Acta*, 51, 2527.

Khare, B. N. *et al.* 1984, *Icarus*, 60, 127

Mennella, V. *et al.* 2002, *ApJ*, 569, 531

MacDonald, G. D., Thompson, W. R., Heinrich, M. *et al.* 1994, *Icarus*, 108, 137

Morrison, D. 1973, *Icarus*, 19, 1

Owen, T. C. *et al.* 2001, *Icarus*, 149, 160

Pendleton, Y. J. & Allamandola, L. J. 2002, *ApJS*, 138, 75

Pendleton, Y. J., *et al.* 1994, *ApJ*, 437, 683

Sagan, C. & Khare, B. N. 1979, *Nature*, 277, 102

Yabuta, H. *et al.* 2007, *LPSC XXXVIII*, abstract 2304.pdf

Zellner, B., Tholen, D. J., & Tedesco, E. F. 1985, *Icarus*, 61, 355

Discussion

KWOK: If these complex organics, or tholin-like materials, are common in small Solar System bodies, where are they on Earth? I mean, were they destroyed during the thermal processing or by shocks, or during the aggregation process? If they are preserved in smaller bodies, some of them must have been here.

CRUIKSHANK: I presume that this complex material has fallen on Earth and continues to fall on Earth in the form of meteorites and IDPs. The method of emplacement on the Saturn satellite is far from clear. This material is probably quite young, by which I mean maybe no more than a few million years old. They may also be only a few tens of centimeters in thickness, suggesting again that it fell on the surface from some external source, but we don't know where. I am not sure any comments about the Earth are totally relevant to the Saturn system, but surely the stuff has fallen on the Earth from whatever source - comets, most likely - and continues to form both comets and meteorites and asteroids. So I think it's here, it's all around us. It's just has been mixed with the biological materials to the extent that we would never recognize it as an independent component.

MATTHEWS: Dale, the main result of your elegant survey of these deep color materials is that there are tholins deposited on many of these bodies. So what are tholins? Tholins are made, as you pointed out, by supplying energy to a mixture of methane and ammonia or methane and nitrogen. And what we have been saying is that what a tholins is essentially two polymers, polymers of HCN and of acetylene that come from methane and ammonia or methane and nitrogen, so you have hydrocarbon polymers but more importantly, you have the HCN polymers. I am suggesting that the main colored material as you're drawing attention to so importantly is HCN polymers. You know when you polymerize the HCN you are starting with colorless liquid, liquid of HCN whatever, and it gets yellow, orange, red, and then finally black, exactly these colors we're talking about. So I'm really so glad, quite thrilled to have you, making this an important issue, but I am really adding a statement there, a very positive statement, that HCN polymers are really important. They have the right colors and they are all over the place. They are important molecules because when you hydrolyze HCN polymers you get amino acids, petites, purines, pyrimidines, it is amazing.

CRUIKSHANK: Yes. Well, I appreciate that comment, Cliff, and you are quite right that HCN is an important component in the laboratory when this material is made in the gas phase and the solid stuff precipitates on the surface of the vessel, there's a residual gas. If we can condense that on a cold finger, we can watch it change in color from a colorless floss to a dark reddish brown glue which is in fact the HCN polymer. So the resulting gas is just the question of the temperature at which the materials is condensed. The residual gas has a lot of HCN which will deposit and polymerize right before your eyes and it is an interesting thing to see. So in the natural setting whatever this tholins is, probably does have an HCN component. Now, why can't we see the nitrile band? We have looked very hard for the nitrile band around 4.6 microns and while there is some sort kind of spectral activity in the data, we cannot identify a nitrile band with any degree of reliability. So we were puzzled by that.

CESSARSKY: How did you date the deposits? Where does this age of ten million year come from?

CRUIKSHANK: The assertion that it is very recent comes from the absence, or the near absence of any bright craters in this black surface. There are about two or three very small craters which apparently have penetrated the surface layer and have spread bright rays of the underlying water ice. Since we think we know something about the cratering frequency, the impact frequency in this part of the Solar System, we can estimate that this number of craters would be produced in something of the order of a million years. There are different models for the cratering rate and some people say it's less than a hundred thousand years and other models suggest maybe as much as ten million years. But that is a good indication of the absence of penetrating impacts that it is a recent deposit. Its thinness, or thickness, is deduced from radar measurements made from Cassini spacecraft which indicated a radar reflectivity, discontinuity, and a depth of a few tens of centimeters. So that's the indication that is a surfacial layer.

Dale Cruikshank (photo by Sze-Leung Cheung).

Organic Matter in Space
Proceedings IAU Symposium No. 251, 2008
S. Kwok & S. Sandford, eds.

Organics in meteorites - Solar or interstellar?

Conel M. O'D. Alexander[1], George D. Cody[2], Marilyn Fogel[2], and Hikaru Yabuta[2]

[1]DTM, Carnegie Institution of Washington,
5241 Broad Branch Road, Washington DC 20015, USA
email: alexande@dtm.ciw.edu

[2]GL, Carnegie Institution of Washington,
5251 Broad Branch Road, Washington DC 20015, USA
email: g.cody@gl.ciw.edu, m.fogel@gl.ciw.edu, hyabuta@gl.ciw.edu

Abstract. The insoluble organic material (IOM) in primitive meteorites is related to the organic material in interplanetary dust particles and comets, and is probably related to the refractory organic material in the diffuse interstellar medium. If the IOM is representative of refractory ISM organics, models for how and from what it formed will have to be revised.

Keywords. Comets, asteroids, ISM: dust

1. Introduction

Refractory organic material is a major component of dust in the interstellar medium (ISM), of comets, of interplanetary dust particles (IDPs) that may come from comets, of at least the surfaces of other outer Solar System bodies and of primitive meteorites (chondrites) derived from the asteroid belt. There are numerous potential mechanisms for making refractory organic material. Here we explore whether the refractory materials in chondrites, IDPs and comets are related, and whether they formed in the ISM or in the early Solar System.

2. Relationship between IOM in meteorites, IDPs and comets

Chondritic porous (CP-) IDPs are generally thought to be the most primitive Solar System objects available for study in the laboratory - it has been argued that they are from comets (Joswiak *et al.* 2000). Similarly, based on large D and ^{15}N excesses, particularly in so-called hotspots (micron to sub-micron regions), the organic material in CP- IDPs has long been considered the most primitive available to us. An apparently macromolecular insoluble organic matter (IOM) makes up the bulk of the organic matter in chondrites (up to $\sim$ 2 wt.% C). The IOM in the most primitive chondrites is enriched in D and ^{15}N in bulk (Alexander *et al.* 2007) and contains hotspots with similar isotopic enrichments and sizes to those in IDPs (Busemann *et al.* 2006). The resolvable hotspots ($>$ 100 nm, $\sim$ 1 vol%) are not sufficiently abundant to contribute significantly to the bulk compositions of the IOM. Some, but not all, hotspots are associated with roughly spherical grains (globules) that are often hollow (Garvie & Buseck 2004, Nakamura-Messenger *et al.* 2006). The bulk H and N isotopic compositions of chondritic IOM with the largest isotopic enrichments are comparable to the average compositions of the most primitive IDPs.

There seems to be a genetic link between the IOM in at least some chondrites and IDPs that may come from comets. Indeed, there is a striking resemblance between the bulk elemental compositions of comet Halley CHON particles, $C_{100}H_{80}N_4O_{20}S_2$ (Kissel & Krueger 1987), and the meteoritic IOM with the most anomalous isotopic compositions ($C_{100}H_{70-79}N_{3-4}O_{11-21}S_{1-5}$). Organic particles from Comet 81P/Wild 2 returned by the Stardust mission have higher N and O contents, as well as more modest D and ^{15}N enrichments, compared to IOM in IDPs and primitive chondrites (Cody *et al.* 2008, Sandford *et al.* 2006). Questions still remain about how much modification (possibly even synthesis) of organics there was during capture, as well as potential sources of contamination. Perhaps more importantly, the silicate mineralogies of Wild 2 particles are not as primitive as that found in IDPs. It seems that the silicates experienced high temperatures, probably prior to formation of the Wild 2 parent body (Ishii *et al.* 2008). The lower abundance of presolar grains in Stardust particles also suggests that the Wild 2 material is not as primitive as IDPs or even the most primitive chondrites. If correct, it is unlikely that the IOM in Wild 2 is as primitive as that in IDPs.

3. A Common Source?

There is a link between the IOM in at least some primitive meteorites, IDPs and comets. Thus, this IOM must have been widespread in the early Solar System. It has been argued for some time that, because of the large D and ^{15}N enrichments, some or all of the IOM must have formed in the ISM. This is supported by the similarity in the 3-4 μm C-H stretch region between chondritic IOM and dust in the diffuse ISM (Pendleton & Allamandola 2002). However, conditions would not have been so very different in the ISM and parts of the solar protoplanetary disk (nebula). Indeed, recently it has been argued that the D enrichments are solar in origin (Gourier *et al.* 2008).

Gourier *et al.* (2008) report an inverse correlation between the D-enrichment of certain functional groups and their C-H bond energies. They propose that this correlation results from the interaction at the surface of the protoplanetary disk of H_2D^+ with organic grains formed near the midplane. Alexander *et al.* (2007) have questioned whether there is a dependence of D-enrichment with functional group. In addition, H_2D^+ is so reactive that it should exchange more-or-less indiscriminately, rather than as a function of C-H bond energy, which seems to be borne out by recent experiments (Thomen *et al.* 2008).

While the Gourier *et al.* model remains controversial, they raise an important issue, namely: how does one distinguish between organic material formed in the ISM and solar nebula? To try to address this issue, we have analyzed the IOM in a wide range of primitive meteorites using a number of techniques (Alexander *et al.* 2007, Cody & Alexander 2005). We have found considerable elemental and isotopic variation in IOM composition between meteorites. If the IOM in all meteorites has an interstellar origin, these variations must be consistent with evolution from a single precursor material. On the other hand, if the IOM is solar, one could expect to see systematic variations in composition and abundance as conditions changed over the 2-3 Ma that different chondrite groups seem to have formed.

In the most primitive chondrites, the ratio of IOM to presolar nanodiamonds is a fairly constant, as are the abundances of both in the low temperature matrices of chondrites (Alexander 2005, Huss & Lewis 1995). The IOM in all chondrites also seems to be associated with trapped noble gases that may be interstellar (Huss *et al.* 1996). These observations are consistent with a common, probably interstellar, origin for IOM. However, even in the most primitive chondrites there is a considerable range in IOM compositions, and the abundance ratios of IOM to presolar grains other than nanodiamonds are more variable.

Much of this variation relative to other presolar grains could simply reflect alteration on the meteorite parent bodies. For instance, presolar silicates are easily destroyed during aqueous alteration and thermal metamorphism. The IOM will also be sensitive to these processes. What are generally considered the most primitive chondrites (CI, CM, CR and Tagish Lake) have all experienced aqueous alteration. The IOM in these meteorites exhibits a wide range of aliphatic/aromatic C ratios (CR>CI>CM>Tagish Lake) that appears to be the result of variable oxidation of the aliphatic fraction of a common precursor at the onset of alteration (Cody & Alexander 2005). In general, the bulk IOM D enrichments decline in the same order as the aliphatic/aromatic ratio. However, there are some notable exceptions. The most dramatic is to be found in the CMs. All CMs have IOM with very similar elemental and functional group chemistries, and most have quite uniform isotopic compositions ($\delta D=800‰$, $\delta^{13}C=-18‰$, and $\delta^{15}N=0‰$). However, the IOM in the CM Bells ($\delta D\approx3000‰$, $\delta^{13}C\approx-34‰$, and $\delta^{15}N\approx400‰$) has a composition that more closely resembles that of the IOM in the CRs, the chondrite group with the generally most isotopically anomalous material. Bells experienced a somewhat different mode of alteration than the other CMs, suggesting that the degree to which IOM is modified during parent body processing can be very sensitive to the conditions.

This sensitivity to conditions can also be seen in those meteorites that experienced thermal metamorphism (CV, CO, O and E chondrites). The metamorphosed chondrites all have lower H/C, O/C and N/C ratios than the aqueously altered ones. This is to be expected as thermal maturation is well known to drive the evolution of carbonaceous materials towards more graphite-like structures, with the associated loss of heteroatoms. The trends in H/C vs. N/C and O/C are all consistent with evolution from a CR/CM-like precursor. Indeed, CMs that have been shock heated roughly parallel the trend in the metamorphosed chondrites, even though the IOM in most metamorphosed chondrites experienced destructive oxidation in addition to thermal maturation. More perplexing are the isotopes. The expectation would be that the D enrichment would decrease with increasing degree of metamorphism. Except in the O chondrites, this is what is seen. In the O chondrites D enrichments increase dramatically with increasing metamorphism. At present, there is no good explanation for this behavior.

The variation in elemental and functional group chemistry of IOM within and between chondrite groups is consistent with parent body modification of a common precursor that looked something like the IOM in the CR chondrites. The isotopes paint a more complicated story that we do not fully understand yet, but it is clear that conditions during parent body processing played a central role. It is not yet possible to see through the parent body processing to the original material in all chondrite groups. However, if one uses the IOM from the most primitive members of each chondrite group, there is no apparent correlation between either IOM composition or its abundance in the matrix and crude estimates for the time and conditions of formation of the chondrite parent bodies.

4. A Solar or ISM Origin?

Based on current evidence, it seems likely that there was a widespread IOM-like precursor material in the formation regions of asteroids and at least some comets. Whether this formed in the ISM or Solar System remains uncertain, although the relatively constant presolar nanodiamond/IOM ratios in chondrites, along with the close similarities in the IR 3-4 μm C-H stretch region between the IOM and diffuse ISM, point towards an interstellar origin. Refractory organic material is abundant in the ISM, and to date it has not been shown that IOM-like material can be made in the solar nebula on the scale required. Nevertheless, the IOM differs in significant ways from what has been inferred for refractory organics in the diffuse ISM. The diffuse ISM material is thought to be composed of large PAHs derived from C-rich AGB stars (Kwok 2004, Pendleton & Allamandola 2002). One way we are pursuing of reconciling this model with an IOM that is dominated by small, highly substituted PAHs is if the large PAHs are heavily damaged by cosmic rays while trapped in ices in molecular clouds. The damaged regions would be hydrogenated, nitrogenated and oxygenated (e.g., Bernstein *et al.* 2003), leaving only small islands of highly substituted PAHs surrounded by short, highly branched aliphatic material. These damaged regions would take on the H, N and O isotopic compositions of the ice, which would be rich in D and ^{15}N. If returned to the diffuse ISM, this process would somehow have to be reversed, but material that was incorporated into the Solar System directly from the protosolar molecular cloud would not have experienced this 'annealing' process.

The C and N isotopes remain problematic to this appealing picture. The C and N isotopes of PAHs formed around C-stars should be far from the solar composition judging from models, astronomical observations and presolar grains. However, the PAHs in IOM are on average essentially solar in their C isotopic compositions (Sephton *et al.* 2000). It cannot be ruled out that on average PAH-producing stars have solar C isotopic compositions, but this seems unlikely at present. The N isotopic compositions of AGB stars are always highly depleted in ^{15}N compared to solar, whereas the bulk and subgrain ^{15}N abundances in the most primitive IOM are enriched. Hence, if the IOM is interstellar in origin and is representative of the refractory organic dust in the ISM, how and from what it formed will have to be rethought.

Acknowledgements

We gratefully acknowledge financial support from NASA's Origins of Solar Systems and Astrobiology programs.

References

Alexander, C. M. O'D. 2005, *Meteorit. Planet. Sci.*, 40, 943

Alexander, C. M. O'D., Fogel, M., Yabuta, H., & Cody, G. D. 2007, *Geochim. Cosmochim. Acta*, 71, 4380

Bernstein, M. P., Moore, M. H., Elsila, J. E., Sandford, S. A., Allamandola, L. J., & Zare, R. N. 2003, *ApJ* (Letters), 582, L25

Busemann, H., Young, A. F., Alexander, C. M. O'D., Hoppe, P., Mukhopadhyay, S., & Nittler, L. R. 2006, *Science*, 314, 727

Cody, G. D. & Alexander, C. M. O'D. 2005, *Geochim. Cosmochim. Acta*, 69, 1085

Cody, G. D., *et al.* 2008, *Meteorit. Planet. Sci.*, 43, 353

Garvie, L. A. J. & Buseck, P. R. 2004, *Earth Planet. Sci. Lett.*, 224, 431

Gourier, D., Robert, F., Delpoux, O., Binet, L., Vezin, H., Moissette, A., & Derenne, S. 2008, *Geochim. Cosmochim. Acta*, 72, 1914

Huss, G. R. & Lewis, R. S. 1995, *Geochim. Cosmochim. Acta*, 59, 115

Huss, G. R., Lewis, R. S., & Hemkin, S. 1996, *Geochim. Cosmochim. Acta*, 60, 3311

Ishii, H. A., *et al.* 2008, *Science*, 319, 447

Joswiak, D. J., Brownlee, D. E., Pepin, R. O., & Schlutter, D. J. 2000, *Lunar Planet. Sci.*, 31, #1500 (abs.)

Kissel, J. & Krueger, F. R. 1987, *Nature*, 326, 755

Kwok, S. 2004, *Nature*, 430, 985

Nakamura-Messenger, K., Messenger, S., Keller, L. P., Clemett, S. J., & Zolensky, M. E. 2006, *Science*, 314, 1439

Pendleton, Y. J. & Allamandola, L. J. 2002, *ApJS*, 138, 75

Sandford, S. A., *et al.* 2006, *Science*, 314, 1720

Sephton, M. A., Pillinger, C. T., & Gilmore, I. 2000, *Geochim. Cosmochim. Acta*, 64, 321
Thomen, A., Robert, F., & Derenne, S. 2008, *Lunar Planet. Sci.*, 39, #2001 (abs.)

Discussion

CRUIKSHANK: Conel, you made a point noting that the bulk of the organic materials is not FTT synthesized. I guess in a large part because it's not associated with mineral grains. Yet in George Flynn's talk, he spoke of an organic rim on mineral grains that may not be the bulk of the organic material in the IDP. But is it possible to distinguish by experimental or other means the difference in the nature of organic material on a mineral grain which may have been FTT synthesized compared to the other material?

ALEXANDER: People have looked at this to some extent. You do see in meteorites and in IDPs carbonaceous material rimming some grains. In the meteorites it's certainly not as pervasive as George Flynn implied. Generally that material, in the TEM at least, looks more graphitic. I don't know if this is true in the IDPs, but certainly in the meteorites the stuff that is rimming the minerals looks more like poorly graphitized carbon, i.e. it exhibits d-spacings, whereas the stuff you see that is unassociated with minerals is essentially featureless.

SANDFORD: You look at all these residues from many, many meteorites and you have done this by using solvents, so presumably you are not necessarily seeing the primordial D to H ratio from anything that is exchangeable. So do you have a sense of what the values that would have been in the original meteorite but were exchanged during the dissolution process.

ALEXANDER: When we've looked at meteorites in situ, generally we see the same range of compositions of hot-spots. But the trouble when you are looking at the meteorites is that much of the hydrogen is in hydrous phases. So it's hard to make a really good comparison. People have done experiments back in the '80s using labeled compounds, e. g., labeled acids, labeled solvents, etc. There's some disagreement, but generally the feeling is that you don't have more than a few percent exchange of hydrogen. You always have to worry about that, but for most purposes I think we don't get a great deal of exchange. All our isolations were done at room temperature, whereas in the old days you would heat things in acids at 60-80 °C.

SANDFORD: In a way, that's a big clue telling you a lot of hydrogen is not in exchangeable forms.

ALEXANDER: Well, that's not entirely true. It is exchangeable. Hikaru Yabuta did hydrothermal experiments and showed that if you heat these things for a couple of days at 300 °C under modest pressure then you entirely exchanged the hydrogen. We don't know what the kinetics of that the process is, and that is something we need to work out.

KWOK: When you say large PAHs, do you mean pure aromatic material or a mixture of aromatic and aliphatic materials, e. g., the kerogen-like compounds that we are seeing in protoplanetary nebulae and planetary nebulae?

ALEXANDER: What I would say is that we have this macromolecular material that seems to be composed of aliphatic material binding together small PAHs. We don't think there's more than a few percent of PAHs with larger than 3 to 4 rings. This material is a macromolecular material with about 40-60% of the carbon in aromatic form, but that material is in small PAHs, not in coronene or larger PAHs.

ZIURYS: Maybe these small PAH aromatic rings are responsible for what we see in the interstellar medium, not the big ones.

ALEXANDER: That's quite possible.

SLOAN: Maybe I missed this, but why do you say the PAHs aren't from the AGB stars?

ALEXANDER: The carbon isotopes of the organic materials are essentially solar, whereas the carbon isotope ratios in AGB star atmospheres and in presolar grains in meteorites from AGB stars are far from solar. Unless you have a magic way of mixing all these material up to make solar composition, there can't be a large fraction of AGB material ending up in the IOM.

Conel Alexander (photo by Sze-Leung Cheung).

Organic Matter in Space
Proceedings IAU Symposium No. 251, 2008
S. Kwok & S. Sandford, eds.

© 2008 International Astronomical Union
doi:10.1017/S1743921308021777

Organics in the samples returned from comet 81P/Wild 2 by the Stardust Spacecraft

Scott A. Sandford

Astrophysics Branch, NASA-Ames Research Center,
Mail Stop 245-6, Moffett Field, California, USA
email: Scott.A.Sandford@nasa.gov

Abstract. The Stardust Mission collected samples from Comet 81P/Wild 2 on 2 Jan 2004 and returned these samples to Earth on 15 Jan 2006. After recovery, a six month preliminary examination was done on a portion of the samples. Studies of the organics in the samples were made by the Organics Preliminary Examination Team (PET) - a worldwide group of over 55 scientists. This paper provides a brief overview of the findings of the Organics PET. Organics in the samples were studied using a multitude of analytical techniques including spatial determination of C and heteroatom elemental abundances (STXM), functional group identification (micro-FTIR/Raman, C,N,O-XANES), and specific molecular identification of certain classes of organics (HPLC-LIF, L2MS, TOF-SIMS). Analyses were also made of spacecraft components and environmental samples collected near the recovered returned capsule to assess contamination issues. The distribution of organics (abundance, functionality, and relative elemental abundances of C,N,O) is heterogeneous both within and between particles. They are an unequilibrated reservoir that experienced little parent body processing after incorporation into the comet. Some organics look like those seen in IDPs (and to a lesser extent, meteorites), while new aromatic-poor and highly labile organics, not seen in meteoritic materials, are also present. The organics are O,N-rich compared to meteoritic organics. Some of the organics have an interstellar heritage, as evidenced by D and ^{15}N enrichments.

Keywords. Stardust Mission, comets, organics, dust, isotopes, astrochemistry, astrobiology

1. Introduction

It is now understood that our Solar System formed from the collapse of a portion of a dense interstellar molecular cloud of gas, ice, and dust (Mannings *et al.* 2000). The collapsing material formed a disk surrounding a central protostar. Much of the material was incorporated into the central protostar, which eventually became our Sun, some was ejected from the system by bipolar jets and gravitational interactions, and most of the remaining material was incorporated into small bodies, called planetesimals, which subsequently accreted to form the planets. However, some of these planetesimals escaped ejection from the system or incorporation into larger bodies and survived in the form of asteroids and comets (Bottke *et al.* 2002, Festou *et al.* 2004).

The importance of these small bodies outweighs their minor contribution to the mass of our Solar System because their materials have undergone less parent-body processing and they contain more pristine early nebular materials. We receive samples of these objects on Earth in the form of meteorites and interplanetary dust particles (IDPs), and these demonstrate that comets and asteroids do contain primitive materials (Kerridge & Matthews 1988, Lauretta & McSween 2006). Since comets are thought to have formed and resided in the outer Solar System, they probably contain a more representative portion of the volatile components of the original protoplanetary disk and have undergone less

Scott Sandford (photo by Dale Cruikshank).

Organic Matter in Space
Proceedings IAU Symposium No. 251, 2008
S. Kwok & S. Sandford, eds.

© 2008 International Astronomical Union
doi:10.1017/S1743921308021789

Chemical diversity of organic volatiles among comets: An emerging taxonomy and implications for processes in the proto-planetary disk

Michael J. Mumma

Solar System Exploration Division, NASA's Goddard Space Flight Center
email: Michael.J.Mumma@nasa.gov

Abstract. As messengers from the early Solar System, comets contain key information from the time of planet formation and even earlier – some may contain material formed in our natal interstellar cloud. Along with water, the cometary nucleus contains ices of natural gases (CH_4, C_2H_6), alcohols (CH_3OH), acids (HCOOH), embalming fluid (H_2CO), and even anti-freeze (ethylene glycol). Comets today contain some ices that vaporize at temperatures near absolute zero (CO, CH_4), demonstrating that their compositions remain largely unchanged after 4.5 billion years. By comparing their chemical diversity, several distinct cometary classes have been identified but their specific relation to chemical gradients in the proto-planetary disk remains murky. How does the compositional diversity of comets relate to nebular processes such as chemical processing, radial migration, and dynamical scattering? No current reservoir holds a unique class, but their fractional abundance can test emerging dynamical models for origins of the scattered Kuiper disk, the Oort cloud, and the (proposed) main-belt comets. I will provide a simplified overview emphasizing what we are learning, current issues, and their relevance to the subject of this Symposium.

Discussion

IRVINE: Could you say a little bit more on the ortho-para ratio? I gathered from your data that you're saying you believe that ratio is primordial in comets.

MUMMA: It seems to be primordial. If you look at Comet Halley over the space of December 1985 to March 1986, the ortho-para ratio did not change. It was around 2.5 on both dates, even though the comet had lost a huge amount of material representing about a 1-meter depletion in the surface. So this is one example where we see it not changing. I showed you an example of how we're trying to test whether this is reset in the coma. Unfortunately, the example I showed you had an ortho-para ratio of 3.0 to begin with, so at rotational temperatures higher than about 25 Kelvin I wouldn't expect it to change and it didn't. We also have other comets with ortho-para ratios of about 2.4 and a graduate student is working with me now on those data to extract the same trend curve. We do think it seems to be primordial, but let's see what happens in the future.

The face-changing (mask illusion) show at the banquet.

Organic Matter in Space
Proceedings IAU Symposium No. 251, 2008
S. Kwok & S. Sandford, eds.

© 2008 International Astronomical Union
doi:10.1017/S1743921308021790

Predominantly left-handed circular polarization in comets: Does it indicate L-enantiomeric excess in cometary organics?

V. Rosenbush[1], N. Kiselev[1], and L. Kolokolova[2]

[1] Main Astronomical Observatory, National Academy of Sciences of Ukraine,
27 Zabolotnoho Str., 03680 Kyiv, Ukraine
email: rosevera@mao.kiev.ua

[2] Department of Astronomy, University of Maryland, College Park, MD, 20742, USA
email: ludmilla@astro.umd.edu

Abstract. Polarimetric observations demonstrated that all comets with significant values of circular polarization show predominantly left–handed circularly polarized light. We discuss the presence of homochiral organics in cometary materials as a source of the observed circular polarization. We have studied the effect of chirality on light–scattering properties of cometary dust considering particles that possess optical activity. Our investigations show that the cometary dust may include optically active materials which can be prebiological homochiral organics.

Keywords. Comets, observations, polarization, light scattering

Recent observations confirmed that the light in the optical continuum of cometary spectra is circularly polarized. There are several mechanisms that may be responsible for the circular polarization (CP) in comets. Among them is scattering of light on particles containing optically active (chiral) materials. Optical activity is concerned with chirality (mirror asymmetry) of molecules and is typical for complex organics. A characteristic property of terrestrial bio–organic molecules is that their amino–acids are left–handed and the sugars are right–handed (called "homochirality"). For a long time it was believed that homochirality, or asymmetry in the number of L and D biomolecules, is of terrestrial origin. But then a significant excess of left–handed amino–acids was found in carbonaceous materials of meteorites (Cronin & Pizzarello 1997, Pizzarello 2004 and references there), suggesting its origin in the pre–solar nebula. Moreover, high values of CP were discovered in star–forming regions and their possible connection to a fundamental problem of astrobiology, homochirality of biomolecules, was widely discussed (Bailey 2000, Bonner & Bean 2000, Meierhenrich & Thiemann 2004, Nuevo *et al.* 2007). It was shown that the photolytic processes, which involve circularly polarized light, may provide a viable mechanism for chiral selection of organic molecules. It is well known that primitive meteorites, interplanetary dust, and comets contain complex organics. This allows us to suppose that primitive solar–system bodies, including comets, may be reservoirs of homochiral organics. In our investigation we focus on a search for chiral organics, which can be detected remotely, by studying CP of the light scattered by cometary dust.

The results of available polarimetric measurements of CP are summarized in Figure 1. As one can see, the observations indicate noticeable circular pllarization in comets and clearly show its systematic trend with the phase angle. Furthermore, CP in the observed comets is predominantly left–handed.

We simulated results of the observations considering light–scattering by particles whose material contained some excess of chiral organics, and, as a result, possessed some optical

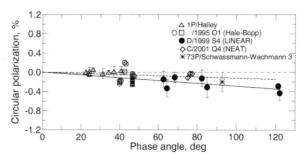

Figure 1. Composite phase–angle dependence of circular polarization for comets 1P/Halley (Dollfus & Suchail 1987), C/1995 O1 (Hale–Bopp) (Rosenbush, Shakhovskoy & Rosenbush 1997, Manset & Bastien 1997), D/1999 S4 (LINEAR) (Rosenbush *et al.* 2007b), C/2001 Q4 (NEAT) (Rosenbush *et al.* 2007a), and 73P/Schwassmann–Wachmann 3 (Tozzi *et al.* 2006). The solid line is the linear fit to the observed data, dashed line represents results of our calculations for optically active (chiral) spherical particles.

activity (for details see Rosenbush *et al.* 2007b). The simulations were done using the theoretical solution for optically active spheres (Bohren & Huffman 1983). To make the calculations more realistic, we considered particles with the power–law size distribution measured *in situ* in comet Halley. We used the optical constants typical for the amino–acids discovered in the Murchison meteorite (Cronin & Pizzarello 1986). Under these conditions we could obtain the observed phase–angle trend of CP, although it does not reach values larger than 0.15% at the phase angle equal to 120°. More realistic model of cometary dust as aggregates of submicron particles can significantly increase the value of CP as multiple–scattering effects get involved. This, together with the spectroscopically detected enriched organic composition of comets, allows us to reasonably speculate that CP in comets indicates the presence of prebiotic homochiral organics.

Acknowledgements

V. Rosenbush thanks the IAU, SOC, and LOC for support to participate in IAU Symposium 251 "Organic Matter in Space" in Hong Kong, China.

References

Bailey, J. 2000, in: G. Lemarchand & K. Meech (eds.), *ASP–CS*, 213, p. 349
Bohren, C. & Huffman, D. 1983, *Absorption and Scattering of Light by Small Particles*, (New York: John Wiley & Sons)
Bonner, W. & Bean, B. 2000, *Origins of Life and Evolution of the Biosphere*, 30, 513
Cronin, J. R. & Pizzarello, S. 1986, *Geochimica et Cosmochimica Acta*, 50, 2419
Cronin, J. R. & Pizzarello, S. 1997, *Science*, 275, 951
Dollfus, A. & Suchail, J. L. 1987, *A&A*, 187, 669
Manset, N. & Bastien, P. 1997, *Icarus*, 145, 203
Meierhenrich, U. & Thiemann, W. 2004, *Origins of Life and Evolution of the Biosphere*, 34, 111
Nuevo, M., Meierhenrich, U. J., D'Hendecourt, L., *et al.* 2007, *Adv. Sp. Res.*, 39, 400
Pizzarello, S. 2004, *Origins of Life and Evolution of the Biosphere*, 34, 25
Rosenbush, V., Kiselev, N., Shakhovskoy, N., *et al.* 2007a, in: G. Videen, M. Mishchenko, M. P. Mengüc & N. Zakharova (eds.), *Tenth Conference on Electromagnetic & Light Scattering*, (New York: Sigma Space Partners), p. 181
Rosenbush, V., Kolokolova, L., Lazarian, A., *et al.* 2007b, *Icarus*, 186, 317
Rosenbush, V. K., Shakhovskoy, N. M. & Rosenbush, A. E. 1997, *Earth, Moon, Planets*, 78, 381
Tozzi, G. P., Bagnulo, S., Boehnhardt, H., *et al.* 2006, *1st European Planetary. Science Congress (EPSC), September 18-22, 2006, Berlin, Germany*

Organic Matter in Space
Proceedings IAU Symposium No. 251, 2008
S. Kwok & S. Sandford, eds.

© 2008 International Astronomical Union
doi:10.1017/S1743921308021807

Solid organics in cometary comae

Gian Paolo Tozzi[1] and Ludmilla Kolokolova[2]

[1]INAF .. Osservatorio Astrofisico di Arcetri, Firenze, Italy
email: tozzi@arcetri.astro.it

[2]University of Maryland, College Park, MD 20742, USA
email: ludmilla@astro.umd.edu

Abstract. It is known since long ago that in comets a large quantity of organic matter exists in form of grains or is embedded in silicate grains. This was detected in situ by cometary space missions as well as inferred as a distributed source of some molecules observed in comets. Since organic matter is rather volatile, finding slow sublimating grains in comets can be good evidence of organics as a constituent of such grains. Here we describe a method to detect sublimating grains in comets. It consists of specific observations, specific data analysis, and some light-scattering modeling. We detect sublimating grains by measuring the quantity of grains as a function of the nucleocentric distance. Once detected, it is possible to get their photometric characteristics and compare them with the results of light-scattering modeling. The method has been applied to several comets. Sublimating grains were reliably identified for two of them.

Keywords. Astrobiology, comets: general, methods: data analysis

1. Introduction

Comets may have played an important role in depositing the organic matter that, between 4.6 to 3.6 billion years ago, allowed the formation of life on the primordial Earth. Radio and near-IR observations have detected many complex organic molecules in the gaseous component of comet atmospheres (for a review, see Bockelée-Morvan *et al.* 2005). However, a large quantity of organic matter may be preserved in cometary solids in the form of organic grains or organics embedded in silicate grains. The Giotto mission to Comet 1P/Halley provided the first evidence of the presence of organic grains in the coma. It was evaluated that they accounted for almost 50% of the mass of the solid component present in the coma of comet Halley (see, e.g., Fomenkova 1999). Remote detection of solid organics is very difficult since their spectroscopic signatures are hidden in more prominent spectroscopic features of other components of cometary dust and since they usually rapidly sublimate under solar radiation. However, the presence of organic matter in solid form has been inferred in several comets indirectly as a distributed source of gas in comae. It has been observed that in some comets the spatial profiles of some molecules cannot be fitted assuming that they are produced directly from the nucleus. Such a distributed source, which cannot be just a more complex parent molecule, is likely constituted by organic grains. For example, Cottin *et al.* (2004) suggested that the distributed source of Formaldehyde observed in comet C/1995 O1 (Hale-Bopp) could be a polymer: Polyoxymethylene. Here we describe another method that not only provides a non-spectroscopic remote search for organic solids in coma, but also allows one to figure out their properties.

313

2. Method & Results

Our method assumes that organics in the grains contain some semivolatile matter, i.e., the grains sublimate under the solar radiation with lifetimes variable from a few hours to tens of hours. By measuring the total cross section of the grains, it is possible then to evaluate, from its change with time, the possible presence of sublimating grains. In the case of a cometary outburst, this can be made by measuring the cross section of the grains in the coma as a function of time. Usually this is difficult to perform because the time elapsed between the onset of the outburst and the first observation is too long and most of the organic grains have already sublimated. However, the method is perfectly applicable to *man-made* outbursts, such as that produced by the *Deep Impact* mission. In case of regular comets, the presence of sublimating grains can be checked by measuring their cross section as a function of the projected nucleocentric distance (ρ), since the grains move away from the nucleus with a constant velocity. Although we integrate along the line of sight, i.e., include different nucleocentric distances, the grains at the corresponding projected distance provide the most significant contribution to the integral because their density is changing as ρ^{-2}. Our method uses the so called ΣAf function, that is Albedo multiplied by the total area covered by the grains in an annulus of radius ρ and unitary depth. For a regular, "quiet" comet, i.e., with constant outflow velocity of the grains and no changes in the production rate, in the absence of fragmentation and/or sublimation this function is constant with ρ. Grain sublimation or the onset of an outburst will be seen as an increase of the function at small nucleocentric distances. To disentangle between the two phenomena it is necessary to repeat the observations about 24 hours later. In the case of an outburst, which is a time dependent phenomenon, the nucleocentric profile of ΣAf will change. In the case of sublimation, the profile will be independent of time. Once the sublimating component is detected, it is possible to figure out its characteristics by measuring ΣAf at ρ=0 and then study the scale length of its change. If the measurements are made with narrowband filters in the visible or/and the near-IR (i.e., in the spectral region where the emission mechanism is the scattering of solar radiation), it is possible to find the scattering efficiency of this sublimating component as a function of the wavelength. Comparison of these efficiencies with the ones computed with some light-scattering model (e.g., Kolokolova *et al.* 2007), allows one to obtain the optical constants of the sublimating component, and hence information on its composition. Several comets have already been observed and analyzed with this method. Some did not show any presence of sublimating components, while others were too active to permit this kind of analysis (C/1995 O1 (Hale-Bopp), C/2001 Q4 (NEAT), 73P/S-W3). However, two of the observed comets, C/1999 WM1 and 9P/Tempel 1, allowed a positive detection of sublimating grains (Tozzi *et al.* 2004, 2007). For the latter comet, the target of the *Deep Impact* mission, a sublimating component was detected pre-impact as well as in the impact cloud.

References

Bockelée-Morvan, D., Crovisier, J., Mumma, M. J., & Weaver, H. A., 2005, *Comet II, Univ. Arizona press,* 391
Cottin, H., Bénilan, Y., Gazeau, M.-C., & Raulin, F., 2004, *Icarus,* 167, 397
Fomenkova, M. N. 1999, *Space Sci. Revs,* 90, 109
Kolokolova, L. Kimura, H., Kiselev, N., & Rosenbush, V., 2007, *A&A,* 463, 1189
Tozzi, G. P., Lara, L. M., Kolokolova, L., Boehnhardt, H., Licandro, J., & Schulz, R. 2004, *A&A,* 424, 325
Tozzi, G. P., Boehnhardt, H., Kolokolova, L., *et al.,* 2007 *A&A,*476, 979

Organic Matter in Space
Proceedings IAU Symposium No. 251, 2008
S. Kwok & S. Sandford, eds.

Organics in cometary and interplanetary dust

A. Chantal Levasseur-Regourd[1] and Jeremie Lasue[2]

[1] Aeronomie IPSL, Univ. Pierre et Marie Curie, Paris, France
email: aclr@aerov.jussieu.fr

[2] Aeronomie IPSL and LPG

Abstract. While gaseous carbon-rich species in cometary comae (coming from the nuclei icy component) are extensively studied by spectroscopic remote observations, so-called CHONs dust particles, i. e. organic compounds coming from the nuclei refractory component, have mostly been studied by dust mass spectrometers flying through the comae of comets 1P/Halley and 81P/Wild 2. However, remote observations of the light scattered by dust in cometary comae and in the interplanetary medium, coupled with both numerical and experimental simulations, have recently allowed us to confirm that such particles harbor a significant fraction of absorbing material, presumably consisting of organic compounds (Levasseur-Regourd *et al.* PSS 2007, Lasue *et al.* A&A 2007).

We estimate the fraction of absorbing material present in cometary dust for extensively observed comets (e.g., 1P/Halley, C/1995 O1 Hale-Bopp) and in the interplanetary dust (from zodiacal light observations). We also establish that, besides compact particles, fluffy aggregates are definitely present in these media. The properties (e.g., size distribution, morphology, composition) of the cometary and interplanetary dust particles, as inferred from light scattering data analysis, are compared with those of the IDPS collected in the upper Earth atmosphere and of the unique samples returned by the Stardust mission at Wild 2. The results are discussed in terms of the formation of comets in the protosolar nebula, and of the possible survival, at the epoch of late early bombardment, of cometary organics embedded in fluffy aggregates.

Discussion

ZINNER: What can you say about the variations between different comets in terms of your measurements?

LEVASSEUR-REGOURD: It's something I did not address here. We have mainly been making comparisons between Halley dust and Hale-Bopp dust. Using light scattering measurements we find that that Hale-Bopp dust has larger amounts of aggregates and tiny grains. Also the drop off in Comet Halley dust as a function of the distance from the nucleus was definitely different for main coma dust particles and dust particles that were observed in the jets, particles which are most likely coming from the subsurface of the nucleus.

From left to right: Joe Nuth, Anny-Chantal Levasseur-Regourd, Max Bernstein (photo by Dale Cruikshank).

Organic Matter in Space
Proceedings IAU Symposium No. 251, 2008
S. Kwok & S. Sandford, eds.

Organic molecules in saturnian E-ring particles. Probing subsurface oceans of Enceladus?

Frank Postberg[1], S. Kempf[2], R. Srama[3], E. Grün[3], J. K. Hillier[4], S. F. Green[4], and N. McBride[4]

[1]MPI für Kernphysik, Heidelberg, Germany
email: Frank.Postberg@mpi-hd.mpg.de

[2]MPI für Kernphysik, Heidelberg, Germany and Institut für Geophysik und extraterrestrische Physik Universität Braunschweig, Germany

[3]MPI für Kernphysik, Heidelberg, Germany

[4]Planetary and Space Science Research Institute, The Open University, Milton Keynes, U. K.

Abstract. The population of Saturn's outermost tenuous E-ring is dominated by tiny water ice particles, some of which contain organic or mineral impurities. Active cryo-volcanism on the moon Enceladus, embedded in the E-ring, has been known to be a major source of particles replenishing the ring since late 2005. Therefore, particles in the vicinity of Enceladus provide crucial information about the dynamic and chemical processes occurring far below the moon's icy surface.

We present a compositional analysis of thousands of impact ionisation mass spectra of Saturn's E-ring particles, with sizes predominantly below 1 μm, detected by the Cosmic Dust Analyser onboard the Cassini spacecraft. Our findings imply that organic compounds are a significant component of icy particles ejected by Enceladus plumes. Our in situ measurements are supported by detections of other Cassini instruments. They hint at a dynamic interaction of a hot rocky core with liquid water below the icy surface, where the organic molecules are generated. Further insights are expected from two close Enceladus flybys to be performed by Cassini in 2008. Then, for the first time, we will obtain spectra of freshly ejected particles at the traversals through the cryo-volcanic plumes.

Discussion

MUMMA: I gather that you did not detect any nitrogen in your samples. Is that correct?

POSTBERG: That is correct, yes.

MUMMA: That's surprising considering the strength of the cyanogen bond. You have 600 K temperatures in the interior of Saturn processing other materials. I can't imagine you wouldn't make triple bonded CN and then get some kind of nitriles.

POSTBERG: Keep in mind that we are analyzing this in solid phase and we are not very sensitive to nitrogen. We are only seeing the cations, and nitrogen doesn't like to form any cations. So we cannot exclude the possibility that nitrogen could be part of the particles, but we don't detect it.

ZIURYS: How unique is your interpretation of your time-of-flight mass spectral data that everything is entirely due to water and hydrocarbons? Could you have oxygen and nitrogen mixed in?

POSTBERG: I wouldn't say that there is only hydrogen and carbon. There could be other functional groups as well, but so far our analysis is just not precise enough. With the upcoming plume crossings, we hope to get spectra of freshly ejected plume particles of different impact speeds, and then we probably will be able to work out a more precise picture.

From left to right: Henry Chan, Cliff Matthews, Sandra Matthews, Bill Irvine, Hans Olofsson, Åke Hjalmarson, Masatoshi Ohishi, Daisy Mak, Gloria Cheung, Selina Chong.

Organic Matter in Space
Proceedings IAU Symposium No. 251, 2008
S. Kwok & S. Sandford, eds.

Laboratory experiments as support to the built up of Titan's theoretical models and interpretation of Cassini-Huygens data

Marie-Claire Gazeau[1], Yves Benilan[1], Et Touhami Es-sebbar[1], Thomas Ferradaz[1], Eric Hébrard[1], Antoine Jolly[1], François Raulin[1], Claire Romanzin[1], Jean-Claude Guillemin[2], Coralie Berteloite[3], André Canosa[3], Sébastien D. Le Picard[3], and Ian R. Sims[3]

[1]LISA (Laboratoire Interuniversitaire des Systèmes atmosphériques), groupe de Physico-chimie Organique Spatiale, UMR CNRS 7583, Universités Paris 12 et Paris 7, Créteil, 94010, France
email: gazeau@lisa.univ-paris12.fr

[2]Sciences chimiques de Rennes, UMR 6226 CNRS-Ecole Nationale Supérieure de Chimie de Rennes, Rennes, F-35700, France

[3]Équipe d'Astrochimie Expérimentale, Université de Rennes 1, France

Abstract. To interpret the concentrations of the products measured in Titan's atmosphere and to better understand the associated chemistry, many theoretical models have been developed so far. Unfortunately, large discrepancies are still found between theoretical and observational data. A critical examination of the chemical scheme included in these models points out some problems regarding the reliability of the description of critical reaction pathways as well as the accuracy of kinetic parameters. Laboratory experiments can be used to reduce these two sources of uncertainty. It can be:
i) experimental simulations: in our laboratory (LISA), representative Titan's simulation experiments are planned to be carried out in a reactor where the initial gas mixture will be exposed, for the first time, to both electrons and photons. Thus, the chemistry between N atoms and CH_3, CH_2, CH fragments, issued from electron dissociation of N_2 and photo-dissociation of CH_4 respectively, will be initiated. Thank to a time resolved technique, we will be able to analyse "in situ", qualitatively and quantitatively, the stable species as well as the short life intermediates. Then, the implied chemistry will be determined precisely, and consequently, its description will be refined in theoretical models. The current status of this program will be given.
ii) specific experiments: they are devoted, for example, to determine kinetic rate constants and low temperature VUV spectra that will be used to feed models and to interpret observational data. Such experiments performed in LISA and in Rennes' laboratory concern polyynes and cyanopolyynes as these compounds could link the gaseous and the solid phase in planetary atmosphere. Results concerning C_4H^+ hydrocarbons kinetic rate constants and VUV cross section of HC_3N and HC_5N will be detailed.

Discussion

ZIURYS: This is a very nice measurement of hydrocarbon radical chain reactions. The rate measurements are very difficult experiments and they also have interstellar applications. What do you think is more important for the formation of these poly-ions in Titan's atmosphere, radical-radical reactions or ion-molecule reactions?

GAZEAU: In fact, it depends on what you are interested in. There is chemistry that is important in the ionosphere, but this is not the same chemistry as that occurring lower in the atmosphere. I am interested in the stratospheric chemistry.

Ralph Lorenz (left) and J. Hunter Waite (right) (photo by Dale Cruikshank).

Organic Matter in Space
Proceedings IAU Symposium No. 251, 2008
S. Kwok & S. Sandford, eds.

The source of heavy organics and aerosols in Titan's atmosphere

J. H. Waite, Jr.[1], D. T. Young[1], A. J. Coates[2], F. J. Crary[1], B. A. Magee[1], K. E. Mandt[1], and J. H. Westlake[1,3]

[1]Southwest Research Institute, San Antonio, TX. 78228,
email: hwaite@swri.edu

[2]Mullard Space Science Laboratory, University College London,

[3]Unversity of Texas at San Antonio, San Antonio, TX. 78249

Abstract. Ion-neutral chemistry in Titan's upper atmosphere (~ 1000 km altitude) is an unexpectedly prodigious source of hydrocarbon-nitrile compounds. We report observations from the Cassini Ion Neutral Mass Spectrometer (INMS; Waite *et al.* 2004) and Cassini Plasma Spectrometer (CAPS; Young *et al.* 2004) that allow us to follow the formation of the organic material from the initial ionization and dissociation of nitrogen and methane driven by several free energy sources (extreme ultraviolet radiation and energetic ions and electrons) to the formation of negative ions with masses exceeding 10,000 amu.

Keywords. Atmospheric effects, planets and satellites: individual (Titan)

1. Introduction

Titan has been recognized as a source of organic aerosols since the 1970's as a result of ground based observations (Kuiper 1944), laboratory simulations (Sagan *et al.* 1992), and Voyager observations (Conrath 1985). However, it was not until imaging of Titan's atmosphere at infrared and centimeter wavelengths by Cassini-Huygens that the impact of the methane cycle on Titan became evident (e.g., Lopez *et al.* 2005). Titan's atmosphere consists of 98% molecular nitrogen, 2% methane, and traces of hydrocarbons such as ethane, acetylene, and propane. A surface temperature of 94 K and pressure of 1.5 bars puts methane near its triple point. This leads to a well-developed methane-based hydrological cycle evidenced by numerous river beds and an extensive high-latitude lake system that appears to operate on seasonal or longer time scales. Over periods on the order of tens of millions of years methane is thought to be irreversibly converted to complex hydrocarbon-nitrile compounds with loss of dihydrogen to space and a corresponding creation of organic aerosols in the atmosphere. It is this longer methane conversion cycle that we focus on in this paper.

Earlier atmospheric models (i.e., Wilson & Atreya 2004) suggested that the primary altitude of aromatic hydrocarbon formation (e.g., benzene is one representative of complex hydrocarbon formation) occurs in the well-mixed portion of the atmosphere (pressure ~ 20 nanobars near 750 km). The proposed process is hydrocarbon radical reactions with a much weaker source due to ion neutral reactions near the ionospheric peak (pressure 0.05 nanobars at 1100 km). Therefore, expectations prior to Cassini were that benzene would be barely measureable at the lowest altitudes observable by Cassini INMS (~ 950 km). However, instead we observe an unexpected abundance of hydrocarbon and nitrile species throughout the INMS mass range (1-100 Daltons) (see Figure 1a). Furthermore, positive ions are measured up to 350 Daltons (Figure 1b) by the CAPS Ion Beam Spectrometer. Even more surprising are observations by the CAPS Electron

Spectrometer of an appreciable concentration of negative ions extending to over 10,000 Daltons (see Figure 1c) (Waite *et al.* 2007, Coates *et al.* 2007).

2. INMS and CAPS Observations and Analysis

Figure 2 shows a neutral mass spectrum from the Cassini INMS averaged over 16 Titan encounters from 2004 to 2008. The data are restricted to encounters with appropriate sampling geometry and to spectra that can be appropriately calibrated and deconvolved (Kasprzak *et al.*, in preparation; Magee *et al.*, in preparation). The rich spectrum shows a progression of hydrocarbon and nitrile species starting from methane and nitrogen and proceeding through toluene. These data were averaged over the altitude range from 1000 to 1100 km, which is below the ionospheric peak. Deconvolved mixing ratio values are shown in Table 1 with error bars that are representative of the atmospheric variance over the 16 flybys. Although the precision of the measurements is high, there is a 20% systematic calibration uncertainty. Furthermore, the C_2 species C_2H_2, C_2H_4, C_2H_6, and the primary nitrile HCN have significant overlap at the mass resolution of INMS and are subject to additional uncertainty.

Analysis of the INMS ion mass spectrum (Figure 1a) allows identification of a host of protonated hydrocarbon and nitrile species (see Table 2) that demonstrate the importance of ion-neutral chemistry in the development of complex organic molecules (Waite *et al.* 2007, Vuitton *et al.* 2007). The dominant ion-neutral reaction is protonation in which a neutral molecule adds a proton, thus producing a positive ion with a mass one

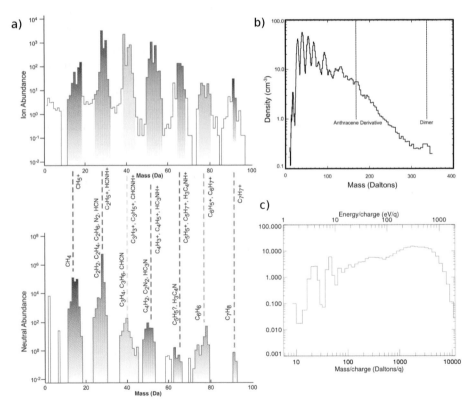

Figure 1. 1a (left panel) INMS ion and neutral spectra. Note the regular mass peak spacing of 12-14 Daltons. 1b (upper right panel) CAPS positive ion spectra, and 1c (lower right panel) CAPS negative ion spectra.

Table 1. Deconvolved average neutral species mixing ratios between 1000 and 1100 km.

Species	Mole fraction	Variance	Species	Mole fraction	Variance
N_2	0.942	6.39×10^{-4}	Ar	1.25×10^{-5}	9.71×10^{-7}
$^{14}N^{15}N$	1.15×10^{-2}	3.56×10^{-5}	CH_3CCH	3.90×10^{-6}	5.01×10^{-7}
CH_4	2.73×10^{-2}	6.24×10^{-5}	C_3H_6	8.94×10^{-7}	1.48×10^{-7}
$^{13}CH_4$	3.18×10^{-4}	6.58×10^{-6}	C_3H_8	3.54×10^{-7}	1.48×10^{-7}
H_2	1.72×10^{-2}	1.18×10^{-4}	C_4H_2	2.60×10^{-6}	3.89×10^{-7}
C_2H_2	4.72×10^{-4}	6.05×10^{-6}	C_2N_2	2.65×10^{-6}	6.14×10^{-7}
C_2H_4	2.46×10^{-4}	4.70×10^{-6}	C_6H_6	1.06×10^{-6}	2.05×10^{-7}
C_2H_6	9.75×10^{-6}	8.48×10^{-7}	C_2HCN	1.68×10^{-6}	4.79×10^{-7}
HCN	6.19×10^{-4}	1.14×10^{-6}			

Table 2. Neutral correspondence in the INMS data set

C or N#	Ions	Neutrals
C1	CH_5^+, NH_4^+	CH_4, NH_3
C2	$HCNH^+$, $C_2H_5^+$	C_2H_2, C_2H_4, C_2H_6, HCN
C3	$C_3H_5^+$, CH_3CNH^+	C_3H_4, CH_3CN
C4	$C_4H_5^+$, HC_3NH^+	C_4H_2, HC_3N, C_2H_3CN
C5	$C_5H_5^+$, $CH_2CCHCNH^+$, $C_5H_7^+$	CH_2CCHCN
C6	$C_6H_7^+$, HC_5NH^+	C_6H_6, C_6H_2, HC_5N
C7	$C_7H_7^+$	$CH_3C_6H_5$, CH_3C_5N
C8	$C_8H_3^+$	C_8H_2

amu greater than the parent neutral. The ion-neutral chemistry thus proceeds as in any reducing environment: from ions whose parent neutrals have a lesser proton affinity (PA) to those whose parent neutrals have a greater PA (Fox & Yelle 1997).

The CAPS positive ion spectrum is obtained with a high resolution (1.7%) electrostatic energy analyzer (Young *et al.* 2004). In the cold (150K) environment of Titan's upper atmosphere the 6 km s^{-1} orbital velocity of the Cassini spacecraft causes the ions to arrive at the sensor in a supersonic beam in which the ion energy spectrum is dominated by the spacecraft ram velocity. This allows us to relate the observed energy/charge peaks in the spectrum to the mass/charge of the observed ion ($m \sim 2E/V^2_{\mathrm{spacecraft}}$) However, the spacecraft potential, which ranges from -0.5 to -2.0 V, introduces an uncertainty that must be measured and taken into account. Failure to incorporate the spacecraft potential

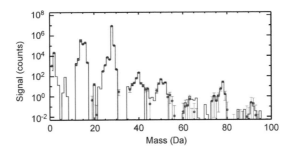

Figure 2. INMS average neutral mass spectrum between 1000 and 1100 km. The dots are the total simulated/fit spectrum including all of the species mentioned in Table 1. The histogram is the actual signal combined over Titan flybys T16-T40 of all viable data between 1000 and 1100 km.

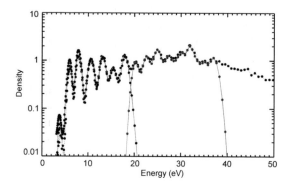

Figure 3. CAPS positive ion spectrum (black) with INMS model fit (blue) and model deconvolution in the extended mass range (red).

Table 3. neutral correspondence in the INMS data set

M/Z Group	Candidates
103-105	$C_8H_7^+$, $C_7H_6N^+$, $C_8H_9^+$
117-119	$C_9H_9^+$, $C_8H_8N^+$, $C_9H_{11}^+$
127-129	$C_{10}H_7^+$, $C_{10}H_8^+$, $C_{10}H_9^+$
141-144	$C_{11}H_9^+$, $C_{11}H_{11}^+$, $C_{10}H_{10}N^+$
153-155	$C_{12}H_9^+$, $C_{11}H_8N^+$, $C_{12}H_{11}^+$
166-169	$C_{13}H_{10}^+$, $C_{13}H_{11}^+$, $C_{12}H_{10}N^+$, $C_{13}H_{13}^+$
178-182	$C_{14}H_{10}^+$, $C_{14}H_{11}^+$, $C_{13}H_{10}N^+$, $C_{14}H_{13}^+$, $C_{13}H_{12}N^+$

introduces uncertainties of $2 \sim 10$ in the ion mass number. This uncertainty is removed by using the INMS mass spectrum in which masses between 1 and 100 Daltons (Da) have been determined independently. The CAPS positive ion spectrum can then be fit using the spacecraft potential and the ion temperatures as free parameters. One such fit is shown in Figure 3 where the INMS fit is shown in blue and the CAPS positive ion data in black. A more complete description of the analysis techniques can be found in (Crary *et al.*, in preparation). Ion spectra measured with CAPS below 1100 km in Titan's ionosphere extend well above 100 Daltons. The spectra can be modeled with an extension of the INMS technique to determine the observed masses (see Figure 3 - the red line fit to the data in black above 100 Daltons). Analysis of several such spectra has been used to identify a series of consistent mass peaks that can be used to identify complex ions which may be present at high mass numbers (see Table 3). The observed complex positive ions are apparently dominated by polyaromatics, heterogeneous aromatics, and nitrile-substituted aromatics throughout the mass range from 100 to 200 Daltons. Candidate heavier aromatics are listed in Table 3. Note that at altitudes around and below 1050 km the heavy positive ions begin to dominate the ion mass spectrum (see Figure 3).

The CAPS negative ion spectrum (see Figure 1c) was obtained with the CAPS electron spectrometer with an energy resolution of 16.7% in an analogous manner to the CAPS positive ion spectrum (Coates *et al.* 2007). The lack of distinctive peaks at higher energies is a result of the complexity of the spectrum and the lower energy resolution of the electron spectrometer relative to the positive ion analyzer. Candidate negative ion species identified with the various spectral peaks are shown in Table 4. The existence and survival of the negative ions in the solar illuminated ionosphere of Titan appears to be caused by the large electron affinities of nitrile substituted polyaromatics. Nitrile end members may also be responsible for polymeric cross links that aid in the formation of the complex negative ion macromolecules. These in turn may have masses up to 50,000 amu due to their suspected multiple charges picked up in Titan's ionospheric environment. Depending on their density, these massive particles may reach a few nanometers in diameter; the size of small aerosols (Waite *et al.* 2007). The integrated density of these negative ions is >200 cm^{-3} or roughly 20% of the free electron density at these altitudes. This surprising observation suggests that reaction of the abundant heavy positive ions with the large negative ions may be an appreciable source of organic macromolecule growth, given the anticipated rapid infrared radiative stabilization of the macromolecule complex formed (due to the large number of vibrational modes of freedom available).

3. Conclusion

We conclude that Titan's upper atmosphere (> 950 km altitude) is the source of a large population of organic molecules and aerosols of greatly varying complexity. We propose that this material starts very simply with dissociation and ionization of molecular nitrogen and methane by solar extreme ultraviolet photons and energetic ions and electrons from Saturn's magnetosphere. Macromolecules then grow to masses exceeding 10,000 Da. via ion-neutral chemistry. We propose that these macromoleules ($\sim$ few nanometers diameter) serve as condensation nuclei for more abundant, less complex, but supersaturated counterparts (e.g., hydrogen cyanide, diacetylene, benzene, etc.) as they precipitate through the stratosphere. The decrease in mixing ratios of the same species of nitriles and hydrocarbons in the stratosphere observed by the Cassini Composite Infrared Spectrometer (Vinatier *et al.* 2007) suggests that this upper atmospheric source may be the primary origin for the abundant atmospheric aerosols in Titan's atmosphere (Tomasko et al 2007, Liang *et al.* 2007).

The ultimate fate of aerosols is precipitation onto the surface where they appear to form extensive dunes (Lorenz *et al.* 2008) and perhaps other surface features made of organic material. The process of aerosol formation thus results in the irreversible conversion of methane into higher order organics on the time scale of a few million years according to the carbon isotopic evidence observed by Cassini INMS (Waite *et al.*, in preparation). The H$_2$ formed in this process is lost to space (Cui *et al.*, in preparation).

References

Coates A. J., Crary, F. J., Lewis, G. R., Young, D. T., Waite, J. H. Jr., & Sittler, E. C. Jr. 2007, *Geophys. Res. Lett.*, 34, L22103, doi:10.1029/2007GL030978.
Conrath, B. J. 1985, ESA-SP-241, ESTEC.
Crary, *et al.*, in preparation
Cui, *et al.*, in preparation
Fox J. & R. Yelle 1997, *Geophys. Res. Lett.*, 24(17), 2179
Kasprzak, *et al.*, in preparation
Kuiper, G. P. 1944, *ApJ*, 100, 378
Liang, M. C., Heays, A. N., Lewis, B. R., Gibson, S. T., & Yung, Y. L. 2007, *ApJ* (Letters), 664, L115
Lopez, R. M. C., Elachi, C., Paganelli, F., Mitchell, K., Stofan, E., Wood, C., Kirk, R., Lorenz, R., Lunine, J., Wall, S. and Cassini RADAR Team, 2005, Flows on the surface of Titan as revealed by the Cassini RADAR, *Bull. Am. Astron. Soc.*, 37, 739
Lorenz, R. D., *et al.* 2008, *Geophys. Res. Lett.*, 35, 2206
Magee, *et al.*, in preparation
Sagan, C., Thompson, W. R., & Khare, B. N. 1992, *Accounts Chem. Res.*, 25, 286
Tomasko, M. G., Doose, L., Engel, S., Dafoe, L. E., West, R., Lemmon, M., Karkoschka, E., & See, C. 2007, *Planetary and Space Science*, doi:10.1016/j.pss.2007.10.012

Vinatier, S., Bézard, B., Fouchet, T., Teanby, N. A., de Kok, R., Irwin, P. G. J., Conrath, B. J., Nixon, C. A., Romani, P. N., Flasar, F. M., & Coustenis, A. 2007, *Icarus*, 188, 120

Vuitton , V., Yelle, R. V., & McEwan, M. J. 2007, *Icarus*, 191, 722

Waite, J. H., Jr., Lewis, W. S., Kasprzak, W. T., *et al.* 2004, *Space Sci. Rev.*, 114, 113

Waite, J. H. Jr., Young, D. T., Cravens, T. E., Coates, A. J., Crary, F. J., Magee, B., & Westlake, J. 2007, *Science*, 316, 870

Waite, J. H. Jr., *et al.*, in preparation (isotope paper)

Wilson, E. H. & Atreya, S. K. 2004, *J. Geophys. Res.*, 109, E06002

Young, D. T., Berthelier, J. J., Blanc, M. *et al.* 2004, *Space Sci. Rev.*, 114, 1

Discussion

KWOK: From what you have described, these tholins are formed almost completely in the atmosphere of Titan. We heard yesterday that tholins are probably present in comets, asteroids, and other primitive bodies in the Solar System. On Tuesday we also heard that macromolecules and kerogen-like materials are being made by stars. Is there any possibility that the tholins that you see in Titan are inherited from the early Solar System rather than being made in situ in the atmosphere?

WAITE: We see the chemistry going on in the atmosphere and we can pretty well put together the pieces. It's consistent with what is going on in the ionosphere and the upper atmosphere, so we see a pathway that can lead to these aerosols in the atmosphere without invoking anything else. Now, what happens on the surface that turns it into grains of organics, we don't know. That's a process that we don't fully understand. Right now I would go with Occam's Razor - If you have everything you need right there at Titan, there's no reason to invoke something from the outside to produce this material. I'm sure the same types of processes go on in other bodies.

MATTHEWS: I would just repeat my point that tholins are polymers of acetylene and polymers of HCN together, and HCN polymers must be a large part of the products you are seeing. I wrote a paper 20 years ago suggesting that the orange haze of Titan could be due to HCN polymers because they are often orange colored too. I take these as evidence that HCN polymers are there.

WAITE: I'd like to comment on that. First of all, the nitrogen compounds we see are small compared with the hydrocarbons. In most of the things we've identified so far, we seen nitriles as terminal groups. Now, complexity is something else. The way I view HCN in this whole process of building bigger molecules is that it may be the 'glue' that holds it all together. In that sense, yes, there are HCN polymers, but it is much more complex. It is a very diverse set of hydrocarbon species and nitriles that are coming together in a very complex way. Some of them are heterocyclics, as well as aromatics with nitrogen in them.

MUMMA: I noticed that you indicated an isotopic ratio of ^{14}N/^{15}N that happens to be the same number one obtains from cometary CN in numerous comets.

WAITE: That's an important point. We've used this ratio to argue that atmosphere has escaped, that we've lost four or five times the current atmosphere. I'd want a different reference standard because losing 80% of your atmosphere is not a good way to start.

Organic Matter in Space
Proceedings IAU Symposium No. 251, 2008
S. Kwok & S. Sandford, eds.

The composition of Europa's near-surface atmosphere

Mau C. Wong[1], Tim Cassidy[2], and Robert E. Johnson[2]

[1]Jet Propulsion Laboratory, California Institute of Technology, Pasadena CA, USA
email: mau.c.wong@jpl.nasa.gov

[2]University of Virginia, Charlottesville VA, USA

Abstract. The presence of an undersurface ocean renders Europa as one of the few planetary bodies in our Solar System that has been conjectured to have possibly harbored life. Some of the organic and inorganic species present in the ocean underneath are expected to transport upwards through the relatively thin ice crust and manifest themselves as impurities of the water ice surface. For this reason, together with its unique dynamic atmosphere and geological features, Europa has attracted strong scientific interests in past decades.

Europa is imbedded inside the Jovian magnetosphere, and, therefore, is constantly subjected to the immerse surrounding radiations, similar to the other three Galilean satellites. The magnetosphere-atmosphere-surface interactions form a complex system that provides a multitude of interesting geophysical phenomenon that is unique in the Solar System. The atmosphere of Europa is thought to have created by, mostly, charged particles sputtering of surface materials. Consequently, the study of Europa's atmosphere can be used as a tool to infer the surface composition. In this paper, we will discuss our recent model studies of Europa's near-surface atmosphere. In particular, the abundances and distributions of the dominant O_2 and H_2O species, and of other organic and inorganic minor species will be addressed.

Discussion

MUMMA: I did my thesis work on electron scattering from simple molecular gases, including O_2, CO_2, and so forth, in the vacuum ultraviolet. The line that Melissa McGrath reported at 1356 Å is the quintet S triple S transition in atomic oxygen. It doesn't have to be produced from O_2, because H_2O can produce atomic oxygen quite easily through UV photolysis.

WONG: I think the way they conclude it is O_2 is from the ratio between the 1356 Å and 1304 Å lines. They look at the ratio of these two lines and conclude it's most likely from the O_2 dissociation.

Participants applauding the singing of the prisoner of war song by Cliff Matthews during the banquet.

Organic Matter in Space
Proceedings IAU Symposium No. 251, 2008
S. Kwok & S. Sandford, eds.

Titan's surface inventory of organic materials estimated from Cassini RADAR observations

Ralph D. Lorenz

Space Department, Johns Hopkins University Applied Physics Laboratory, USA
email: `Ralph.lorenz@jhuapl.edu`

Abstract. Cassini RADAR observations now permit an initial assessment of the inventory of two classes, presumed to be organic, of Titan surface materials: polar lake liquids and equatorial dune sands. Several hundred lakes or seas have been observed, of which dozens are each estimated to contain more hydrocarbon liquid than the entire known oil and gas reserves on Earth. Dark dunes cover some 20% of Titan's surface, and comprise a volume of material several hundred times larger than Earth's coal reserves. Overall, however, the identified surface inventories ($> 3 \times 10^4$ km^3 of liquid, and $> 2 \times 10^5$ km^3 of dune sands) are small compared with estimated photochemical production on Titan over the age of the Solar System. The sand volume is too large to be accounted for simply by erosion in observed river channels or ejecta from observed impact craters. The lakes are adequate in extent to buffer atmospheric methane against photolysis in the short term, but do not contain enough methane to sustain the atmosphere over geologic time. Thus, unless frequent resupply from the interior buffers this greenhouse gas at exactly the right rate, dramatic climate change on Titan is likely in its past, present and future.

Discussion

NITTLER: Does the similarity in the dune morphologies between Titan and the Earth imply similar grain size?

LORENZ: It does in the sense that there is an optimum size for the transport of granular materials in any given environment. If particles are very large, they have a low area-to-mass ratio, so it is difficult for wind to pick them up against gravity. If they're very small, then they stick together effectively. If you play with sugar with different particle sizes you can see that the very fine grain stuff actually has some cohesion. As it turns out, the optimum particle size is more or less the same for Titan and the Earth. The wind speeds required to transport sandy material on Titan are quite low, maybe about a meter per second, which is comparable with what is expected for the tidal currents in the atmosphere created by Titan's eccentric orbit around Saturn. There are some outstanding issues with the dunes, and what they indicate about the prevailing wind direction, that challenges global circulation models right now. But there's a whole new field of aeolian geomorphology to be explored with Titan now and what it means for the meteorology.

SLOAN: Would you please define the word "smust" that appeared on one of your slides?

LORENZ: It's not a word I like; it wasn't even a paper I liked. Don (Donald) Hunten came up with this idea that the solid aerosol particles somehow absorb liquid ethane onto them and stop it from acting like a liquid. I can't remember the etymology of the term - I think it was a way to refer to something like "moist dust". You have to look it up - it was in Nature last year I believe.

ZIURYS: You described some nice experiments where tholins and water lead to amino acids and pyrimidines and pyrroles, but how did you get rid of the nitrogen and produce sugars? Has anyone done experiments on that?

LORENZ: To be honest, this whole story has not been explored in anywhere near the depth in the laboratory that it needs to be. I know Catherine Neish in Arizona is dabbling with some hydrolysis rate coefficients. But to really explore the complexity here - and what stuff is removed and what stuff precipitates out - somebody needs to do that. I can't imagine these are particularly difficult experiments to at least have a stab at, but they may be difficult to do right.

ZIURYS: There may be some kind of reactions where you get N_2 and that gets lost in the gas phase so you can make things without nitrogen ...

LORENZ: Right. There is the whole question of how these hydrolysis processes depend on the pH of the solution - whether you are hydrolyzing them with water, or perhaps, as is more likely in the Titan environment, a strong ammonia solution. Ammonia acts as an antifreeze in the water and brings the melting point down. There's a lot of stuff to explore.

Scott Sandford performing the cutting of the pig ceremony, a Cantonese custom of ensuring the success at the opening of a major event.

Organic Matter in Space
Proceedings IAU Symposium No. 251, 2008
S. Kwok & S. Sandford, eds.

Possibilities of lightning-induced processes in gas-dusty atmosphere of water-containing bodies of Solar System

Yuriy Serozhkin

Institute of Semiconductor Physics
41, Prospekt Nauky, 03028, Kyiv, Ukraine
email: yuriy.serozhkin@zeos.net

Abstract. Lightning is considered as one of energy sources for synthesis of biochemical compounds. Numerous theoretical and experimental researches of gas-grain chemistry show that chemical reactions on the gas - ice boundary play a considerable role in such synthesis. In this connection the greatest interest represents studying lightning in gas-dusty atmospheres of water-containing bodies (comets, ice satellites of Jupiter and Saturn).

Keywords. Astrobiology, plasmas, comets: general, planets and satellites: individual (ice satellites)

1. Lightning in water-containing bodies of the Solar system?

Modern theories of charge accumulation and separation in thunderstorm clouds are based on the study of the microphysics of ice (Rakov & Uman 2003). The size and the sign on a charge formed during the collision of ice crystals and snowflakes does not depend on electric fields and is defined by a profile of temperatures and physical characteristics of the colliding grains. At the altitude 6 – 8 km and temperatures of about $-15°$, there exists a negatively charged layer (so-called charge reverse by thickness some hundreds meters). The separation of charges occurs at collisions between ice crystals falling downwards and snowflakes rising upwards. The ice crystals are charged positively below the charge reverse layer, and negatively above. Above this layer the snow pellets are charged positively, and below negatively. Thus, the negatively charged layer consists of negatively charged ice crystals and snow pellets. Positively charged snow pellets form a charge at the top of the cloud, and positively charged ice crystals form positive charge in the bottom of the cloud.

The majority of terrestrial lightning discharges are connected to participation of water (see Table 1). Another, significantly smaller group, includes discharges caused by the accumulation of charges on dust (discharges in dust storms) or ashes (volcano lightning).

So, there are two reasons for considering lightening in waterless atmospheres. It follows from Table 1 that the huge volumes and amounts of material in sprites gives a reason to speak about their large possibilities for synthesis of organic compounds.

2. Lightning-induced processes in gas-dusty atmospheres

It is possible that conditions exist for electrostatic charging of grains in atmospheres of two bodies: jets from the surfaces of comets, and geysers on Enceladus. The atmosphere of comets are complicated dynamic objects in which there is a directional motion of particles, convection, melting of ice, and fracture of large particles. The temperature profile of the Enceladus geysers is very similar to a profile of temperatures in thunderstorm clouds near their charge reverse layer. In these conditions charging of ice grains is

Table 1. Properties of the terrestrial lightnings.

Type of discharge	Number (in year)	Energy MJ	Volume km^3	Matter in volume, ton	Matter in year, ton
Cloud-cloud,cloud-ground lightning	$\approx 3 \cdot 10^9$	$\approx 5 \cdot 10^9$	≈ 0.001	$\approx 1 \cdot 10^3$	$\approx 3 \cdot 10^{12}$
Sprites* (lightning in mesosphere)	$\approx 3 \cdot 10^7$	≈ 10	$\approx 1 \cdot 10^4$	$\approx 2 \cdot 10^6$	$\approx 6 \cdot 10^{13}$
Lightning at tornado and volcano	$\approx 1 \cdot 10^3$?	≈ 10	≈ 0.0001	≈ 100	$\approx 1 \cdot 10^5$

Notes:
* Sprites are the most frequent and powerful lightning-induced events in the mesosphere and occur above thunderstorms after about one percent of lightning (Pasko *et al.* 1997). The duration is several ms. Sprites can extend about 50 km in horizontal and vertical directions. The brightest region takes place in the altitude of 65-75 km.

possible, and conditions for separation of charges which can lead to discharges in area of geysers. For estimates of parameters of plasma of hypothetical discharges we will assume that physical characteristics of the atmosphere over an active spot of comets and in the area of geysers are close to characteristics of terrestrial atmosphere at heights of 90 – 100 km. The electric discharges in rarefied gas-dusty atmospheres will serve not only as energy sources for synthesis of biochemical compounds. It is possible that the more important consequences for gas-grain organic chemistry will be the presence the properties of dusty plasma ordered structures (plasma crystals) (Tomas *et al.* 1994). Under certain conditions in dusty plasmas it is possible to form helical dust structures (Tsytovich *et al.* 2007). The plasma of sprites and hypothetical lightning's discharge in gas-dusty atmospheres on Mars and comets can also be a medium for the formation of ordered structures (Serozhkin 2005b).

The criterion of the first step to formation of ordered structure in dusty plasma is the Coulomb coupling parameter. It is determined as the relation of a potential energy of interaction of charged particles to a kinetic energy of particles in a center-of-mass system of dust particles. For example, the estimation of the requirement to energy of micron-size grains with concentration 10^3 cm^{-3} and charge 1000e (in case of an ingress of a dusty component into an area of sprites) for transition in a Coulomb liquid (Coulomb coupling parameter greater than 2) gives a value of kinetic energy less than 10^{-12} erg, which corresponds to a velocity dispersion of less than 0.5 cm s^{-1} (Serozhkin 2005a).

Can such ordered structures play the role of matrices for the formation of pre-biological compounds? Can conditions exist on the Earth or in the Solar System for the formation of plasma crystals? The research of the synthesis of organic compounds in dusty plasmas of discharges would submit additional capabilities for understanding of processes of formation and self-organizing of pre-biological compounds.

References

Pasko, V. P., Inan, U. S., Bell, T. F., & Taranenko, Y. N. 1997, *J. Geophys. Res.*, A102, 4529
Rakov, V. A. & Uman M.A. 2003, *Lightning: Physics and Effects* (Cambridge University Press)
Serozhkin, Y. 2005a, in L. Boufendi, M. Mikkian & P. K. Shukla (eds), *New Vistas in Dusty Plasmas, AIP Conference Proceedings, 799* (Melville, New York), p. 383
Serozhkin Y. 2005b, in R. Hoover, A. Rozanov & R. Paepe (eds), *Perspectives in Astrobiology, NATO Science Series: Life and Behavioral Sciences, 366* (IOS press, Amsterdam), p. 170
Thomas, H. M. *et al.* 1994, *Phys. Rev. Lett.*, 73, p. 652
Tsytovich, V. N. Morfill, G. E. Fortov, V. E., Gusein-Zade, N. G., Klumov, B. A., & Vladimirov, S. V. 2007, *New Journal of Phys.*, 9, p. 263

Organic Matter in Space
Proceedings IAU Symposium No. 251, 2008
S. Kwok & S. Sandford, eds.

© 2008 International Astronomical Union
doi:10.1017/S1743921308021881

Structural, chemical and isotopic examinations of interstellar organic matter extracted from meteorites and interstellar dust particles

Henner Busemann[1], Conel M. O'D. Alexander[2], Larry R. Nittler[2], Rhonda M. Stroud[3], Tom J. Zega[3], George D. Cody[4], Hikaru Yabuta[4], and A.L. David Kilcoyne[5]

[1] Planetary and Space Sciences Research Institute, The Open University, U. K.
email: h.busemann@open.ac.uk

[2] Department of Terrestrial Magnetism, Carnegie Institution of Washington

[3] Materials Science and Technology Division, Naval Research Laboratory Washington

[4] Geophysical Laboratory, Carnegie Institution of Washington

[5] Chemical Science Division, Berkeley National Laboratory

Abstract. Meteorites and Interplanetary Dust Particles (IDPs) are supposed to originate from asteroids and comets, sampling the most primitive bodies in the Solar System. They contain abundant carbonaceous material. Some of this, mostly insoluble organic matter (IOM), likely originated in the protosolar molecular cloud, based on spectral properties and H and N isotope characteristics. Together with cometary material returned with the Stardust mission, these samples provide a benchmark for models aiming to understand organic chemistry in the interstellar medium, as well as for mechanisms that secured the survival of these fragile molecules during Solar System formation. The carrier molecules of the isotope anomalies are largely unknown, although amorphous carbonaceous spheres, so-called nanoglobules, have been identified as carriers. We are using Secondary Ion Mass Spectrometry to identify isotopically anomalous material in meteoritic IOM and IDPs at a ~100-200 nm scale. Organics of most likely interstellar origin are then extracted with the Focused-Ion-Beam technique and prepared for synchrotron X-ray and Transmission Electron Microscopy. These experiments yield information on the character of the H- and N-bearing interstellar molecules: While the association of H and N isotope anomalies with nanoglobules could be confirmed, we have also identified amorphous, micron-sized monolithic grains. D-enrichments in meteoritic IOM appear not to be systematically associated with any specific functional groups, whereas ^{15}N-rich material can be related to imine and nitrile functionality. The large ^{15}N- enrichments observed here (δ^{15}N >1000 ‰) cannot be reconciled with models using interstellar ammonia ice reactions, and hence, provide new constraints for understanding the chemistry in cold interstellar clouds.

Discussion

IRVINE: What is a 'nano-globule'?

BUSEMANN: It's a hollow piece of amorphous carbon that has a roundish shape, which is why it's called a 'globule.' We see this shape in the TEM and these show isotopic anomalies.

BERNSTEIN: Could you be more specific about the correlation between the ^{15}N isotopic anomalies and the functional groups? In your abstract you mention that the isotopic anomalies are associated in imines and nitriles. However, in one of your slides you showed it associated with amorphous carbon and talked about the ^{15}N being associated with NH, amino, and then in another slide it was nitrile and amide.

BUSEMANN: The regions examined with the different techniques are not the same. The infrared information is from an enriched region shows only OH and NH. If you look at the XANES spectra you see that the imine and the nitrile are both definitely enhanced. I see the most pronounced enrichment in the nitrile, but the imine group is very close to it in frequency space, so it is likely that there is some contribution from that as well.

MUMMA: Could you clarify the changes in the Charnley and Rogers model that now produces predictions of enriched ^{15}N even larger than the values you observed?

BUSEMANN: I don't understand it fully, but I think they freeze CO out of the gas phase relative to the nitrogen and this results in higher ^{15}N enrichments in the gas phase chemistry.

From left to right: Henner Busemann, George Flynn, Ernst Zinner, Conel Alexander, and Larry Nittler (photo by Dale Cruikshank).

Organic Matter in Space
Proceedings IAU Symposium No. 251, 2008
S. Kwok & S. Sandford, eds.

Micro-Raman study of nanodiamonds from Allende meteorite

Arnold Gucsik[1], Ulrich Ott[1], Edit Marosits[1], Anna Karczemska[2], Marcin Kozanecki[3], and Marian Szurgot[4]

[1]Max Planck Institute for Chemistry, Department of Geochemistry, Mainz, Germany
email: gucsik@mpch-mainz.mpg.de

[2]Technical University of Lodz, Institute of Turbomachinery, Lodz, Poland

[3]Technical University of Lodz, Department of Molecular Physics, Lodz, Poland

[4]Technical University of Lodz, Center for Mathematics and Physics, Lodz, Poland

Abstract. We have studied the Raman spectroscopic signatures of nanodiamonds from the Allende meteorite in which some portions must be of presolar origin as indicated by the isotopic compositions of various trace elements. The spectra of the meteoritic nanodiamond show a narrow peak at 1326 cm^{-1} and a broad band at 1590 cm^{-1}. Compared to the intensities of these peaks, the background fluorescence is relatively high. A significant frequency shift from 1332 to 1326 cm^{-1}, peak broadening, and appearance of a new peak at 1590 cm^{-1} might be due to shock effects during formation of the diamond grains. Such changes may have several origins: an increase in bond length, a change in the electron density function or charge transfer, or a combination of these factors. However, Raman spectroscopy alone does not allow distinguishing between a shock origin of the nanodiamonds and formation by a CVD process as is favored by most workers.

Keywords. (Stars:) supernovae: general, shock waves, ISM: evolution, meteors, meteoroids

1. Introduction

Primitive meteorites contain abundant (up to 1500 ppm) amounts of nanodiamonds (Huss 1990). At least some subpopulation must be of pre-solar (stardust?) origin, as indicated by the isotopic composition of trace elements the diamonds carry, in particular noble gases (e.g., Huss & Lewis 1994, 1995) and tellurium (Richter *et al.* 1998). On the other hand, the isotopic composition of the major element, carbon, is unremarkable, i. e., within the range reasonably expected for Solar System materials (e.g., Russell *et al.* 1991). As a consequence many workers believe that the majority of the diamonds is of local, i. e., Solar System, origin and that the fraction that is pre-solar is relatively small (e.g., Zinner 1998). Two main theories exist for the formation process of the meteoritic nanodiamonds (e.g., Daulton *et al.* 1996, and references therein): (1) Chemical vapor deposition (CVD), and (2) shock origin. TEM investigations, in particular, seem to suggest that formation by a CVD process is most likely (Daulton *et al.* 1996).

In this study, we present results of the study of meteoritic nanodiamonds from the Allende meteorite by means of Raman spectroscopy in an attempt to obtain further constraints with regard to the formation process.

2. Results

The individual grain sizes of meteoritic nanodiamonds vary between 2 and 7 nm (e.g., Daulton *et al.* 1996), which is small compared to the $\sim$ 1 μm diameter of the laser excitation beam on the surface of the sample. The Raman spectra of the nanodiamonds exhibit two broad bands centered at $\sim$1326, and $\sim$1590 cm^{-1} (Figure 1). In general,

336 Arnold Gucsik *et al.*

peak intensities of these bands are relatively low, which indicates the strong background fluorescence. This may be due to lack of well-crystalline parts of the sample; alternatively, it may (also) be related to the small grain size. Following the data correction, the band at 1326 cm^{-1} shows a 14.4 cm^{-1} FWHM and a 6 cm^{-1} peak shift from the 1332 cm^{-1} peak position of standard or reference nanodiamond (Zhang & Zhang 2005, Karmenyan *et al.* 2007).

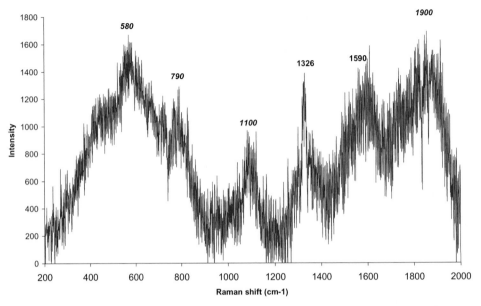

Figure 1. Raman spectral features of nanodiamonds from Allende meteorite (1326 and 1590 cm^{-1}), and glass holders at 580, 790, 1100, and 1900 cm^{-1}. Raman spectra were recorded using the confocal Raman micro-spectrometer T-64000 (Jobin-Yvon) equipped with a BX-40 (Olympus) microscope at Technical University of Lodz, Poland (Argon line λ=514.5 nm).

3. Discussion and Conclusions

The Raman spectra of single crystal diamond is dominated by a Brillouin zone-center point as *T2g* mode at 1332 cm^{-1} (strong or very strong peak for carbon sp^3 bonding) with approximately 5-10 cm^{-1} FWHM (Ferrari & Robertson 2004, Zhang & Zhang 2005, Karmenyan *et al.* 2007, Karczemska *et al.* 2008). This relatively sharp and single peak is frequently used as a signature of high crystalline quality. In previous studies, several additional peaks have been described in the Raman spectra of micro-and nanodiamond samples, as follows. The two most typical and significant ones in the spectra of artificially produced chemical vapor deposited (CVD) nanocrystalline diamonds are at 1150 and 1450 cm^{-1} (trans-polyacethylenes; Ferrari & Robertson 2001). It has been proposed that these peaks are related to phonon modes with q_0, which are activated by the disorder induced by small grain sizes in nanocrystalline or amorphous diamond (Filik *et al.* 2006). Additional medium or strong bands in the Raman spectra of nanocrystalline diamond samples are usually seen at 1350 cm^{-1} and at 1590 cm^{-1}. The 1350 cm^{-1} feature is related to the D-band, which is a normally Raman inactive A1g mode. It is activated due to the finite crystal size (Filik *et al.* 2006, and references therein). The G-band at 1590 cm^{-1} is assigned to carbon sp^2 bonding graphitic structures (Nasdala *et al.* 2004).

Frequency shifts of the 1332 cm^{-1} band by a few cm^{-1} (6 cm^{-1} in this case) may be due to strained nanodiamond caused by shock waves or high-pressure-induced deformation,

Table 1. Micro-Raman properties of nanodiamonds separated from CVD, meteoritic, and shock metamorphic samples.

Origin	Raman Bands (cm^{-1})							References
CVD	500	1150	1240-1280	1332	1350	1450	1590	Ferrari & Robertson (2001) Ferrari & Robertson (2004) Filik *et al.* (2006) Karmenyan *et al.* (2007) Karczemska *et al.* (2008)
Meteoritic				1332[1]			1580	Greshake et al (2000) Karczemska *et al.* (2008)
Shock metamorphic				1332[1]			1590	El Goresy *et al.* (2001) Kenkmann *et al.* (2002) Chen *et al.* (2004, 2006) Dunlop *et al.* (2007)

Note
[1] Peak shift to the lower frequency modes

but may also be due to disorder in the nanodiamond structure. Similar frequency shifts, broad bands (10-20 cm^{-1} at FWHM), and relatively high background fluorescence were observed in impact-induced diamond as well as in nanodiamond samples from different shock metamorphic environments such as terrestrial impact structures (El Goresy *et al.* 2001) and meteorites (Greshake *et al.* 2000, Mostefaoui *et al.* 2002) as well as in shock recovery experiments (Kenkmann *et al.* 2002). We note, however, that the frequency shift of the band at 1332 cm^{-1} and the peak broadening from higher modes to the lower ones may also be assigned to the effects of phonon / quantum confinement (Yoshikawa *et al.* 1995, Chen *et al.* 1999, Sun *et al.* 2000, Berg *et al.* 2008). Shifts were also observed after irradiation by neutrons (Guo *et al.* 2004). An additional peak at 1590 cm^{-1} was also described in the previous studies, which is probably related to the amorphous carbon phase present as the result of shock metamorphism (El Goresy *et al.* 2001, Kenkmann *et al.* 2002). These observations on shock-produced diamonds are in good agreement with the Raman spectral properties of nanodiamonds from our Allende meteorite sample. The frequency shift may be associated with the modification of the local configuration of the sample by means of the transformation of graphite into nanodiamond (Dunlop *et al.* 2007). In particular, the difference in frequency values for the Raman C-C bonding vibrations between 1332 and 1326 cm^{-1} in nanodiamond could indicate a change of the C-C bond strength caused by the phase transition at high pressure and temperature.

In conclusion, our results from Raman spectroscopy alone are not conclusive, especially since it is currently difficult to distinguish between the effects of shock transformation and small grain size. They leave open, however, the possibility that a significant fraction of the nanodiamonds in primitive meteorites were formed by shock transformation from graphite / amorphous carbon in the interstellar medium. As noted above, this possibility was immediately recognized after their discovery (Tielens *et al.* 1987), but more recent work has mostly concluded that a CVD-like process is more likely (e.g., Daulton *et al.* 1996, Le Guillou *et al.* 2006, Le Guillou & Rouzaud 2007). Other processes that are possible in principle (see also Anders & Zinner 1993), but have received less attention, are photolysis of hydrocarbons (Buerki & Leutwyler 1991), annealing by UV photons (Nuth & Allen 1992) and transformation by energetic particle irradiation (e.g., Ozima & Tatsumoto 1997). Note that Raman peak shifts were observed after irradiation by neutrons (Guo *et al.* 2004), but otherwise information is lacking concerning the effects of these processes on the Raman spectral properties of nanodiamonds.

References

Anders, A. & Zinner, E. 1993, *Meteoritics*, 28, 490

Berg, T., Marosits, E., Maul, J., Nagel, P., Ott, U., Schertz, F., Schuppler, S., Sudek, Ch., & Schönhense, G. 2008, *J. Appl. Phys.*, (submitted)

Buerki, P. R. & Leutwyler, S. 1991, *J. Appl. Phys.*, 69, 3739

Chen, J., Deng, S. Z., Chen, J., Yu, Z. X., & Xu, N. S. 1999, *Appl. Phys. Lett.*, 74, 3651

Chen, P., Huang, F., & Yun, S. 2004, *Mater. Res. Bull.*, 39, 1583

Chen, P., Huang, F., & Yun, S. 2006, *Diam. Rel. Mat.*, 15, 1400

Daulton, T. L., Eisenhour, D. D., Bernatowitz, T. J., Lewis, R. S., & Buseck, P. R. 1996, *Geochim. Cosmochim. Acta*, 60, 4853

Dunlop, A., Jaskierowicz, G., Ossi, P. M., & Della-Negra, S. 2007, *Phys. Rev. B*, 76, 155403

El Goresy A., Gillet, P., Chen, M., Künstler, F., Graup, G., & Stähle, V. 2001, *Am. Min.*, 86, 611

Filik, J., Harvey, N., Allan, N. L., May, P. W., Dahl, J. E. P., Liu, S., & Carlson, R. M. K. 2006, *Phys. Rev. B*, 035423

Ferrari, A. C. & Robertson, J. 2001, *Phys. Rev. B.*, 63, 121405

Ferrari, A. C. & Robertson, J. 2004, *Phil. Trans. R. Soc. Lond. A*, 362, 2477

Guo, Y., Zheng, Z., Feng, Y., & Li, Y. 2004, *J. Peking Univ. (Science Edition)*, 40, 212

Greshake, A., Kenkmann, T., & Scmitt, R. T. 2000, *Meteorit. Planet. Sci.*, 35, A65

Huss, G. R. 1990, *Nature*, 347, 159

Huss, G. R. & Lewis, R. S. 1994, *Meteoritics*, 29, 791

Huss, G. R. & Lewis, R. S. 1995, *Geochim. Cosmochim. Acta*, 59, 115

Karczemska, A. T., Szurgot, M., Kozanecki, M., Szynkowska, M. I., Ralchenko, V., Danilenko, V. V., Louda, P., & Mitura, S. 2008, *Diam. Real Mat.* (submitted)

Karmenyan, A., Perevedentseva, E., Chiou, A., & Cheng, C-L. 2007, *J. Phys:Conf. Series*, 61, 517

Kenkmann, T., Hornemann, U., & Stöffler, D. 2002, *LPSC XXXIII*, abs.#1052

Le Guillou, C. & Rouzaud, J. N. 2007, *LPSC XXXVIII*, abs.#1578

Le Guillou, C., Rouzaud, J. N., & Brunet, F. 2006, *LPSC XXXVII*, abs.#1635

Mostefaoui, S., El Goresy, A., Hoppe, P., Gillet, Ph., & Ott, U. 2002, *Earth Planet. Sci. Lett.*, 204, 89

Nasdala, L., Smith, D. C., Kaindl, R., & Zieman, M. A. 2004, In: A. Beran and E. Libowitzky (Eds.) Spectroscopic methods in mineralogy. EMU Notes In Mineralogy 6, Eötvös University Press, p. 281-343.

Nuth, J. A. III & Allen, J. E. Jr. 1992, *Astrophys. Space Sci.*, 196, 117

Ozima, M. & Tatsumoto, M. 1997, *Geochim. Cosmochim. Acta*, 61, 369

Richter, S., Ott, U., & Begemann, F. 1998, *Nature*, 391, 261

Russel, S. S., Arden, J. W., & Pillinger, C. T. 1991, *Science*, 254, 1188

Sun, Z., Shi, J. R., Tay, B. K., & Lau, S. P. 2000, *Diam. Relat. Mater*, 9, 1979

Tielens, A. G. G. M., Seab, C. G., Hollenbach, D. J., & McKee, C. F. 1987, *ApJ* (Letters), 319, L109

Yoshikawa, M., Mori, Y., Obata, H., Maegawa, M., Katagiri, G., Ishida, H., & Ishitani, A. 1995, *Appl. Phys. Lett.*, 67, 694

Zhang, D. & Zhang, R. Q. 2005, *J. Phys. Chem. B*, 109, 9006

Zinner, E. 1998, *Annu. Rev. Earth Planet. Sci.*, 26, 147

Discussion

FLYNN: We looked at the carbon XANES of diamonds from Allende a few years ago and also we were unable to find the diamond exciton feature which is characteristic of CVD diamond but is not characteristic of shock diamonds, so we also have some indications that there are not well crystallized diamonds.

GUCSIK: This is a good comment because we selected the Allende meteorite on the basis of the phase equilibria diagram by Carl Agee. He estimated that the maximum pressure for the internal pressure of the parent body of the Allende meteorite wasn't more than 30 GPA. This is kind of a low shock regime of the meteoritic background because you can see here that we need at least 40 GPA for the phase transformation between graphite and diamond. So I think there are 10 GPA of pressure that might be missing in this model.

NITTLER: For a contrasting view, there was a very detailed study by Tyrone Dalton some years ago who did a very detailed high resolution transmission electron microscopy study of the meteoritic nanodiamonds and compared the microstructures of diamonds produced by chemical vapor deposition and diamonds produced by shock using the same technique in the same instrument. He found that the microstructures looked much more like the CVD diamonds then the shock diamonds and argued the exact opposite of what you are arguing.

GUCSIK: I have a surprise for you. I was playing around with the shock wave calculator of the normal shock wave front coming from a supernova explosion. This shock wave calculator can be found on the net. And I managed to get 40 GPA pressure circumstances required for having the phase transformation between graphite and diamond, and the shock wave calculator found 2400 Kelvin as a temperature for the post shock effects. I think both processes, the shock wave front and post shock temperature effects, might both be leading processes for the formation of the meteoritic diamonds. On the basis of very preliminary data I can conclude this kind of behavior of nanodiamonds.

GUCSIK: It is important to note that, in general, although CVD diamonds show Raman shift of a peak at 1331 cm^{-1}, they do not exhibit a coexistent peak at around 1600 cm^{-1} and the peak shifting. Moreover, compared to CVD diamonds, our Allende meteoritic nanodiamonds do no not contain some major Raman bands at around 1150 and 1450 cm^{-1} (sp^2, transpolyacetylene) peaks.

HENNING: Actually I have a comment. There is a another mechanism that is able to produce nanodiamonds - ion radiation of small particles. This may actually be a mechanism that is operating in the environment of young stars because these nanodiamonds seem to be different from the meteoritic nanodiamonds. It would be worth while to study nanodiamonds that are produced by ion irradiation as well.

GUCSIK: Yes. But the coexistence of peak shifting of the peaks at 1331 cm^{-1} and 1600 cm^{-1} have only been known from the shock metamorphic nanodiamonds. This can aid (or even support) understanding more about the supernova-driven shock wave front and its effect on the graphite-diamond high pressure transition in the interstellar medium.

Catherine Cesarsky being asked to put red paint on the the eye of the lion with a brush in order to liven up the lion.

Organic Matter in Space
Proceedings IAU Symposium No. 251, 2008
S. Kwok & S. Sandford, eds.

Stardust in meteorites: A link between stars and the Solar System

Ernst Zinner

Laboratory for Space Sciences and the Physics Department, Washington University, St. Louis, MO USA
email: ekz@wustl.edu

Abstract. Ultimately, all of the solids in the Solar System, including ourselves, consist of elements that were made in stars by stellar nucelosynthesis. However, most of the material from many different stellar sources that went into the making of the Solar System was thoroughly mixed, obliterating any information about its origin. An exception are tiny grains of preserved stardust found in primitive meteorites, micrometeorites, and interplanetary dust particles. These μm- and sub-μm-sized presolar grains are recognized as stardust by their isotopic compositions, which are completely different from those of the Solar System. They condensed in outflows from late-type stars and in SN ejecta and were included in meteorites, from which they can be isolated and studied for their isotopic compositions in the laboratory. Thus these grains constitute a link between us and our stellar ancestors. They provide new information on stellar evolution, nucleosynthesis, mixing processes in asymptotic giant branch (AGB) stars and supernovae, and galactic chemical evolution. Red giants, AGB stars, Type II supernovae, and possibly novae have been identified as stellar sources of the grains. Stardust phases identified so far include silicates, oxides such as corundum, spinel, and hibonite, graphite, silicon carbide, silicon nitride, titanium carbide, and Fe-Ni metal.

Discussion

HENNING: Is there any hope to do the same work to iron sulphide grains? As you know, it's strongly debated whether iron sulphide is produced in the protoplanetary disk or in the molecular cloud.

ZINNER: Well, one question is of course how stable sulphide grains are. One type of non-solar grain has been found because it could be extracted from meteorites and then studied, and the other type is actually studied in situ. The only hope is to find some sulphide grains in situ and then measure their sulphur isotopic compositions. This is not too hard to do in principle, but not much has been done yet. It's still possible that they exist.

CRUIKSHANK: Scott Sandford tells me that sulphides are well represented in the Stardust samples. Is that still true?

ZINNER: There is also in a lot of iron sulphide in meteorites, but most of it is not of stellar origin. So it's the same as the silicate grains. The Solar System makes silicate grains and so you have to measure thousands of them in order to find the stellar ones. The same thing might be necessary for the sulphide grains, but not much has been done.

ZIURYS: What do you expect for the magnesium isotope ratio between AGB stars and supernova?

ZINNER: There are differences, but it's a little complicated. There are always large anomalies in ^{26}Mg, but this is from the decay of ^{26}Al, so you cannot say anything about the Mg itself. You have to examine the ^{25}Mg:^{24}Mg ratio. There are very large variations that we think are from AGB stars because oxygen isotopes in the same grains suggest they are from AGB stars. There are a few grains which we think are of supernova origin, and Mg is also anomalous in 25:24 in them. Making these measurements also depend on how much Mg is in the grains. Mg is an element that doesn't like to go into silicon carbide. However, Al goes there and then you find almost mono-isotopic ^{26}Mg from the decay of ^{26}Al. Mg is more volatile and doesn't like to go into silicon carbide or graphite, so it's much more difficult to do such measurements.

D.B. Vaidya asking a question.

Organic Matter in Space
Proceedings IAU Symposium No. 251, 2008
S. Kwok & S. Sandford, eds.

© 2008 International Astronomical Union
doi:10.1017/S1743921308021911

Presolar grains in the Solar System: Connections to stellar and interstellar organics

Larry R. Nittler

Department of Terrestrial Magnetism, Carnegie Institution of Washington
email: `lnittler@ciw.edu`

Abstract. A small fraction of primitive meteorites and interplanetary dust particles (IDPs) consists of grains of presolar stardust. These grains have extremely unusual isotopic compositions, relative to all other planetary materials, indicating that they condensed in the outflows and explosions of prior generations of stars (Clayton & Nittler 2004). Identified presolar grain types include silicate, oxide and carbonaceous phases. The latter include graphitic carbon, diamond and SiC. Although many of these phases do not have a direct connection to organic chemistry, this is not true of the graphitic spherules. Many of these, with isotopic compositions indicating an origin in C-rich asymptotic giant branch (AGB) star outflows, have a structure consisting of naonocrystalline cores surrounded by well-graphitized C (Bernatowicz *et al.* 1996). The cores include isotopically anomalous polycyclic aromatic hydrocarbons (Messenger *et al.* 1998) and represent a link between molecular chemistry and dust condensation in stellar outflows. Meteorites and IDPs also contain abundant isotopically anomalous organic matter, including distinct organic grains, some of which probably formed in stellar outflows and/or the interstellar medium (ISM) (Busemann *et al.* 2006, Floss *et al.* 2004). In some IDPs, deuterium- and ^{15}N-enriched organic matter is closely associated with presolar silicate grains (Messenger *et al.* 2005, Nguyen *et al.* 2007), suggesting an association in the ISM prior to Solar System formation.

References

Bernatowicz, T. J., *et al.* 1996, *ApJ*, 472, 760
Busemann, H., *et al.* 2006, *Science*, 312, 727
Clayton, D. D. & Nittler, L. R. 2004, *ARA&A*, 42, 39
Messenger, S., *et al.* 1998, *ApJ*, 502, 284
Messenger, S., *et al.* 2005, *Science*, 309, 737
Nguyen, A. N., *et al.* 2007, *LPSC*, 38, Abstract #2332

Discussion

ZIURYS: You make the point that you have to find an anomalous isotope ratio to say a material is presolar. However, if you look at ^{12}C/^{13}C ratios in AGB stars, you find ratios from 4 to 100 in the molecular material - UU Omega has a ^{12}C/^{13}C ratio of 89, and D Hydra is seventy something. So when you find an odd ratio you can definitely say it comes from outside the Solar System, but just because you see a semi-normal ratio doesn't necessarily mean that it doesn't come from an AGB star.

NITTLER: Well I disagree with you and the point is that it's not 'sort of normal'. The fact is this organic matter is normal with 'normal' meaning the same as terrestrial carbon standards within 1 or 2%. And you are actually right that you do have values that overlap the terrestrial value, but they are rare in AGB stars. Most of the stuff is very highly non-solar and that means the average is nowhere close to solar. So if you have a mixture of materials, it's going to be incredibly coincidental that it has a solar $^{12}C/^{13}C$ ratio.

ZIURYS: That's Dave Lambert's study in 1986 and there's far more data these days.

NITTLER: But the plot you showed in your talk was the same. Every object on that plot was non-solar.

ZIURYS: That is starting with red giants and going into AGBs. I am just saying be cautious since there are carbon AGB stars that have a greater enrichment of ^{12}C to ^{13}C.

NITTLER: I agree. There is a huge range. And the thrust of my argument is that because of the huge range, it's extraordinarily unlikely that you would end up with something the same as terrestrial to within 1% in any sort of average sense, but you are right that you cannot rule the possibility out. However, if you measure a hundred samples, and they are all terrestrial, it is highly unlikely all one hundred came from an AGB star.

Eric Quirico and Cecelia Ceccarelli (photo by Dale Cruikshank).

Organic Matter in Space
Proceedings IAU Symposium No. 251, 2008
S. Kwok & S. Sandford, eds.

© 2008 International Astronomical Union
doi:10.1017/S1743921308021923

The flux of interstellar particles detected in the Solar System

Mária Hajduková Jr.

Astronomical Institute of the Slovak Academy of Sciences
Dúbravská cesta 9, 845 05 Bratislava, The Slovak Republic
email: astromia@savba.sk

Abstract. The present work shows the proportion of interstellar meteors from different mass ranges (exceeding 20 orders of mass scale) detected by various observational techniques. Having analysed the IAU Meteor Data Center, we find that the mass index of interstellar particles continuously increases towards higher masses, but there is a significant change between $10^{-10} - 10^{-11}$ kg. This break is possibly caused by different physical processes leading to different populations of interstellar particles and might be connected with their origin.

Keywords. Meteors, meteoroids, interplanetary medium

1. Observations of interstellar particles by different techniques

A search for interstellar meteoroids entering the Solar System is currently going on using different techniques, covering a mass scale of more than 20 orders, from faint particles detected by cosmic detectors up to the range of bolides detected by photographic methods. The substantial question of how many interstellar particles there are in comparison with those belonging to our interplanetary cloud has led to many searches. Reports from the Advanced Meteor Orbit Radar (AMOR) in New Zealand and space borne observations, mainly from the Ulysses and Galileo spacecrafts, contradict, in some way, the most precise classical photographic observations as well as the broader photographic data included in the IAU Meteor Data Center (MDC), where only a very few meteors slightly exceeding the hyperbolic velocity limit in the mass range of large particles corresponding to $10^{-4} - 10^{1}$ kg (Hajduková 1994, 2008).

The orbital and geophysical data of the latest version of MDC, containing 4,581 photographic and 62,906 radar meteors (Lindblad *et al.* 2005), has been analysed. As detailed analysis showed, the vast majority of the hyperbolic orbits are a consequence of measurement errors in the determination of meteor velocity. We obtained values for interstellar meteor fluxes of 6×10^{-14} m^{-2}s^{-1} for radar data ($m > 7 \times 10^{-8}$ kg) and 7×10^{-19} m^{-2}s^{-1} for photographic meteors $m > 10^{-3}$ kg.

Hawkes & Woodworth (1997a) obtained a contribution of hyperbolic meteors at the level of 1 - 2% of the mass range between $10^{-4} - 10^{-9}$ kg using a video detector technique. Mathews *et al.* (1999), using an ultra high frequency (UHF) radar technique, detected a small level of hyperbolic particles in the mass range of $10^{-9} - 10^{-12}$ kg, reaching the upper mass limit of space-born particles. Weryk & Brown (2005) using the Canadian Meteor Orbit Radar (CMOR) orbital data obtained the flux of interstellar meteoroids arriving at the Earth of 6×10^{-6} meteoroids km^{-2} h^{-1} for the mass of 1×10^{-8} kg.

The question arising from the above results is whether it is possible to bring all the above mentioned data to a common view of the real contribution of interstellar particles across the broad scale of mass, exceeding 20 orders of magnitude, or whether we should search for errors in the results or methods leading to them?

2. The flux and mass distribution of interstellar particles

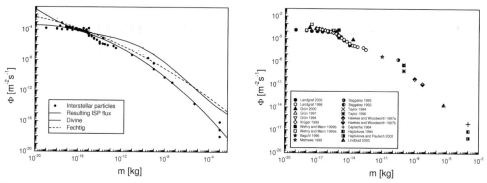

Figure 1. Flux of interstellar particles along the mass range (right panel), from observations by different techniques, in comparison with the flux of interplanetary particles according to Divine (solid line) and Fechtig (dashed line) (left panel).

Figure 1 is constructed from the results of different authors, as well as from our obtained values of interstellar meteor flux from the MDC. The sources of the data are given in the right side of the figure. The heavy line in Figure 1 (left) represents a second order polynomial of interstellar fluxes, over the whole mass scale from all given data. The fainter lines represent the second order polynomial for the interplanetary flux data given by Divine *et al.* (1993) (solid line) and Fechtig (1976) (dashed line). As is seen in Figure 1, the flux of interstellar particles is more than 2 orders of magnitude lower than the flux of interplanetary ones in the mass range of large particles, but it increases towards the smaller particles. Their mass distribution is steeper, however the critical value of the mass index $s_{is} = 2$ corresponds to almost the same mass interval, between 10^{-10}–10^{-11} kg. The mass index changes continuously, increasing towards higher masses. The data implies 3 distinct mass intervals, however a substantial change of s values occurs between 10^{-11}–10^{-10} kg as a break in the interstellar flux distribution. This break is possible caused by different physical processes leading to population of interstellar particles and might be connected with their origin (Hajduková & Hajduk 2006).

Acknowledgements

This work was supported by the Slovak Scientific Grant Agency VEGA, grant No 3067.

References

Divine, N., Grün, E., & Staubach, P. 1993, in: W. Flury, Darmstadt (eds.), *Proceeding of the First European Conference on Space Debris*, ESA SD - 01, p. 245

Fechtig, H. 1976, in: C. L. Hemenvay, P. M. Millman and A. F. Cook (eds.), *Evolutionary and Physical Properties of Meteoroids*, NASA SP, 319, p. 218

Hajduková, M. Jr. 1994, *A&A*, 288, 330

Hajduková, M. Jr. 2008, *Earth, Moon, Planets*, 102, 67

Hajduková, M. Jr. & Hajduk, A. 2006, *Contri. Astron. Obs. Skalnaté Pleso*, 36, 15

Hawkes, R. L. & Woodworth, S. C. 1997a, *J. R. Astron. Soc. Can.*, 91, 68

Hawkes, R. L. & Woodworth, S. C. 1997b, *J. R. Astron. Soc. Can.*, 91, 218

Lindblad, B. A., Neslušan, L., Porubčan, V., & Svoreň, J. 2005, *Earth, Moon, Planets*, 93, 249

Mathews, J. D. *et al.* 1999, in: W. J. Baggaley and V. Porubčan (eds.), *Meteoroids 1998*, (Slovak Academy of Sciences, Bratislava), p. 79

Weryk, R. J. & Brown, P. 2005, *Earth, Moon, Planets* 95, 221

Organic Matter in Space
Proceedings IAU Symposium No. 251, 2008
S. Kwok & S. Sandford, eds.

The occurence of interstellar meteoroids in the vicinity of the Earth

Mária Hajduková Jr.

Astronomical Institute of the Slovak Academy of Sciences
Dúbravská cesta 9, 845 05 Bratislava, The Slovak Republic
email: `astromia@savba.sk`

Abstract. If interstellar meteors are present among the registered meteor orbits, the distribution of the excesses of their heliocentric velocities should correspond to the distribution of radial velocities of close stars. Hence, for the velocity $v_i = 20$ kms^{-1} of an interstellar meteor (with respect to the Sun) we obtain a heliocentric velocity $v_H = 46.6$ kms^{-1} of an interstellar meteor arriving at the Earth. Moreover, a concentration of radiants to the Sun's apex should be observed. An analysis of the hyperbolic meteors among the 4581 photographic orbits of the IAU Meteor Data Center showed that the identification of the vast majority of the hyperbolic orbits in these catalogues has been caused by an erroneous determination of their heliocentric velocity and/or other parameters. Any error in the determination of v_H, especially near the parabolic limit, can create an artificial hyperbolic orbit that does not really exist. On the basis of photographic meteors from the IAU MDC, the proportion of possible interstellar meteors decreased significantly (greater than 1 order of magnitude) after error analysis and does not exceed the value 2.5×10^{-4}. Neither any concentration of radiants to the Sun's apex, nor any distribution following the motion of interstellar material has been found.

Keywords. Meteors, meteoroids, interplanetary medium

1. Introduction

From the position of the whole Solar System in the Galaxy and the studies of processes in the interstellar medium, it is clear that the Solar System is not an isolated system. Its interaction with the interstellar medium should lead to the presence of interstellar particles.

The present work is based on the analysis of 4581 meteor orbits collected in the photographic catalogues of the IAU Meteor Data Center (MDC) (Lindblad *et al.* 2005). Among them, 527 orbits (11.5%) are hyperbolic. It was shown (Hajduková 2008) that the vast majority of hyperbolic orbits has been caused by the erroneous determination of their heliocentric velocity or other parameters and that the number of hyperbolic meteors in the investigated catalogues of MDC does not represent in any case the number of interstellar meteors in observational data. Earlier (Hajduková 1994), from an analysis of the 2910 photographic orbits from the older version of the MDC, it was shown that there is a lack of statistical argument for the presence of real hyperbolic orbits; hence, it is necessary to analyse all special cases individually. This paper deals with the individual approach to the photographic orbits showing extremely high hyperbolic excesses found in 6 of 17 catalogues.

2. Interstellar meteoroids and meteoroids with hyperbolic orbits

Meteoroids entering the atmosphere of the Earth and coming from the interstellar medium have heliocentric velocities $v_H > v_p$, where v_p is the parabolic velocity with respect to the Sun, and the hyperbolicity of their orbits (with semimajor axis $a < 0$ and eccentricity $e > 1$) is not caused by planetary perturbations inside the Solar System. The difference between hyperbolic and parabolic heliocentric velocity is the hyperbolic excess of heliocentric velocity defined as $\Delta v_H = v_0[(2/r - 1/a)^{1/2} - (2/r)^{1/2}]$, where a is measured in AU and v_H in kms^{-1}; v_0 is the mean heliocentric velocity of the Earth, 29.8 kms^{-1} and $r = 1$. For this approximation we have selected and analysed individual meteors from the photographic catalogues of the IAU MDC.

The value of semimajor axis, a, is very sensitive to the value of the heliocentric velocity v_H, especially near the parabolic limit. Hence, any error in the determination of v_H can create an artificial hyperbolic population that does not really exist. If interstellar meteors are present among the hyperbolic orbits, the distribution of the hyperbolic excesses of their heliocentric velocities should correspond to the distribution of radial velocities of close stars. For the velocity of an interstellar meteor $v_i = 20$ kms^{-1}, which represent the relative velocity of the Sun in the nearby stellar environment and taking into account the equation $v_i^2 = v_H^2 - v_p^2$, with $v_p = 42.1$ kms^{-1}, we obtain a heliocentric velocity $v_H = 46.6$ kms^{-1} of an interstellar meteor arriving at the Earth.

3. Extreme hyperbolic orbits in the IAU MDC

There are 527 formally hyperbolic orbits (with eccentricity $e > 1$) among the total of 4581 photographic meteoroid orbits in the MDC catalogues. However, almost half of them (247 orbits) have been identified with the major meteor showers. We have found 59 orbits with $e > 1$ and $v_H > 46.6$ kms^{-1} in the photographic catalogues of MDC, of which 54 are from the Dushanbe, Odessa and Kiev catalogues, 3 from the European Fireball Network, 1 from the Nippon Meteor Society and 1 in the McCrosky - Shao catalogue. There is no meteoroid with $v_H > 46.6$ kms^{-1} in the 11 other photographic catalogues, containing more than 2/3 of all photographic orbits. In the most precise 413 Superschmidt orbits from the catalogue of Jacchia and Whipple the largest value of Δv_H is 0.7 kms^{-1}. The presence of 23 - 26 shower meteors among this sample of 59 extremely hyperbolic meteors points to very large observational or measurement errors.

From this analysis it can be concluded that many apparent hyperbolic orbits presented in the low quality data are not true hyperbolic meteors. In more precise data, there may be an argument for the presence of true hyperbolic orbits. However, the hyperbolic excesses of the velocities in these cases are very low, about one order less than required from the velocity distribution of neighbouring stars, which argues against the presence of interstellar meteors even in the most precise data. Summarizing the above data, the proportion of real hyperbolic meteors decreased significantly after the error analysis, from 11.5 $\times 10^{-2}$ to 2.5 $\times 10^{-4}$ of the total number of photographic meteors in the database. Hence, the flux of interstellar meteors is much lower than was declared by the authors of the catalogues, or was believed in some analyses of these observations. It is at least 1 order of magnitude less.

Acknowledgements

This work was supported by the Slovak Scientific Grant Agency VEGA, grant No 3067.

References

Lindblad, B. A., Neslušan, L., Porubčan, V., & Svoreň, J. 2005, *Earth, Moon, Planets*, 93, 249
Hajduková, M. Jr. 1994, *A&A*, 288, 330
Hajduková, M. Jr. 2008, *Earth, Moon, Planets*, 102, 67

Organic Matter in Space
Proceedings IAU Symposium No. 251, 2008
S. Kwok & S. Sandford, eds.

Formation of biomolecule precursors in space?

Wolf D. Geppert[1], Erik Vigren[1], Mathias Hamberg[1],
Vitali Zhaunerchyk[1], Magdalena Kaminska[2], Richard D. Thomas[1],
Fabian Österdahl[1], Fredrik Hellberg[1], Anneli Ehlerding[1],
Mathias Danielsson[1], and Mats Larsson[1]

[1]Physics Department, Stockholm University
Roslagstullsbacken 21, S-10691, Stockholm, Sweden
email: `wgeppert@physto.se`

[2]Swietokrzyska Akademy, Swietokrzyska, PL-25406 Kielce,Poland

Abstract. The possibility of an extraterrestrial origin of biomolecule building blocks has been a subject of intense discussions for many years. The detection of amino acids in meteorites opens the possibility of a delivery of biomolecules synthesized in the interstellar medium or star-forming regions to the primeval Earth. Whereas it can be doubted if more complex species like amino acids can survive the strong UV radiation in the early Solar System, this does not necessary hold for more primitive precursor molecules like nitriles. These compounds can also be synthesized very efficiently in methane-nitrogen dominated atmospheres like the one present on Titan and the early ages of Earth. This Contribution focuses on the formation and degradation processes of nitriles in interstellar clouds and planetary atmospheres and on their possible role in the generation of biomolecules.

Keywords. Astrochemistry, molecular processes, plasmas, ISM: molecules, astrobiology

1. Introduction

One of the most interesting questions in astrobiology is to what extent the basic building blocks of life can be formed in the interstellar medium and from what level of complexity planetary conditions (dense atmospheres, oceans) are necessary. For a long time it was taken as certain that biomolecules were formed in a "primordial soup" under an atmosphere of water, ammonia and methane that was subject to high-voltage discharges, an environment mimicked by the famous Urey-Miller experiment (Miller 1953). There are two problems with that pathway of biomolecule formation: Firstly, amino acids are formed as racemates in the Urey-Miller reactor and, secondly, their synthesis is hindered by the presence of oxygen, whose abundance in the atmosphere of early Earth is still a subject of controversial discussion (Rosing & Frei 2004). On the other hand, analysis of certain meteorites revealed the presence of several amino acids and investigations into the hydrogen isotope distribution suggested an interstellar origin of these compounds (Bernstein *et al.* 2002). Furthermore, some of the amino acids detected in the Murchison and Murray meteorite showed quite substantial enantiomeric excesses of the (biological) L-form (Pizarrello & Cooper 2002, Cronin & Pizarrello 1997), which raises the question if the first amino acids on Earth, and thus homochirality of life, could have an interstellar origin. The problem with that scenario is that amino acids cannot be expected to survive the strong UV field present in the early Solar System (Ehrenfreund & Sephton 2006). This, however, does not hold for nitriles, which possess a much higher photostability than their corresponding acids (Bernstein *et al.* 2004) and can, after their arrival at planetary surfaces, hydrolyze to form amino acids or polymerize to tholines. Nitriles have been detected in several different interstellar objects, like circumstellar envelopes (Matthews & Sears 1986) and dark clouds (Johansson *et al.* 1984). Interestingly, the largest molecule hitherto detected in the interstellar medium, $HC_{11}N$, is a nitrile (Bell *et al.* 1982).

Nitriles are especially important in nitrogen-methane dominated atmospheres like those of Titan and the early Earth. In one of the flybys of the Cassini probe, much higher abundances of protonated nitriles than predicted by previous models have been discovered (Vuitton *et al.* 2006). It is very likely that nitriles are involved in the formation of the haze in Titan's ionosphere (Lavvas *et al.* 2008a), since copolymers of acetylene and HC_3N have been found to exhibit optical properties similar to that haze (Clarke & Ferris 1997). In preparation for projected space research programs like the Titan Explorer mission it is of pivotal importance to elucidate the formation and destruction mechanisms of these species to understand the chemical processes in the ionosphere of Titan.

If one assumes that nitriles can be delivered to planets from the interstellar medium, the question of their formation pathway arises. In the gas phase, a synthesis involving ion-neutral reactions (e.g., radiative associations) leading to a protonated nitrile followed by a dissociative recombination (DR) that yields the unprotonated species and a hydrogen atom, is often invoked, as in the case of acetonitrile (Huntress & Mitchell 1979).

$$CH_3^+ + HCN \rightarrow CH_3CNH^+ + h\nu \quad (1)$$

$$CH_3CNH^+ + e^- \rightarrow CH_3CN + H \quad (2)$$

Another important mechanism for nitrile formation is the reaction of CN radicals with hydrocarbons (Herbst & Leung 1990). In the case of acetylene and ethylene (Balucani *et al.* 2000a), such processes have experimentally been shown to lead to cyanoethylene (HC_3N) and acrylonitrile (C_2H_3CN), respectively.

$$C_2H_2 + CN \rightarrow HC_3N + H \quad (3)$$

$$C_2H_4 + CN \rightarrow C_2H_3CN + H \quad (4)$$

In planetary atmospheres insertion reactions of nitrogen atoms in the long-lived excited (^2D) state are crucial to the formation of nitriles (Balucani *et al.* 2000b):

$$C_2H_4 + N(^2D) \rightarrow CH_3CN + H \quad (5)$$

Due to their high proton affinity, nitriles can easily be protonated (in the interstellar medium by H_3^+, HCO^+ and N_2H^+, in Titan's ionosphere by reactions with hydrocarbon ions), which explains the high nitrile abundances detected by the Ion Neutral Mass Spectrometer (INMS) on board the Cassini spacecraft. Since, according to models, abundances of many nitriles in Titan's atmosphere peak at an altitude of around 1000 km (Lavvas *et al.* 2008b), i.e., well into the ionosphere, protonation of nitriles is very probably an important decay process of nitriles there and could compete with polymerization and ionization by solar UV photons and magnetospheric electrons. DR is one of the most crucial destruction pathways of these protonated nitriles in interstellar clouds and planetary ionospheres, but, unfortunately, little experimental data are available on the rates and product branching ratios of these reactions. This Contribution sums up recent work performed by the Stockholm Group on the DR of protonated nitriles to address this lack of fundamental data.

2. Experimental details

The DR experiments have been performed at the heavy-ion storage ring CRYRING at the Manne Siegbahn Laboratory, Stockholm University. The experimental procedure has been described in detail elsewhere (Neau *et al.* 2000) and is therefore summarized only briefly here. For technical reasons related to the data analysis (better resolution on the detector) fully deuterated compounds were used in the case of protonated acetonitrile and protonated methylacetylene. After creation of the ions in the source they were further accelerated by a RF field to translational energies of 1.78 MeV ($C_2H_3CNH^+$ and $DCCCND^+$) and 2.09 MeV (CD_3CND^+), respectively. The ions were stored in the ring for several seconds to allow for cooling by radiative deactivation and superelastic collisions with electrons. The ion beam was then merged with a mono-energetic electron beam in an electron cooler, the length of the interaction region being 0.85 m. Neutral products generated by DR reactions in the electron cooler left the ring tangentially and were detected by an energy-sensitive ion implanted silicon detector with a radius of 17 mm mounted at a distance of 3.85 m from the centre of the interaction region. A background signal due to neutral products emerging from collisions of the ions with residual gas was also present: this was measured with the relative kinetic energy between the ions and electrons tuned to 1 or 2 eV, where the DR cross section is very low and the measured neutral fragments are therefore almost exclusively produced by rest gas collisions. This background was subsequently subtracted from the total detector signal.

3. Results and Discussion

Cross sections and reaction rate constants - For the cross section measurements, the kinetic energy in the centre-of-mass frame between the ions and the electrons was changed between 1 and 0 eV by ramping the cathode voltage of the electron cooler over a certain time interval (1-2 s) from a high value corresponding to a centre-of-mass energy of 1 eV, the electrons being faster than the ions, down to a low-value also corresponding to 1 eV where the electrons were slower than the ions. Thus, a voltage corresponding to a centre-of-mass energy of 0 eV is reached during the scan. The signal from the detector was monitored by a single channel analyzer, selecting signals only when all the fragments reach the detector simultaneously, and thereafter recorded by a multichannel scaler, yielding the number of counts vs. storage time and, therefore, at a given relative kinetic energy. After correction for space charge, toroidal effects, and the energy spread of the electrons (see, e. g., Geppert *et al.* 2006 for details), the cross sections of the investigated DR reactions could be determined. Under assumption of a Maxwell-Boltzmann distribution of kinetic energies of the reactants, their thermal rate constants were calculated. Both cross sections and rate constants are listed in Table 1.

Branching fractions - The fragments produced by a DR event reached the detector within a very short time interval compared with its response time. The pulse height of the detector signal was therefore proportional to the total mass. To measure the branching fractions a metal grid with a transmission $T = 0.297 \pm 0.015$ was inserted in front of the detector (Neau *et al.* 2000). In this case the registered DR spectrum splits into a series of peaks with different energies corresponding to the sum of the masses of the particles passing the grid. Unfortunately, for the investigated ions, the resolution of the detector did not allow a separation of fragment masses differing only by a the mass of a D or H atom, so only conclusions about the probabilities of ruptures of bonds between the heavy atoms could be obtained. The branching fraction for the different break-up

Wolf D. Geppert *et al.*

Table 1. Reactive cross sections and thermal rate constants of the DR of investigated protonated nitriles.

Ion	Cross section/cm^2 (E in eV)	Rate constant/cm^{3}s^{-1} (T in K)	Reference
DCCCND$^+$	$2.3 \pm 0.4 \times 10^{-15}$ E$^{-1.10 \pm 0.02}$	$1.5 \pm 0.3 \times 10^{-6}$ (T/300)$^{-0.58 \pm 0.02}$	Geppert *et al.* (2004)
CD$_3$CND$^+$	$7.4 \pm 1.5 \times 10^{-16}$ E$^{-1.23 \pm 0.02}$	$8.1 \pm 1.6 \times 10^{-7}$ (T/300)$^{-0.69 \pm 0.02}$	Vigren *et al.* (2008)
C$_2$H$_3$CNH$^+$	$1.1 \pm 0.2 \times 10^{-15}$ E$^{-1.29 \pm 0.02}$	$1.8 \pm 0.4 \times 10^{-6}$ (T/300)$^{-0.80 \pm 0.02}$	Preliminary data

channels have been computed using a simple Gaussian elimination process (see Geppert *et al.* 2004 for details). In the case of DCCCND$^+$, the probability of the retention of the carbon chain is $52 \pm 5\%$, and the rest of DR processes leads to fragmentation of the central C-C bond. For C$_2$H$_3$CNH$^+$ the branchings fractions of these two pathways are equal (50%). Interestingly, no rupture of the C-N or the terminal C-C bond is observed for both ions. In CD$_3$CND$^+$ DR leads to conservation of the C-CN chain in 65% of all events. The remainder result in the break-up of one C-N or C-C bond. Fragmentation of more than one heavy atom bond was not observed in any cases.

These results would imply that the sequence of Reactions (1) and (2) is a feasible pathway for synthesis of acetonitrile in dark interstellar clouds. However, it has been revealed in a FT-ICR experiment that the product of the first process is not protonated acetonitrile but a highly excited state of protonated methyl isonitrile (CH$_3$NCH$^+$), which can either radiatively decay or convert to CH$_3$CNH$^+$ through collisions (Anicich *et al.* 1995). Nevertheless, the latter process is too slow to be of any significance in the interstellar medium or planetary ionospheres. A way out could be the reaction of methyl cations with HNC, of which considerable densities are found in dark clouds, especially in TMC-1 where it is more abundant than HCN (Pratap *et al.* 1997):

$$CH_3^+ + HNC \rightarrow CH_3CNH^+ + h\nu \quad (6)$$

The degradation of nitriles through the protonation - DR sequence partly leads to recovery of the nitriles, but also results in formation of reactive products, which can undergo further chemical reactions. Since the reaction rates obtained for the investigated nitriles are in the upper range of those usually found in DR reactions [in the case of protonated acetonitrile they are about a factor of 2.5 higher than previous flowing afterglow measurements (Geoghegan *et al.* 1991)], they could have a far-reaching influence on the chemistry of nitrogen compounds in the interstellar medium and planetary ionospheres. Future model calculations incorporating these new data will show what impact our results have on the predictions of molecular abundances emerging from such computations.

Acknowledgements

The authors want to thank the staff at Manne Siegbahn Laboratory for excellent technical support. Funding from the Swedish Space Board and the Swedish Research Council is gratefully acknowledged.

References

Anicich, V. G., Sen, A. D., Huntress, W. T., & McEwan, M. 1995, *J. Chem. Phys.*, 102, 3256
Balucani, N., Asvany, O., Huang, L. C. L., Lee, Y. T., Kaiser, R. I., Osamura, Y., & Bettinger, H. F. 2000a, *ApJ*, 545, 892

Balucani, N., Cartechini, L., Alagia, M., Casavecchia, P., & Volpi, G. G. 2000b, *J. Phys. Chem. A.*, 104, 5655

Bell, M. B., Feldman, P. A., Kwok, S., & Matthews, H. E. 1982, *Nature*, 295, 389

Bernstein, M. P., Dworkin, J. P., Sandford, S. A., Cooper, G. W., & Allamandola, L. J. 2002, *Nature*, 416, 401

Bernstein, M. P., Ashbourn, S. F. M. Sandford, S. A., & Allamandola, L. J. 2004, *ApJ*, 601, 365

Clarke, D. W. & Ferris, J. P. 1997, *Icarus*, 127, 158.

Cronin, J. R. & Pizarrello, S. 1997, *Science*, 295, 951

Ehrenfreund, P. & Sephton, M. A. 2006, *Faraday Disc.*, 133, 277

Geoghegan, M., Adams, N. G., & Smith, D. 1991, *J. Phys. B.*, 24, 2589

Geppert, W. D., Ehlerding, A., Hellberg, F., Semaniak, J., Österdahl, F., Kaminska, M., Al-Khalili, A., Zhaunerchyk, V., Thomas, R., af Ugglas, M., Källberg, A., Simonsson, A., & Larsson, M. 2004, *ApJ*, 613, 1302

Geppert, W. D., Hamberg, M., Thomas, R. D., Österdahl, F., Hellberg, F., Zhaunerchyk, V., Ehlerding, A., Millar, T. J., Roberts, H. Semaniak, J., af Ugglas, M., Djuric, N., Källberg, A., Simonsson, A., Kaminska, M., & Larsson, M. 2006, *Faraday Disc.*, 133, 177

Herbst, E. & Leung, C. M. 1990, *A&A*, 233, 177

Huntress, W. T. & Mitchell, G. F. 1979, *ApJ*, 231, 456

Johansson, L. E. B., Andersson, C., Elldér, J., Friberg, P., Hjalmarsson, A., Höglund, B., Irvine, W. M., Olofsson, H., & Rydbeck, G. 1984, *A&A*, 130, 227

Lavvas, P. P., Coustenis, A., & Vardavas, I. M., 2008a, *Planet. Space Sci.*, 56, 27

Lavvas, P. P., Coustenis, A., & Vardavas, I. M., 2008b, *Planet. Space Sci.*, 56, 67

Matthews, P. & Sears, T. J. 1986, *ApJ*, 300, 766

Miller, S. L. 1953, *Science*, 117, 528

Neau, A., Al-Khalili, A., Rosén, S., Le Padellec, A., Derkatch, A. M, Shi, W., Vikor, L., Larsson, M., Nagard, M. B., Andersson, K., Danared, H., & af Ugglas, M. 2000, *J. Chem. Phys.*, 113, 1762

Pizarrello, S. & Cooper, G. W. 2001, *Meteoritics and Planetary Science*, 36, 897

Pratap, P., Dickens, J. E., Snell, R. L., Miralles, M. P. Bergin, E. A., Irwine, W. M., & Schloerb, F. P. 1997, *ApJ*, 486, 862

Rosing, M. T. & Frei, R. 2004, *Earth Planet. Sci. Lett.*, 217, 234

Vigren, E., Kaminska, M., Hamberg, M., Zhaunerchyk, V. Thomas, R. D., Danielsson, M. Semaniak, J. Andersson, P. U., Larsson, M, & Geppert, W. D., 2008 *Phys. Chem. Chem. Phys.*, 10, 4014

Vuitton, V., Yelle, R. V., & Anicich, V. G. 2006, *ApJ* (Letters), 647, L175

Discussion

BERNSTEIN: It's interesting that when you have the protonated acrylonitrile you get the acrylonitrile product half of the time and half of the time you get carbon bond fragmentation. When the protonated acrylonitrile fragments, what other products do you get? Does that give you acetonitrile ultimately? That's a very interesting pathway as well.

GEPPERT: I think that is an important point. You unfortunately cannot work out the complete reaction scheme, so you cannot really make any conclusions about what products you get. The upper limit of recovery of acrylonitrile is, as said, about 50%. Unfortunately there are many different exoergic pathways for breaking up after recombination. Amongst the events which involve a break up of a bond between two heavy atoms I would assume that a lot goes into HNC, and C_2H_3 (which is a comparatively stable radical) but I would not bet my (comparatively mundane) salary on it.

Dancing on board the Bauhinia during the harbour cruise (from left to right: Lucy Ziurys, Peter Strittmatter, Catherine Cesarsky, and Sun Kwok) (photo by Bruce Hrivnak).

Session III

Laboratory analogs of organic
compounds in space

Organic Matter in Space
Proceedings IAU Symposium No. 251, 2008
S. Kwok & S. Sandford, eds.

PAHs in Astronomy - A Review

Farid Salama

NASA Ames Research Center, Space Science Division,
Mail Stop: 245-6, Moffett Field, California 94035-1000, USA
email: farid.salama@nasa.gov

Abstract. Carbonaceous materials play an important role in space. Polycyclic Aromatic Hydrocarbons (PAHs) are a ubiquitous component of organic matter in space. Their contribution is invoked in a broad spectrum of astronomical observations that range from the ultraviolet to the far-infrared and cover a wide variety of objects and environments from meteorites and interplanetary dust particles to outer Solar System bodies to the interstellar medium in the local Milky Way and in other galaxies. Extensive efforts have been devoted in the past two decades to experimental, theoretical, and observational studies of PAHs. A brief review is given here of the evidence obtained so far for the contribution of PAHs to the phenomena aforementioned. An attempt is made to distinguish the cases where solid evidence is available from cases where reasonable assumptions can be made to the cases where the presence - or the absence - of PAHs is purely speculative at this point.

Keywords. Infrared: ISM, ultraviolet: ISM, ISM: molecules, ISM: dust, extinction, methods: laboratory, techniques: spectroscopic, line: identification, line: profiles, surveys, molecular data

1. Overview

Carbon molecules and ions play an important role in space. Polycyclic Aromatic Hydrocarbons (PAHs) are an important and ubiquitous component of the organic materials. Twenty years have passed since the PAH model was put forward to account for the ubiquitous infrared (IR) emission bands associated with a wide range of interstellar environments in our local galaxy and in other galaxies (Leger & Puget 1984, Allamandola *et al.* 1985). The infrared bands are observed at 3.29, 6.2, 7.7, 8.7, 11.3, and 12.7 μm and are often accompanied by minor, weaker, bands and underlying broad structures in the 3.1 - 3.7, 6.0 - 6.9, and 11 - 15 μm ranges. In the model dealing with the interstellar spectral features, PAHs are present as a mixture of radicals, ions, and neutral species (Allamandola *et al.* 1999). The ionization states reflect the ionization balance of the medium while the size, composition, and structure reflect the energetic and chemical history of the medium. The proposed excitation (pumping) mechanism of the IR bands is a one-photon mechanism that leads to the transient heating of the PAH molecules and ions by stellar ultraviolet (UV), visible, and/or NIR photons. The IR bands are associated with the molecular vibrations of PAH structures present either as free molecules and ions (for the discrete bands) or as subunits of larger carbonaceous grains (for the broad, continuum-like structures). PAHs thus constitute the building blocks of interstellar dust grains and play an important role in mediating energetic and chemical processes in the interstellar medium (ISM). The PAH model has considerably evolved over the years (see Figure 1), thanks in large part to the extensive laboratory and theoretical efforts that have been devoted to this issue over the years. There is a wide consensus now, in the post-Spitzer era, that PAHs are the best candidates to account for the IR emission bands (Draine & Li 2007). The bands that were once dubbed the "unidentified" IR bands (UIR) are now

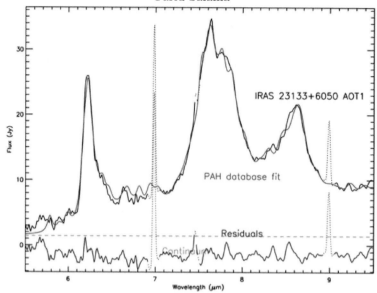

Figure 1. Least squares fit of the PAH spectra in the Ames' database to the ISO SWS
spectrum of the H II Region IRAS 23133. From Cami *et al.* (2008)

routinely called the "PAH bands" and are used as probes of the ISM in extra-galactic
environments (Smith *et al.* 2007).

PAHs are also thought to contribute to the interstellar extinction and to be among
the carriers of the diffuse interstellar absorption bands (DIBs). The DIBs are ubiquitous
spectral absorption features observed in the line of sight to stars that are obscured by
diffuse or translucent interstellar clouds. Close to 500 bands have been reported to date
spanning from the near UV to the near IR with bandwidths ranging from 0.4 to 40 Å
(Tielens & Snow 1995, Snow & McCall 2006). DIBs are also detected in extragalactic
environments (Cox *et al.* 2007, Cordiner *et al.* 2008). The present consensus is that the
DIBs arise from gas-phase, organic molecules and ions that are abundant under the
typical conditions reigning in the diffuse ISM. The PAH hypothesis is consistent with
the cosmic abundance of carbon and hydrogen and with the required photostability of
the DIB carriers against the strong VUV radiation field in the diffuse interstellar clouds
(Salama *et al.* 1996). It should be noted, however, that DIBs that are unambiguously
associated with specific neutral or ionized PAHs are yet to be found.

Photo-luminescence from charged PAH clusters has been advanced as the possible
source of the extended red emission (ERE), a broad emission feature ranging from 540
to beyond 900 nm with a peak wavelength longward of 600 nm to beyond 800 nm, that
is observed in many environments where both dust and UV photons are present (photodissociation regions found in the diffuse interstellar medium, reflection nebulae, planetary
nebulae, and even in other galaxies). The carriers of the ERE are widespread throughout
the diffuse interstellar medium of the Milky Way Galaxy and other galaxies (Rhee *et al.*
2007, Berné *et al.* 2008, Witt *et al.* 2008).

PAHs have been proposed as the carriers of the unknown fluorescence bands that were
detected in the 280-400 nm range by the TKS spectrometer on board Vega in the coma
of comet Halley (Moreels *et al.* 1994, Clairemidi *et al.* 2007). Although specific PAHs
[anthracene ($C_{14}H_{10}$), phenanthrene ($C_{14}H_{10}$), and pyrene ($C_{16}H_{10}$)] were singled out in
this case, the observed bands are too broad for unambiguous assignments and further
confirmation should await higher resolution data. If verified, however, this finding would
be the first detection of the signature of *specific* PAHs in space.

PAHs have also recently been tentatively invoked as trace components in the complex refractory organics that best model the low albedo surface materials of the moons of Saturn, Iapetus, and Hyperion (Cruikshank *et al.* 2007), and as a component of the complex organic materials that form in Saturn's rings (Cuzzi 2008).

Finally, PAHs have been unambiguously characterized in meteoritic samples (cf. Plows *et al.* 2003) and in interplanetary dust particles (cf. Clemett *et al.* 1993).

2. Laboratory and Theoretical Studies of PAHs

The major challenges that face the study of cosmic PAHs are of two orders. First, these studies must meet the general basic key requirements for laboratory astrophysics, i. e., insure that the laboratory measurements of the physical/chemical phenomena under consideration are relevant and that the experimental conditions accurately mimic (in a realistic way) the known physical and chemical conditions that exist in the specific space environments under study. Second, the nature of the samples must be properly reproduced. PAHs are refractory materials with low vapor pressure, are often toxic, and represent a particularly difficult challenge for laboratory studies. Theoretical studies are also challenging. Improved quantum chemistry models and programs must be developed to account for such large polyatomic molecular structures (Weisman *et al.* 2003).

For example, the harsh physical conditions reigning in IS clouds -low temperature, collisionless, strong UV radiation fields – are simulated in the laboratory by isolating the molecular PAH entities (neutral and ions) in cold inert-gas matrices (He, Ne, and Ar) or, in the most recent experiments, by forming molecular beams seeded with PAH molecules (Figure 2). Cold PAH ions and radicals are formed *insitu* from the neutral precursors in an isolated environment either through one-photon ionization or through soft penning ionization in a cold plasma expansion or "glow discharge" and probed with high-sensitivity cavity ringdown spectroscopy in the NUV-NIR range (Romanini *et al.* 1999, Biennier *et al.* 2003, 2004, Sukhorukov *et al.* 2004, Tan & Salama 2005a, b, 2006) or with laser depletion mass spectroscopy measurements (Bréchignac & Pino 1999, Pino *et al.* 1999). Carbon nanoparticles are also formed during the short residence time of the precursors in the plasma and are characterized with time-of-flight mass spectrometry. These experiments provide unique information on the spectra of large carbonaceous molecules and ions in the gas phase that can now be directly compared to interstellar and circumstellar observations (DIBs, extinction curve). These findings, when combined with detailed investigations of the flow dynamics in the PDN (Remy *et al.* 2005, Biennier *et al.* 2006) and fed into a plasma modeling program (Broks *et al.* 2005a, b) that describes the electron density and energy, as well as the argon ion and metastable atom number density, permit a full characterization of the plasma and hold great potential for understanding the formation process of interstellar carbonaceous grains.

Almost ten years have passed since our last review of the status of the PAH model (Salama 1999). Substantial progress has been made in this past decade both in experimental and in theoretical studies of PAHs (neutrals and ions), allowing for more decisive comparisons with astronomical observations. In this chapter, we briefly review the advances made on the various fronts.

PAHs in the infrared: The current consensus is that the interstellar IR emission spectra correspond to the composite emission of vibrational modes of a complex mixture of molecular PAHs of different sizes (i.e., number of carbon atoms), structures (compact, linear, branched) and charge states (neutrals, positive and negative ions). The vibrational spectra of neutral and ionized PAHs were measured in solid inert-gas matrices of argon at low temperature (10 K) for an extended set of molecular structures containing up to 50 carbon atoms (Hudgins & Allamandola 2004, Mattioda *et al.* 2005). This extended

360 Farid Salama

spectral database, combined with theoretical calculations, has allowed major progress in fitting the IR observational data that now permits one to take into account the subtle variations that are observed among the various regions of space probed by these bands as illustrated in Figure 1.

Despite their success in explaining the IR emission bands, PAH IR spectra alone cannot - by definition - lead to the identification of *specific* individual molecular PAH structures. The C-C and C-H vibrational frequencies measured in the mid-infrared mostly characterize chemical bonds that are present in almost all PAHs and have a weak dependence on the size and structure of the molecule, making it impossible to characterize specific PAH molecules and/or ions. The identification of individual PAH structures can only come from the detection of their skeleton modes in the far-infrared (FIR) range, their rotational spectra in the microwave, and/or from the detection of their electronic spectra in the UV and visible, and this, assuming that individual contributions can be deconvolved from the global distribution. The realization of these limitations has led to the development of new innovative approaches in laboratory astrophysics.

PAHs in the far-infrared and microwave ranges: FIR and sub-millimeter transitions probe skeleton motions and are a direct function of the size and the shape of the molecular structure, making them very interesting candidates for the detection of specific PAH structures in space. Interstellar PAHs should produce a forest of lines over a broad spectral range and the signature of specific PAHs might be detectable in the near future with instruments such as Herschel and SOFIA. This domain has been largely unexplored until recently when FIR spectra of a set of neutral PAHs have been obtained in absorption in solid inert-gas matrices of argon at low temperature (Mattioda *et al.* 2008) and in the gas phase via thermal emission (Pirali *et al.* 2006). The FIR spectra of cold neutral and ionized PAHs in the gas phase have recently been reported from resonantly enhanced multiphoton ionization (REMPI) and zero kinetic energy photoelectron (ZEKE) spectroscopy measurements (Zhang *et al.* 2008) opening the way for *direct* comparison with astronomical data when available.

Advances have also been observed in the microwave range, where the rotational spectra of cold neutral PAHs have recently been reported (Thorwirth *et al.* 2007). Although PAHs with only carbon and hydrogen atoms have weak dipole moments, nitrogen-containing PAHs have larger dipole moments that might be observable in this range, opening new prospects for the detection of specific PAHs.

PAHs in the visible and ultraviolet:

PAH ions: The electronic spectra of selected PAH ions were also measured in order to derive their *intrinsic* characteristics for comparison with interstellar spectra. Since

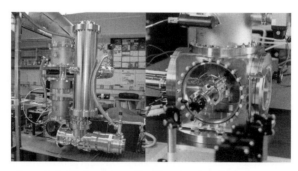

Figure 2. Configuration of the Laboratory Facility. *Right:* close-up view of the Chamber consisting of a Pulsed Discharge Nozzle coupled to a Cavity Ringdown Spectrometer apparatus and Reflectron time-of-flight mass spectrometer.

the discharge does not affect the vibrational temperature of the aromatic ions formed in cold plasma expansion (Remy *et al.* 2005, Biennier *et al.* 2006), detailed analysis of band profiles indicates that the vibronic bands are typically broad (FWHMs of the order of a few tens of cm^{-1}), lack substructure and exhibit ultra fast relaxation lifetimes (a few tens to a few hundred femtoseconds). These features are characteristic of non-radiative intramolecular relaxation processes and explain the UV photon pumping mechanism that occurs in the ISM and the observations of the IR emission bands by radiative cascade. The characteristics of the PAH ion bands measured to date are recapitulated in Table 1 where they are compared to the strong broad DIB at 4428 Å that exhibits very similar characteristics. The 4428 Å DIB is an averaged Lorentzian profile resulting from measurements in the lines-of-sight of 35 highly-reddened O and early B stars with reddening, E(B-V) ranging from 1.0 to 2.5 mag, and located in the Cyg OB2 association (Snow *et al.* 2002). The preliminary conclusion that can be derived from this comparison is that if PAH ions contribute to the DIBs they contribute to a class of *broad* DIBs. The search for weaker broad DIBs has been inconclusive so far. This would tend to indicate the absence of these specific ions in the line of sights that were probed. The non-observation might also be due, however, to the difficulty of identifying weak broad features in the astronomical spectra. The case will be settled when dedicated surveys of broad DIBs with low error bars will be available for comparison with the laboratory data.

Neutral PAHs: The electronic spectra of cold (50 K rotational temperature) neutral methylnaphthalene ($C_{11}H_{10}$), acenaphthene ($C_{12}H_{10}$), phenanthrene ($C_{14}H_{10}$), pyrene ($C_{16}H_{10}$), perylene ($C_{20}H_{12}$), pentacene ($C_{22}H_{14}$), and benzoperylene ($C_{22}H_{12}$) were measured in the gas-phase in order to compare with astronomical spectra (Tan & Salama

Table 1. Electronic state peak positions and band widths of cold gas-phase PAH ions measured in the laboratory are compared to the characteristics of the strong broad 4428 Å DIB

Molecular ion	Electronic State	λ (Å)	$\Delta\lambda$ (Å)
Naphthalene$^+$ [$C_{10}H_8^+$][a,1]	D_2	6707.7	10
Naphthalene$^+$ [$C_{10}H_8^+$][b,2]	D_3	4548.5	19
Acenaphtene$^+$ [$C_{12}H_{14}^+$][a,1]	D_2	6462.7	22
Fluorene$^+$ [$C_{13}H_{10}^+$][c,2]	D_3	6201.7	53
Phenanthrene$^+$ [$C_{14}H_{10}^+$][d,2]	D_2	8919.0	12
Anthracene$^+$ [$C_{14}H_{10}^+$][e,1]	D_2	7087.6	47
Pyrene$^+$ [$C_{16}H_{10}^+$][f,1]	D_5	4362.0	28
Pyrene$^+$ [$C_{16}H_{10}^+$][g,2]	D_4	4803.3	30
Pyrene$^+$ [$C_{16}H_{10}^+$][g,2]	D_2	7786.6	97
Methylpyrene$^+$ [$C_{17}H_{12}^+$][h,1]	D_5	4411.3 (4413.3; 4409.3)	10
Pyrene(COH)$^+$ [$C_{17}H_{10}O^+$][h,1]	D_8	4457.8	20
Pyrene(COH)$^+$ [$C_{17}H_{10}O^+$][h,1]		4442.7	
Pyrene(COH)$^+$ [$C_{17}H_{10}O^+$][h,1]		4431.4	
4428 Å DIB^i	*4428.4 ± 1.4*	*17.3 ± 1.64*	

Notes: [1]CRDS, [2]Laser Depletion MS, [a]Biennier *et al.* (2003), [b]Pino *et al.* (1999), [c]Bréchignac *et al.* (2001), [d]Bréchignac & Pino (1999), [e]Sukhorukov *et al.* (2004), [f]Biennier *et al.* (2004), [g]Pino (1999), [h]Tan & Salama (2006), [i]Snow *et al.* (2002)

2005a, b, Rouillé *et al.* 2004, 2007). Typical spectra are shown in Figures 3 & 4. Neutral PAHs exhibit narrower bands than the ions (FWHMs are of the order of only a few cm^{-1}) with a profile that is *closely* similar to the profile of the narrower DIBs. The case is strikingly illustrated in the comparison of the 5363 Å band of neutral pentacene, $C_{22}H_{14}$, with the narrow (< 2 Å FWHM) 5364 Å DIB detected with the echelle spectrograph of the OHP 2m-telescope (Figure 4). Note that $C_{22}H_{14}$ exhibits another band of similar strength at 5340 Å where only a weaker feature is found in the astronomical spectra. Additional comparisons are being performed between laboratory spectra and the astronomical spectra of extensive sets of reddened O and B stars observed with the ELODIE/OHP and the UVES/VLT telescopes in the 3100 Å to 5400 Å range. The stars have reddening E(B-V) ranging from 0.3 to 1.35 mag. Features can be detected at the 0.2% level when co-adding all spectra. From these comparisons, we can derive upper limits to the abundances of individual PAHs in the observed lines of sight. Values of the order of 10^{-4} to 10^{-6} are derived for the fraction of cosmic carbon locked up in these PAHs (Cami *et al.* 2005, Galazutdinov *et al.* 2008). These preliminary results await comparison with improved modelling of stellar lines in the NUV.

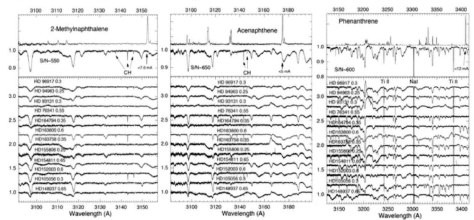

Figure 3. Comparison of jet-cooled neutral PAH spectra in the NUV-Visible to the ESO/VLT/UVES spectra of reddened O and B stars. (Galazutdinov *et al.* 2008)

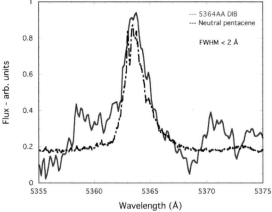

Figure 4. Comparison of band profiles of jet-cooled pentacene ($C_{22}H_{14}$) with CRDS and the 5363.8 Å DIB measured with the OHP echelle spectrograph (see text). The laboratory band (dotted line) was slightly shifted in wavelength for illustration purposes. Taken from Salama (2008).

Discharge products and formation of carbon particles: We have identified the formation of CH as one of the high-energy plasma fragmentation products formed in the PDN through the measurement of its (A-X) (0-0) absorption band. We have simultaneously observed an increase in the extinction of the CRD signal caused by the formation of carbon nanoparticles in the plasma and the formation of soot on the electrodes of the PDN source. Analysis of the soot with $\mu L^2 MS$ has shown the formation of larger PAH particles in the discharge (Biennier *et al.* 2005). These preliminary findings hold some potential for exploring the formation (and destruction) processes of carbon-bearing molecules in space and are the subject of investigations in this and in other laboratories as evidenced by presentations at this symposium (contributions by Pino *et al.*, Brunetto *et al.*, Saito *et al.*).

3. Implications

This brief overview of Laboratory Astrophysics studies applied to the characterization of PAHs in space indicates that tremendous progress has been made in the past decade leading to laboratory data that can now be *directly* compared to the astronomical observations. These preliminary results demonstrate the power of the laboratory approach associating cold molecular beams and plasma sources with sensitive high-resolution spectroscopy (CRDS, REMPI, ZEKE) and mass spectrometry (e.g. RETOF-MS) techniques for generating and characterizing laboratory analogs of large interstellar organic molecules. The laboratory spectra of cold PAHs are decongested, making it possible *for the first time* to perform unambiguous searches for specific PAH molecules in astronomical spectra and derive meaningful upper limits for their abundances. These experiments also provide first hand data on the spectroscopy and on the molecular dynamics of *free*, cold, large complex organic molecules and ions in the gas phase. We are now, for the first time, in the position to directly search for *individual* PAH molecules and ions in astronomical spectra.

This new generation of laboratory experiments opens a new page and offers tremendous opportunities for the data analysis of upcoming space missions that have the potential to lead to the identification of *specific* PAHs in space. Herschel and SOFIA will soon pioneer observations in the FIR. The Cosmic Origins Spectrograph (COS) due to be installed on the Hubble Space Telescope in the next servicing mission will search for the signature of large aromatic molecules and ions in the UV extinction of highly reddened stars. Finally, the new generation of laboratory experiments that couples mass spectrometry instruments to molecular beam and high-sensitivity spectroscopy instruments also provide a powerful tool for a better understanding of the formation mechanisms of carbonaceous dust nanoparticles from molecular precursors in the circumstellar shells of carbon-rich stars. This is the topic of future exciting research.

Acknowledgements

This work is supported by the NASA APRA, Cosmochemistry, and Planetary Atmosphere Programs of the Science Mission Directorate. I wish to acknowledge L. Biennier, J. Cami, J. Remy, and X. Tan for their contribution to this work and the outstanding technical support of R. Walker. I also thank L. Allamandola, A. Mattioda, D. Cruikshank, A. Witt, and J. Cuzzi for stimulating discussions and for providing support materials for this review.

References

Allamandola, L. J., Hudgins, D., & Sandford S.A. 1999, *ApJ* (Letter), 511, L115
Allamandola L.J., Tielens A.G., & Barker J.R. 1985, *ApJ* (Letter), 290, L25

Group picture at the University of Hong Kong.

Organic Matter in Space
Proceedings IAU Symposium No. 251, 2008
S. Kwok & S. Sandford, eds.

© 2008 International Astronomical Union
doi:10.1017/S1743921308021972

Theoretical PAH emission models for aromatic infrared bands

Amit Pathak† and Shantanu Rastogi

Physics Department, D. D. U. Gorakhpur University, Gorakhpur - 273009, INDIA.
email: amitpathak1234@rediffmail.com

Abstract. Aromatic Infrared Bands (AIBs) show significant profile variations in different astrophysical environments. Theoretical IR data is used to develop emission models to understand these variations. A good match with the observed "7.7" μm feature from different regions is obtained.

Keywords. Astrochemistry, stars: circumstellar matter, infrared: ISM

1. Introduction

ISO data reveals important variations in peak positions and band shapes of the AIBs (ISO results 1996). Lack of extensive experimental data on PAHs enhances the importance of theoretical data that may be used to model AIBs. We report models of composite emission from different PAH size groups using calculated IR data (Pathak 2006).

2. Emission models and astrophysical implications

The emission spectrum for individual PAHs is computed using the absorption spectrum as input assuming canonical ensembles as suggested in the cascade emission model (Schutte *et al.* 1993, Cook & Saykally 1998, Pech *et al.* 2002). Complete details of the emission modeling may be found elsewhere (Pathak & Rastogi 2008). Obtained emission spectra of individual PAHs are co-added assuming equal number of species and are plotted using a Lorentzian profile with FWHM of 20 cm^{-1}.

The model spectrum of neutrals is dominated by C−H stretch and C−H out-of-plane bend vibrations. The observed C−H stretch (3.3 μm) and C−H out-of-plane bend (11.2 μm) intensity ratio is a suitable parameter constraining the size of PAHs. The absorption intensities of C−H stretch vibrations reduce drastically upon ionization while small intensity variations are observed for the C−H wag modes. Among the AIBs the 6.2 and 7.7 μm emission features are most intense. The C−C stretch vibrations set up in ionized PAH molecules give rise to these bands. Profile variations observed in these bands are a direct measure of the background environments that excite the PAHs.

The spectral models are specifically used to correlate with the "7.7" μm band. The profiles of two different models are shown in Figure 1. Model 'a', comprising medium sized compact PAHs, has a stronger lower wavelength component as observed in UV-rich environments of H II regions and reflection nebulae and have been classified as A' profiles (Peeters *et al.* 2002). The spectral model comprising large PAHs has the 7.7 μm band dominated by the higher wavelength component at 7.8 μm (1285 cm^{-1}) with a shoulder at 7.6 μm (1315 cm^{-1}). Such profiles conform with observations of relatively benign

† Present address: Inorganic and Physical Chemistry Department, I. I. Sc., Bangalore, India.

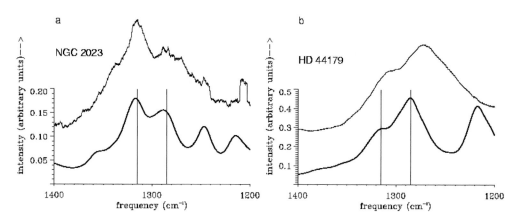

Figure 1. PAH cations emission model in the 1200 to 1400 cm^{-1} region compared with ISO spectra of PAHs with (a) less than 40 C atoms in varying proportions, and (b) more than 40 C atoms. The 7.7 μm band lies within the two vertical lines.

Table 1. Band position and strength ratios of the two components of 7.7 μm composite.

PAH model	7.6	7.8	Diff. (cm^{-1})[1]	Diff.$^2_{obs}$ (cm^{-1})	$I_{7.6}/I_{7.8}$[1]	Obs $I_{7.6}/I_{7.8}$[2]
with < 40 C atoms	1347	1285	62	~30	1.06	1.56 (NGC 2023)
						1.35 (Orion peak 2)
						1.20 (IRAS 23133)
with > 40 C atoms	1315	1285	30	~45	0.65	0.64 (NGC 7027)
						0.52 (IRAS 17047)
						0.42 (HD 44179)

[1]Difference of the band position and intensity ratio of the two components of the 7.7 μm composite as calculated from the theoretical spectral models.;
[2]The observed results are from Peeters *et al.* (2002), Table 2.

astrophysical regions classified as B' profiles (Peeters *et al.* 2002). Intensity ratios of the two components of the 7.7 μm complex and their wavenumber separation in the three models is presented in Table 1.

3. Conclusion

The 7.8 μm component, observed in benign regions, correlates with the model spectra of large PAH cations. The 7.6 μm sub-feature dominant in UV rich regions matches the model spectra of medium sized PAH cations. This indicates formation of large PAHs in outflows of post-AGB stars that transform to medium sized ones in strong UV sources.

References

Cook, D. J. & Saykally, R. J. 1998, *ApJ*, 493, 793
ISO Results 1996, *A&A* (Letters), 315, L26
Pathak, A. 2006, *Ph.D. Thesis, D. D. U. Gorakhpur University, Gorakhpur (India)*
Pathak, A. & Rastogi, S. 2008, *A&A*, 485, 735
Pech, C., Joblin, C., & Boissel, P. 2002, *A&A* 388, 639
Peeters, E. *et al.* 2002, *A&A* 390, 1089
Schutte, W. A., Tielens, A. G. G. M., & Allamandola, L. J., 1993, *ApJ*, 415, 397

Organic Matter in Space
Proceedings IAU Symposium No. 251, 2008
S. Kwok & S. Sandford, eds.

© 2008 International Astronomical Union
doi:10.1017/S1743921308021984

Photoproduction of H_3^+ from gaseous methanol inside dense molecular clouds

S. Pilling[1], D. P. P. Andrade[2], A. C. F. Santos[2], and H. M. Boechat-Roberty[2]

[1]LNLS, Laboratório Nacional de Luz Síncrotron, São Paulo, Brazil.
email: sergiopilling@yahoo.com.br

[2]UFRJ, Universidade Federal do Rio de Janeiro, Rio de Janeiro, Brazil.
email: diana_andrade@ufrj.br, toni@if.ufrj.br, heloisa@ov.ufrj.br

Abstract. We present experimental results obtained from photoionization and photodissociation processes of abundant interstellar methanol (CH_3OH) as an alternative route for the production of H_3^+ in dense clouds. The measurements were taken at the Brazilian Synchrotron Light Laboratory (LNLS) employing soft X-ray and time-of-flight mass spectrometry. Mass spectra were obtained using the photoelectron-photoion coincidence techniques. Absolute averaged cross sections for the production of H_3^+ due to molecular dissociation of methanol by soft X-rays (C1s edge) were determined. The H_3^+'s photoproduction rate and column density were been estimated adopting a typical soft X-ray luminosity inside dense molecular and the observed column density of methanol. Assuming a steady state scenario, the highest column density value for the photoproduced H_3^+ was about 10^{11} cm^2, which gives the ratio photoproduced/observed of about 0.05%, as in the case of dense molecular cloud AFGL 2591. Despite the small value, this represent a new and alternative source of H_3^+ into dense molecular clouds and it is not been considered as yet in interstellar chemistry models.

Keywords. Methods: laboratory, molecular data, ISM: molecules, astrochemistry

1. Introduction

The H_3^+ ion plays an important role in diverse fields from chemistry to astronomy, for example in the chains of reaction that lead to the production of many of complex molecular species observed in the interstellar medium (Herbst & Klemperer 1973, Dalgarno & Black 1976, McCall *et al.* 1998 and references therein). A detailed review about this simplest stable interstellar polyatomic molecule could be found in Oka (2006).

In interstellar regions its main formation pathway occurs via ionization of molecular hydrogen by ubiquitous cosmic rays or local X-rays, followed by the efficient ion-neutral reaction, $H_2^+ + H_2 \rightarrow H_3^+ + H$. Its dominant destruction pathway occurs via proton-hop reaction with the abundant interstellar carbon monoxide, $H_3^+ + CO \rightarrow HCO^+ + H_2$. However, as pointed out by Maloney *et al.* (1996) and Koyama *et al.* (1996), dense clouds, mainly the ones with embedded protostars, the soft X-ray field may represents the dominant excitation/ionization source, penetrating great depths into molecular clouds.

2. Experimental methodology and results

In an attempt to simulate the effect of stellar soft X-ray flux on gaseous molecules inside dense clouds we have used synchrotron radiation as a light source. The measurements were taken at the toroidal grating monochromator (TGM) beamline at Brazilian Synchrotron Light Laboratory (LNLS), Brazil, employing soft X-rays photons over the C1s resonance energy range (200-310 eV). The incoming radiation perpendicularly intersects the gas sample inside a high vacuum chamber. Conventional time-of-flight mass spectra were obtained using photoelectron and photoion coincidence (PEPICO) techniques. The complete description of the experimental setup can be found elsewhere (Boechat-Roberty *et al.* 2005, Pilling *et al.* 2006, 2007b).

Methanol is one of the most abundant molecules in the interstellar medium and in dense molecular clouds. Therefore, even despite the reduced production of H_3^+ from X-rays photodissociation process, it is reasonable to expect that at least a fraction of the detected H_3^+ in molecular clouds may be produced from this simple methyl compound molecule.

The averaged cross sections for H_3^+ production by soft X-rays photons at C1s resonance, were determined taking into an account the relative intensities of the H_3^+'s dissociative channels on PEPICOs spectra and the respective simple ionization and double ionization cross section of the parent molecule (see details in Pilling *et al.* 2007a).

The H_3^+'s photoproduction rate and column density were estimated by adopting a typical soft X-ray luminosity inside dense molecular cloud (Stäuber *et al.* 2005) and the observed column density of its most abundant parent ion, methanol. The values for H_3^+'s photoproduction cross section due to the dissociation of methanol by photons over the C1s edge were about 1.4×10^{-18} cm^2.

3. Conclusion

Assuming a steady state scenario and a typical X-ray luminosity of $L_x \gtrsim 10^{31}$ erg s^{-1} as the case for AFGL 2591 (Stäuber *et al.* 2005), the highest column density value for the photoproduced H_3^+ was about 10^{11} cm^2, which gives the ratio photoproduced/observed of about 0.05%. Despite the small value, this represents a new and alternative source of H_3^+ inside dense molecular clouds that has not been considered as yet in interstellar chemistry models. Better estimative for H_3^+ photoproduction rate depends of more accurate soft X-ray radiation field determinations.

Moreover, the energetic ionic products released by dissociation of CH_3X molecules, including the H_3^+ ion, become an alternative and efficient route to complex molecular synthesis, since some ion-molecule reactions do not have an activation barrier and are also very exothermic. We hope that these cross sections give rise to more precise values for some molecular abundances in interstellar clouds and even in planetary atmosphere models.

References

Boechat-Roberty, H. M., Pilling, S., & Santos, A. C. F. 2005, *A&A*, 438, 915.
Dalgarno, A. & Black, J. H. 1976, *Rep. Prog. Phys.*, 39, 573.
Herbst, E., & Klemperer, W. 1973, *ApJ*, 185, 505.
Koyama, K., Hamaguchi, K., Ueno, S., Kobayashi, N., & Feigelson, E. D. 1996, *PASJ* (Letters), 48, L87.
Oka, T. 2006, *PNAS*, 103, 12235.
Maloney, P. R., Hollenbach, D. J., & Tielens, A. G. G. M. 1996, *ApJ*, 466, 561.
McCall, B. J., Geballe, T. R., Hinkle, K. H., & Oka, T. 1998, *Science*, 279, 1910.
Stäuber, P., Doty, S. D., van Dishoeck, E. F., & Benz, A. O. 2005, *A&A*, 440, 949.
Pilling, S., Andrade, D. P. P., Neves, R., Ferreira-Rodrigues, A. M., Santos, A.C.F., & Boechat-Roberty, H. M. 2007a, *MNRAS*, 375, 1488.
Pilling, S., Neves, R., Santos, A. C. F., & Boechat-Roberty, H. M. 2007b, *A&A*, 464, 393.
Pilling, S., Santos, A. C. F., & Boechat-Roberty, H. M. 2006, *A&A*, 449, 1289.

Organic Matter in Space
Proceedings IAU Symposium No. 251, 2008
S. Kwok & S. Sandford, eds.

Survival of gas phase amino acids and nucleobases in space radiation conditions

S. Pilling[1], D. P. P. Andrade[3], R. B. de Castilho[3], R. L. Cavasso-Filho[1], A. F. Lago[1], L. H. Coutinho[2], G. G. B. de Souza[3], H. M. Boechat-Roberty[3], and A. Naves de Brito[1]

[1]LNLS, Laboratório Nacional de Luz Síncrotron, São Paulo, Brazil
email: sergiopilling@yahoo.com.br, cavasso@lnls.br, alago@lnls.br,
arndaldo@lnls.br

[2]UEZO, Centro Universitário Estadual da Zona Oeste, Rio de Janeiro, Brazil
email: coutinholh@yahoo.com

[3]UFRJ, Universidade Federal do Rio de Janeiro, Rio de Janeiro, Brazil
email: diana_andrade@ufrj.br, castilho@iq.ufrj.br, gerson@iq.ufrj.br,
heloisa@ov.ufrj.br

Abstract. We present experimental studies on the photoionization and photodissociation processes (photodestruction) of gaseous amino acids and nucleobases in interstellar and interplanetary radiation analogs conditions. The measurements have been undertaken at the Brazilian Synchrotron Light Laboratory (LNLS), employing vacuum ultraviolet (VUV) and soft X-ray photons. The experimental set up basically consists of a time-of-flight mass spectrometer kept under high vacuum conditions. Mass spectra were obtained using a photoelectron photoion coincidence technique. We have shown that the amino acids are effectively more destroyed (up to 70–80%) by the stellar radiation than the nucleobases, mainly in the VUV. Since polycyclic aromatic hydrocarbons have the same survival capability and seem to be ubiquitous in the ISM, it is not unreasonable to predict that nucleobases could survive in the interstellar medium and/or in comets, even as a stable cation.

Keywords. Methods: laboratory, molecular data, astrochemistry, astrobiology

---

1. Introduction

The search for amino acids and nucleobases (and related compounds) in the interstellar medium/comets has been performed over at least the last 30 years, but unfortunately it has yet to be successful (e.g., Brown *et al.* 1979, Simon & Simon 1973). However, recently some traces (upper limits) of the simplest amino acid, glycine (NH_2CH_2COOH), were observed in the comet Hale-Bopp (Crovisier *et al.* 2004) and in some molecular clouds associated with star forming regions (Kuan *et al.* 2003a), but these identifications have yet to be verified (Snyder *et al.* 2005, Cunningham *et al.* 2007). Despite no direct detection of nucleobases in comets or in molecular clouds, some of their precursor molecules like HCN, pyridines, pyrimidines and imidazole were been reported in the Vega 1 flyby of comet Halley (Kissel & Krueger, 1987) and have been searched for in the interstellar medium (Kuan *et al.* 2003b).

The search for these biomolecules in meteorites, on the contrary, has revealed an amazing number of proteinaceous and non-proteinaceous amino acids, up to 3 parts per million (ppm) (e.g., Cronin 1998 and references therein), and some purine and pyrimidine based nucleobases up to 1.3 ppm (e.g., Stocks & Schwartz 1981 and references therein). This dichotomy between the carbonaceous chondrites meteorites and interstellar medium/comets remains a big puzzle in the field of astrochemistry and in investigations about the origin of life.

The goal of this work is to review some experimental gas-phase photoionization and photodissociation studies of amino acids and nucleobases induced by vacuum ultra-violet (VUV) and soft X-ray photons, obtained recently by our group (Lago *et al.* 2004, Coutinho *et al.* 2005, Marinho *et al.* 2006, Pilling *et al.* 2007b). A possible direction on the different survivability of these biomolecules on astrophysical environments are given.

2. Experimental methodology and results

In an attempt to simulate the stellar/solar VUV and soft X-ray flux we have used synchrotron radiation as a light source. The experiments were performed at the Brazilian Synchrotron Light Laboratory (LNLS), employing harmonic free VUV photons (Cavasso-Filho *et al.* 2007) and soft X-ray photons from the toroidal grating monochromator (TGM) beamline. The emergent photon beam flux was recorded by a light sensitive diode. Briefly, the radiation ($\sim 10^{12}$ photons s^{-1}) from the beamline perpendicularly intersect the vapor-phase sample at the center of the ionization region inside a high vacuum chamber (Boechat-Roberty *et al.* 2005, Pilling *et al.* 2006). Mass spectra were obtained using the photoelectron photoion coincidence (PEPICO) technique (Pilling *et al.* 2007b, c and references therein).

In Figure 1 we have shown the time-of-flight mass spectra of the fragments released from the amino acid glycine and the nucleobase adenine, recorded at different photon energies over the VUV (12-21 eV) and soft X-ray (~ 150 eV) ranges. As a general rule, even at low photon energy, the amino acids have only a small contribution ($\lesssim 10\%$) or even were not detected, a consequence of their high photodestruction degree. The release of the carboxyl group (COOH) as a neutral or cationic species, depending on the amino acid, is one of the most import dissociation channels (see also Jochims *et al.* 2004). The nucleobases have shown a higher molecular stability in comparison with the amino acids. For these molecules, the parent ions remain relatively strong over all the studied VUV photon energy range. In all spectra, as the photon energy increases, the fragmentation profile also increases, as expected. Some minor contamination of water was observed in the spectra, which reflects the high hydrophilic character of the samples.

In previous studies of photodissociation of nucleobases (e.g., Jochims *et al.* 2005, Schwell *et al.* 2006) in the VUV photon energy range the authors have identified some important photodissociation channels as well as HNCO loss from thymine and uracil and HCN loss from adenine. As pointed out by Pilling *et al.* (2007b), in the case of

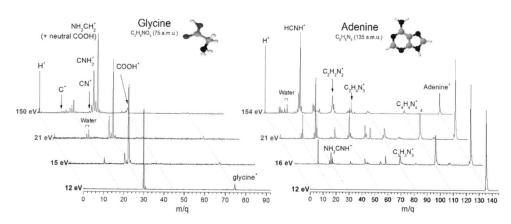

Figure 1. Time-of-flight mass spectra of gaseous amino acid glycine and nucleobase adenine recorded at different photon energies at VUV (12-21 eV) and at soft X-ray (~ 150 eV) ranges.

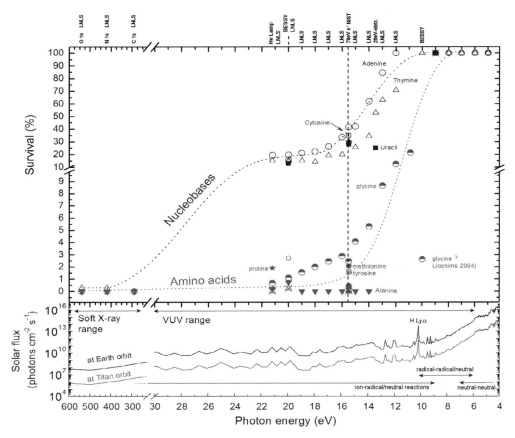

Figure 2. (Top) Comparison between the survival (photoresistence) of different amino acids and nucleobases due to ionizing radiation field in the VUV and soft X-ray. (Bottom) The solar photon flux at Earth and Titan orbit (adapted from Gueymard 2004). See details in text.

adenine, the neutral HCN may represent as much as 40% of its photodissociative channels. The ion $HCNH^+$ is another largely photoproduced fragment from both amino acids and nucleobases.

A comparison between the survival of different nucleobases and amino acids due to ionizing radiation field from 4 to 600 eV is given at Figure 2. Our data is represented by LNLS labels. The literature data were gather mainly from the NIST database and Jochims *et al.* (2004, 2005). The amino acids are effectively more destroyed by VUV stellar/solar radiation than the nucleobases. The higher resistance of nucleobases to the ionizing photons may be associated with the presence of the hetero-cyclic structure and unsaturated bonds.

For comparison, we also present in the bottom panel of Figure 2, the solar photon flux at ultraviolet and soft X-rays at Earth and Titan orbit (adapted from Gueymard 2004). The different photochemical domains are also given. According to the detailed review of Schwell *et al.* (2006), the photoabsorption cross section of these molecules is higher in VUV as compared to the mid-UV (< 6 eV). In particular, all molecules absorb strongly at 10.2 eV, where the intense Lyα (10.2 eV) stellar/solar emission is located. Most of the small biomolecules studied have first ionization energies (IE) below this energy, making photoionization phenomena an important issue to study. Since most of the amino acids has their first AE below 10.2 eV, the stellar/solar hydrogen Lyα has a great influence on their molecular survival. This is not observed in the case of nucleobases, which have their first AE at energies above the hydrogen Lyα energy, where the photon flux is about 2 orders of magnitude lower (in the case of Sun).

3. Conclusion

We have shown that the amino acids are effectively more destroyed by stellar radiation than the nucleobases, mainly in the VUV spectral range where the differences reach up 70–80%, decreasing to high-energy photon range, corroborating other experimental results given in the literature. The nucleobases are able to form a stable cation in the gas phase and since polycyclic aromatic hydrocarbons (PAHs) and polycyclic aromatic nitrogen-rich hydrocarbons (PANHs) have the same capability and seem to be ubiquitous in the ISM (Allamandola *et al.* 1986), it is not unreasonable to predict that aromatic nucleic acid bases could survive in the interplanetary and interstellar media.

These results lead us to ask an interesting question. Why don't we find nucleobases in cometary/molecular clouds radioastronomical observations since they are more resistant to stellar ultraviolet radiation than the detected amino acid (e.g., glycine)? Probably, the answer may be associated with the formation pathway efficiencies rather than with the detection limits and more studies of this subject are need.

Finally, a possible direction to the search of large pre-biotic molecules, as the case of amino acids and nucleobases or even for larger molecules, could be the search not for the molecules themselves, but from their photoproduced daughter species like, for example, the fragments $COOH^+$, $HNCO^+$, HCN^+, and $HCNH^+$. The abundances, derived from the radio-observation of these fragments, combined with the laboratory data (e.g., relative ion yield and photoproduction cross section) may give us a clue about the presence and the amount (upper limit) of their parent molecules.

References

Allamandola, L. J., Tielens, A. G. G. M., & Barrer, J. R. 1986, *ApJS*, 71, 733.
Boechat-Roberty, H. M., Pilling, S., & Santos, A. C. F. 2005, *A&A*, 438, 915.
Brown, R. D., Godfrey, P. D., Storey, J. W. V., Bassez, M.-P., *et al.* 1979, *MNRAS*, 186, 5P.
Cavasso-Filho, R. L., Homem, M. G. P., Lago, A., Pilling, S., & Naves de Brito, A. 2007, *J El. Spect. Rel. Phen.*, 156–158, 168.
Coutinho, L. H., Homem, M. G. P., Cavasso-Filho, R. L., Marinho, R. R. T., Lago, A.F., de Souza, G. G. B., & Naves de Brito, A. 2005, *Braz. J. Phys.*, 35, 940.
Cronin, J. R. 1998, in: A. Brack (eds.), *The Molecular Origins of Life*, (Cambridge University Press, UK)
Crovisier, J., Bockelée-Morvan, D., Colom, P., Biver, N., Despois, D., Lis, D.C., *et al.* 2004, *A&A*, 418, 1141.
Cunningham, M. R., Jones, P. A., Godfrey, P. D., *et al.* 2007, *MNRAS*, 376, 1201.
Gueymard, C. A. 2004, *Solar Energy*, 76, 423.
Jochims, H.-W., Schwell, M., Baumgartel, H., & Leach, S. 2005, *Chem. Phys.*, 314, 263.
Jochims, H.-W., Schwell, M., Chotin, J-.L., Clemino, M., Dulieu, F., Baumgärtel, H., & Leach, S. 2004, *Chem. Phys.*, 298, 279.
Kissel, J. & Krueger, F. R. 1987, *Nature*, 326, 755.
Kuan, Y.-J., Charnley, S. B., Huang, H.-C., Tseng, W.-L., & Kisiel, Z. 2003a, *ApJ*, 593, 848.
Kuan, Y.-J., Yan, C.-H., Charnley, S. B., Kisiel, Z., Ehrenfreund, P., & Huang, H.-C. 2003b, *MNRAS*, 345, 650.
Lago, A. F., Coutinho, L. H., Marinho, R. R. T., Naves de Brito, A., & de Souza, G.G.B. 2004, *Chem. Phys.*, 307, 9.
Marinho, R. R. T., Lago, A. F., Homem, M. G. P., Coutinho, L. H., de Souza, G. G. B., & Naves de Brito, A. 2006, *Chem. Phys.*, 324, 420.
Pilling, S., Andrade, D. P. P., Neves, R., Ferreira-Rodrigues, A. M., Santos, A.C.F., & Boechat-Roberty, H. M. 2007a, *MNRAS*, 375, 1488.

Pilling, S., Lago A. F., Coutinho, L. H., de Castilho R.B., de Souza, G. G. B., & Naves de Brito, A. 2007b, *Rapid Commun. on Mass Spectrom.*, 21, 3646.

Pilling, S., Neves, R., Santos, A. C. F., & Boechat-Roberty, H. M. 2007c, *A&A*, 464, 393.

Pilling, S., Santos, A. C. F., & Boechat-Roberty, H. M. 2006, *A&A*, 449, 1289.

Schwell, M., Jochims, H.-W., Baumgärtel, H., Dulieu, F., & Leach, S. 2006, *P&SS*, 54, 1073.

Simon, M. N., & Simon, M. 1973, *ApJ*, 184, 757.

Snyder, L. E., Lovas, F. J., Hollis, J. M., Friedel, D. N., Jewell, P. R., Remijan, A., Ilyushin, V. V., Alekseev, E. A. & Dyubko, S. F. 2005, ApJ, 619, 914.

Stocks, P. G. & Schartz, A. W. 1981, *Geochim Cosmochim. Acta*, 45, 563.

Discussion

KOBAYASHI: You irradiate with UV and X-rays. How many photons did you expose these molecules to, and how does this correspond to exposure times in space?

PILLING: Our flux is about 10^{12} photons per second in the ultraviolet. At the moment, I don't know how this corresponds to exposure time in the interstellar environment.

ZIURYS: Some of those products you proposed from your fragmentation are seen in the interstellar medium, for example, $COOH^+$ and $HCNH^+$. We have no information on some of the other molecules because we don't know their rest frequencies.

PILLING: We are proposing that if you combine the radio observations for molecules like $COOH^+$, together with the branching ratios from the amino acids from the lab, maybe we can trace some upper limits from these molecules.

ZIURYS: Yes, look at the relative abundances and look at the branching ratios. It's a good idea.

Mia Hajdukova and Sergio Pilling (photo by Mia Hajdukova).

Organic Matter in Space
Proceedings IAU Symposium No. 251, 2008
S. Kwok & S. Sandford, eds.

Formation of alcohols on ice surfaces

H. M. Cuppen, G. W. Fuchs, S. Ioppolo, S. E. Bisschop, K. I. Öberg, E. F. van Dishoeck, and H. Linnartz

Raymond and Beverly Sackler Laboratory for Astrophysics
Leiden Observatory, Leiden University
P.O. Box 9513, 2300 RA Leiden, the Netherlands
email: cuppen@strw.leidenuniv.nl

Abstract. As the number of detections of complex molecules keeps increasing, answering the question about their formation becomes more pressing. Many of the saturated organic molecules are found to have a very low gas phase formation rate and are therefore thought to be formed on the icy surfaces of dust grains. In the Sackler Laboratory for Astrophysics we started a systematic study of the surface reaction routes that have been suggested over the years. Here we present the experimental results on the formation of methanol and ethanol by hydrogenation reactions of carbon monoxide and acetaldehyde ice. Computer simulations of the surface processes under similar conditions using the continuous-time random-walk Monte Carlo technique reveal some of the underlying physical processes. A better understanding of the physical conditions in which these molecules are formed can help in the interpretation of the observational results. The CO hydrogenation results will appear in detail in Fuchs *et al.* (2008). For more details on ethanol formation we refer to Bisschop *et al.* (2007).

Keywords. Astrochemistry, methods: laboratory, techniques: spectroscopic, molecular processes

1. Introduction

Surface processes play an important role in many astrophysical processes. Abundant molecules, like hydrogen, water and methanol, are formed mainly through surface reactions. Complete chemical networks on icy grains have been suggested where atom-bombardment by H, C, O, or N atoms leads to complex organic molecules as depicted in Figure 1. These reaction schemes as proposed by Tielens & Hagen (1982) and Tielens & Charnley (1997) are now in reach of experimental approaches and in the Sackler Laboratory for Astrophysics we have recently started a systematic study of this network by checking the different surface reactions and measuring the corresponding rates. This contribution summarizes the results on the formation of methanol and ethanol by hydrogenation reactions of carbon monoxide and acetaldehyde, indicated by the boxes in Figure 1. For hydrogenation studies of CO_2 and $HCOOH$, we refer to Bisschop *et al.* (2007).

2. CO hydrogenation

Laboratory studies in which CO ice is exposed to atomic hydrogen have been performed independently by two groups (Hiraoka *et al.* 1994, Watanabe & Kouchi 2002). Hiraoka observed only formaldehyde formation, whereas Watanabe also found an effective methanol production. In a series of papers, these conflicting results have been discussed (Hiraoka *et al.* 2002, Watanabe *et al.* 2003, Watanabe *et al.* 2004) and the existing discrepancy

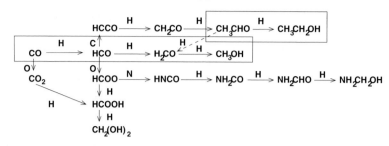

Figure 1. Surface reaction network (solid arrows) based on Tielens & Charnley (1997). The boxes indicate the reaction scheme presented in the present work, the dash arrow the newly found route.

has been explained by different experimental conditions, most noticeable the adopted H-atom flux (Hidaka *et al.* 2004). An experimental verification, however, has been lacking so far. In order to verify the origin of the conflicting results, an additional systematic study has been performed focusing on the physical dependencies that may affect CO ice hydrogenation schemes, in particular: surface temperature and H-atom flux. For this purpose, a specialized UHV set-up has been used in which a CO ice layer of controllable thickness is deposited on a gold substrate. Typical thicknesses of 10 monolayers (ML) are used. These layers are then exposed to an atomic hydrogen beam obtained by a thermal cracking source. The decay of CO and the formation of the products formaldehyde and methanol are monitored *in-situ* by means of Reflection Absorption InfraRed Spectroscopy (RAIRS). After three hours of H-atom exposure, an additional analysis technique, Temperature Programmed Desorption (TPD), is applied in which the sample is heated and the desorption of the species is recorded as a function of temperature using a mass spectrometer. The symbols in Figure 2 show the evolution of CO, H_2CO, and CH_3OH as a function of fluence and time for a surface temperature of 12 K and a relatively high exposure of 5×10^{13} H-atoms cm^{-2}s^{-1}. Here clearly both H_2CO and CH_3OH are formed, in agreement with Watanabe *et al.* (2004). A comparible experiment is performed with a flux similar to the one applied by (Hiraoka *et al.* 1994). At this very low flux and fluence, 1×10^{12} cm^{-2}s^{-1} and 1×10^{16} cm^{-2}, the formation of methanol cannot be confirmed by the RAIR spectra as its peak height is near the detection limit. However, the TPD spectrum shows a peak for mass 32 amu around the desorption temperature of methanol which is attributed to its formation. This leads to the conclusion that methanol is also formed at lower fluxes, although around the detection limit, and that the pricinple mechanism is not strongly dependent on the H flux.

3. Interpretation by Monte Carlo simulations

To interpret the results, a continuous-time, random-walk Monte Carlo simulation method was applied. This technique simulates a sequence of processes that can occur on the grain surface. These processes include hopping and desorption of the species and reactions between two species. It is a powerful tool to translate experimental data to interstellar conditions, since it can handle both the relatively high fluxes used in the laboratory and the low interstellar fluxes. It can also simulate a relatively large system over a long period of time, allowing for multiple processes to occur and to study their relative importance and interaction. In contrast with rate equation methods that are often applied, the method follows the individual atoms during a simulation, in this way one can consider the layering and topology of the system. For a detailed description of the method we refer to previous papers (Chang *et al.* 2005, Cuppen & Herbst 2005, Cuppen & Herbst 2007).

Here, two successive simulations are performed to simulate one experiment. First a CO layer is deposited starting from a bare surface and then during a second simulation this CO layer is exposed to hydrogen atoms. Input parameters for the simulations are hopping and reaction barriers and desorption energies. All processes except reactions are assumed to exhibit Arrhenius-like behavior where the barrier is crossed thermally. Some of the input parameters were varied in order to reproduce the experimental product evolutions. The solid lines in Figure 2 represent the simulated results that match the experiments. From our Monte Carlo analysis we can draw the following conclusions:

• the reaction rates show very little temperature dependence, indicating that tunneling through the reaction barrier is important,

• the production rate of methanol and formaldehyde decreases for increasing temperature, which is due to a higher hopping and desorption rate,

• the penetration depth of the hydrogen atoms into the CO ice is higher for higher surface temperatures, which is probably because of the higher mobility of the CO atoms in the ice.

The strong temperature dependence of the methanol formation rate may explain the large fluctuations in the astronomically observed methanol abundances.

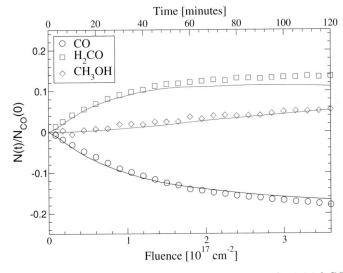

Figure 2. Evolution of CO, H_2CO and CH_3OH with respect to the initial CO-ice signal as a function of fluence and time. The symbols represent the experiment, the solid curves the results from the Monte Carlo simulations.

4. Ethanol formation

An additional series of experiments is performed in which acetaldehyde ice is exposed to H-atoms. According to the reaction scheme in Figure 1, ethanol should be formed in these experiments. However, in the RAIR spectra formaldehyde, methanol and methane were detected, suggesting that acetaldehyde breaks down into formaldehyde and methane upon hydrogen exposure. Formaldehyde is subsequently hydrogenated to methanol. No clear absorption is observed at 1050 cm^{-1}, where the strongest C_2H_5OH band, the C=O stretching mode, is expected. Since this frequency region is particularly problematic in our

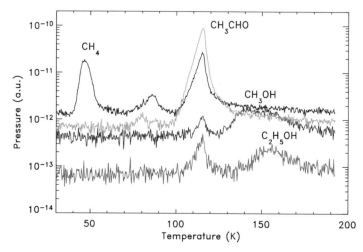

Figure 3. TPD spectra of acetaldehyde ice exposed to H-atoms 5×10^{17} cm^{-2} at 14.5 K. Methane, methanol and ethanol are formed. Taken from Bisschop *et al.* (2007).

detector, the detection sensitivity is low. Another strong band of ethanol is expected at 3.5 μm. Unfortunately, this feature overlaps with a number of methanol modes. Broad, weak features are indeed detected in this range, but due to the complexity of both CH$_3$OH and C$_2$H$_5$OH absorptions and the relatively weak signal this cannot be used to determine whether methanol is present. Fortunately, more information can be obtained by inspection of the TPD spectra as shown in Figure 3. These corroborate the formation of CH$_4$, H$_2$CO, and CH$_3$OH, which peak at 45 K, 100 K, and 140 K, respectively. In addition, a TPD desorption peak is located at $\sim$160 K for masses 45 and 46 amu. This is assigned to ethanol desorption based on a comparison with the TPD of pure non-bombarded acetaldehyde ices. In conclusion, a fraction of CH$_3$CHO, below the infrared detection limit of the 1050 cm^{-1} band, is converted to ethanol and a larger fraction forms formaldehyde, methanol, and methane. This experimentally confirms a pathway in the proposed reaction scheme by Tielens & Charnley (1997) and allows us to add another pathway. The latter is indicated by the dashed arrow in Figure 1: from CH$_3$CHO to CH$_2$O under the influence of atomic hydrogen.

References

Bisschop, S. E., Fuchs, G. W., van Dishoeck, E. F., & Linnartz, H. 2007, *A&A*, 474, 1061
Chang, Q., Cuppen, H. M., & Herbst, E. 2005, *A&A*, 434, 599
Cuppen, H. M. & Herbst, E. 2005, *MNRAS*, 361, 565
Cuppen, H. M. & Herbst, E. 2007, *ApJ*, 668, 294
Fuchs, G. W., Cuppen, H. M., Ioppolo, S., Bisschop, S. E., Andersson, S., van Dishoeck, E. F., & Linnartz, H. 2008, *A&A*, submitted
Hidaka, H., Watanabe, N., Shiraki, T., Nagaoka, A., & Kouchi, A. 2004, *ApJ*, 614, 1124
Hiraoka, K., Ohashi, N., Kihara, Y., Yamamoto, K., Sato, T., & Yamashita, A. 1994, *Chem. Phys. Lett.*, 229, 408
Hiraoka, K., Sato, T., Sato, S., Sogoshi, N., Yokoyama, T., Takashima, H., & Kitagawa, S. 2002, *ApJ*, 577, 265
Tielens, A. G. G. M. & Charnley, S. B. 1997, *Origins of Life and Evolution of the Biosphere*, 27, 23
Tielens, A. G. G. M. & Hagen, W. 1982, *A&A*, 114, 245
Watanabe, N. & Kouchi, A. 2002, *ApJ* (Letters), 571, L173

Watanabe, N., Nagaoka, A., Shiraki, T., & Kouchi, A. 2004, *ApJ*, 616, 638
Watanabe, N., Shiraki, T., & Kouchi, A. 2003, *ApJ* (Letters), 588, L121

Discussion

CECCARELLI: Did you try to see what happens if you change the substrate? In particular, did you try to see whether you get a different result if your substrate is already ices, like water-ices?

CUPPEN: We did some experiments with mixtures of ices. If you want to do this at low flux, I think you will really have a problem with detection. The work we've done this far was mostly in pure ices, but we did some work on mixtures of ices.

CECCARELLI: And did you find any differences?

CUPPEN: I will have to check and tell you later.

From left to right: Peter Sarre, In-Ok Song, Sun Kwok, Angela Speck, Steve Pointing (photo by Dale Cruikshank).

John Maier (left) and Svatopluk Civis (right).

Organic Matter in Space
Proceedings IAU Symposium No. 251, 2008
S. Kwok & S. Sandford, eds.

Simulation of organic interstellar dust in the laboratory

Walt W. Duley

Department of Physics and Astronomy, University of Waterloo,
200 University Ave. West, Waterloo, Ontario, Canada N2L3G1
email: wwduley@uwaterloo.ca

Abstract. New techniques for the generation and analysis of carbon nanoparticles (CNPs) have been developed and have resulted in the production of CNP samples whose infrared spectral properties are essentially identical to those observed in interstellar absorption and emission. These particles are of mixed aromatic/aliphatic composition. Spectra of CNPs containing 10^2–10^4 atoms will be discussed in relation to spectra of interstellar materials. We find that infrared line widths in these samples are typically 10–30 cm^{-1}, but can be as small as 2 cm^{-1}. Simulation of the 3.28 μm feature is shown to yield important new insight into the nature of interstellar CNPs.

Keywords. Infrared: ISM, ISM: molecules, ISM: lines and bands, (stars:) circumstellar matter

1. Introduction

What is "organic interstellar dust"? Although one tends to think of a particular material, the composition of "organic interstellar dust" is undoubtedly not unique as the observed properties of this material vary depending on location in the ISM. In the diffuse interstellar medium (DISM), organic dust appears to have a composition similar to that of hydrogenated amorphous carbon (HAC), while dust seen in emission seems to consist of carbon nanoparticles (CNPs) having mixed polycyclic aromatic hydrocarbon (PAH) and aliphatic composition. The morphology of these materials is indicated to be that of aggregates of CNPs forming larger particles with an enhanced aliphatic composition in the DISM, while material in emission objects consists of individual CNPs having fewer aliphatic chains. In this model, the aliphatic chains act as "glue" between aromatic components in CNPs. A hierarchical model in which HAC is decomposed in the vicinity of emission objects is indicated by observation.

Key information on the composition of organic interstellar material can be obtained from infrared spectra of many circumstellar, interstellar and extra-galactic sources as these are dominated by emission from PAHs (Allamandola *et al.* 1989, Peeters *et al.* 2002, 2004, van Diedenhoven *et al.* 2004, Sloan *et al.* 2005) The strongest emission features (historically called the unidentified infrared (UIR) bands) occur at 3.3, 3.4, 6.2, ∼7.7, ∼8.6, 11.3 and 12.7 μm and have been known for many years, but recent observations indicate that there are numerous other weaker bands (Werner *et al.* 2004, Sloan *et al.* 2005). Observations have shown that there are three basic types of emission spectra as indicated by the relative intensity and characteristic profile of features in the 3.3/3.4, 6-9 and 11-11.4 μm wavelength range (Geballe 1997, Tokunaga 1996, Peeters *et al.* 2002, van Diedenhoven *et al.* 2004). Adopting the notation of Peeters, most sources can be classified as type A, with a pronounced feature at 6.19-6.23 μm (1616-1605 cm^{-1}) together with bands at ∼7.6/7.8 μm (1316 / 1280 cm^{-1}) and ∼8.6 μm (1163 cm^{-1}). Type B sources have bands at 6.24-6.28, 7.6/7.9 and 8.7 μm (1603 / 1592, 1316 / 1265 and 1150 cm^{-1}).

A few type C objects (e.g., IRS 13416) exhibit features at ~6.3 and 8.2 μm (1587 and 1220 cm^{-1}) but do not have a band near 7.7 μm (1300 cm^{-1}). In practice, some sources combine type A and type B characteristics, and the observed 6.2 μm band often contains more than one component.

These emission features are indicative of the presence of aromatic hydrocarbons (Duley & Williams 1981, Leger & Puget 1984, Allamandola et al. 1989) and other bands, notably those in the 3.4–3.6 μm range, are characteristic of CH vibrations in sp^3-hybridized bonded hydrocarbons (Duley & Williams 1983). However, the composition of the compounds responsible for these bands is still uncertain despite numerous theoretical and laboratory studies (Scott et al. 1997, Hudgins et al. 1999, 2004, van Diedenhoven et al. 2004). This difficulty arises because the dominant observational bands at 6.2 and 7.7 μm do not appear at these wavelengths in laboratory or theoretical emission spectra of known PAH molecules. This suggests that the species responsible for the emission bands are either much larger molecules, or are radicals and radical ions derived from fully hydrogenated molecules through dehydrogenation and ionization in interstellar sources.

While the general features of these emission spectra are reproduced in laboratory spectra of hydrogenated amorphous carbon (HAC) (Scott et al. 1997, Grishko & Duley 2000, Duley et al. 2005), detailed assignments are precluded due to broadening in these solids. This broadening occurs because of the overlap of spectral features from a wide variety of molecular components. We have subsequently found a way to significantly reduce this broadening by using surface enhanced Raman spectroscopy (SERS) to sample highly-localized areas on the surface of carbon nanoparticles. Essentially this technique enables spectra to be obtained from molecular-sized units. An important aspect of SERS is that the technique samples both Raman and infrared-active vibrations (Etchegoin et al. 2003). As a result, spectra represent the full density of states and can therefore be directly compared with those emitted by large carbon molecules in astronomical sources.

2. Experimental

Thin films of amorphous carbon (a:C) and HAC were deposited at temperatures between 77 and 573 K on clean fused quartz substrates or polished p-type Si wafers with (100) orientation. A base pressure of 2×10^{-7} Torr (2.67×10^{-5} Pa) was obtained in a high vacuum chamber evacuated by a turbo-molecular pump with a liquid nitrogen trap. One mJ pulses from a Ti:sapphire laser ($\lambda = 800$ nm, pulse duration $\tau = 120$ fs, repetition frequency 500 Hz, incidence angle 45°) were used to ablate high purity pyrolytic graphite (99.99%). Sample microstructure was characterized with a scanning electron microscope equipped with a field emission gun and with a commercial atomic force microscope. In general, these films were found to be assembled from individual nanoparticles having sizes < 100 nm. Further details on the structure and properties of these materials can be found elsewhere (Hu et al. 2006, 2007). Film thickness was measured using a profilometer (AMBios XP-2) after deposition. The SERS technique requires that the sample be coated with a thin silver layer (Dieringer et al. 2006). To avoid possible contamination with organic solvents, the silver layer in these experiments was deposited in vacuum using laser ablation of Ag metal. Raman spectra were obtained using a Renishaw micro-Raman spectrometer with a He-Ne laser source at an excitation wavelength of 633 nm. The spectral resolution was 1 cm^{-1}. The morphology of these samples is as shown in Figure 1. These materials have a nano-assembled structure whereby carbon nano-particles (CNPs) combine to form a range of larger structures (Hu et al. 2007a, b).

Some advantages of the SERS technique are that it can be used to sample molecular-sized units and that the energies of transitions are accurately reproduced for molecules such as naphthalene. In addition, both infrared and Raman transitions appear in SERS

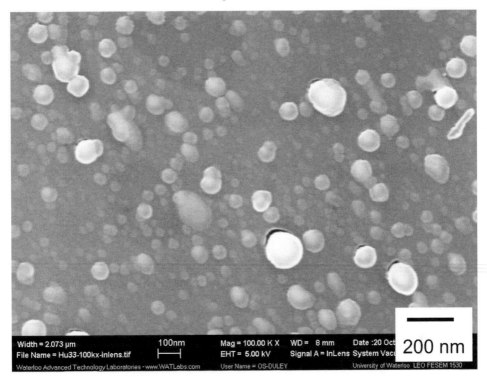

Figure 1. Nano-assembled structure of CNP sample. Individual particles have sizes of several nm. These combine to form larger aggregates.

spectra (Moskovits 1985). The linewidth of these transitions is determined by vibrational delocalization on timescales of 1-10 psec (10 cm^{-1}), similar to that which occurs in ISM spectra.

3. Laboratory spectra of CNPs and astronomical spectra

A summary of several spectral features seen in CNP samples is given in Figure 2. These originate from CH molecular groups present in individual carbon nanoparticles. Apart from the band at 3.26 μm, which corresponds to the stretching vibration of aromatic CH, all features have linewidths in the 10 cm^{-1} range comparable to those observed in interstellar spectra. The larger width of the 3.26 μm feature is consistent with rapid delocalization of excitation in the aromatic CH stretching band (Hu & Duley 2007, 2008). The SERS technique is especially sensitive to the presence of C=C bonds, so spectra tend to be enhanced at wavelengths where the vibrations of these species occur. Two low-resolution survey spectra (Figure 3) illustrate this point. These show strong peaks near 1600 and 1320 cm^{-1} (6.25 and 7.6 μm) that can be identified with vibrational modes of C=C bonds in aromatic ring and other hydrocarbon molecules (Ferrari & Robertson 2000). It is important to note that only the peaks near 7.6/7.8 μm are diagnostic of the presence of aromatic rings since the C=C vibration in other non-aromatic hydrocarbon molecules can also contribute to the 6.2 μm band. The features in Figure 3 all contain structure arising from individual molecular groups and are, themselves, superimposed on a broad continuum extending from ~1700 – 1100 cm^{-1} (5.9–9.1 μm) similar to that commonly observed in astronomical spectra (Allamandola *et al.* 1989, Peeters *et al.* 2002). The relative strength of the two main peaks in Figure 3 is found to be variable, but there is no obvious detailed correlation between the relative amplitude of the 6.2 and

7.6/7.8 μm peaks and preparation conditions. This suggests that both aromatic and non-aromatic molecules contribute to the band near 6.2 μm. There is also little difference in the wavelengths of these peaks in hydrogenated and de-hydrogenated samples. This result is expected since a range of ring compositions are present in all of the samples studied and ring frequencies in large molecules are little affected by hydrogenation. For example, the spectrum of coronene is strongly correlated with that of graphite (Mapelli *et al.* 1999). This suggests that the spectra reported here for molecular groups within clusters will be representative of individual PAH species.

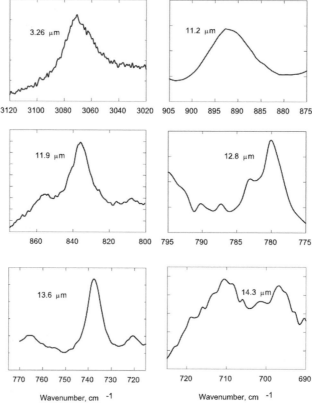

Figure 2. A selection of spectra of CNPs corresponding to the CH stretching and bending modes in PAH groups in these particles. The stretching band of CH appears at 3.26 microns, while other bands correspond to bending modes of CH in various configurations.

Spectra of three samples representative of A, B and C-type features in the 6.2 μm region are shown in Figure 4. Vertical lines show the average wavelength of the observed features in this range. None of these spectra are "pure" in the sense that they all show structure. The appearance of different profiles is then due to variations in the relative strength of these components. For example, the peak of the type C profile in Figure 4 appears as a shoulder in the type B spectrum. Measurements also indicate that the 6.2 μm feature extends from 6.06–6.45 μm (1650–1550 cm^{-1}) implying that the observation of type A, B, or C emission may simply reflect changes in the excitation and/or concentration of a limited number (4–5) of specific molecular structures. There is also evidence for another component at 5.98 μm (1671 cm^{-1}) in the type B spectrum (Figure 4) that could correspond to the weak emission feature observed at 2248 6.0 μm in the spectrum

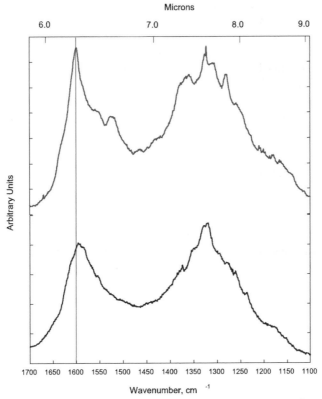

Figure 3. Low-resolution SERS spectra of samples in the 1700–1100 cm^{-1} (5.9–9.0 μm) range. Upper spectrum: nano-assembled film of amorphous carbon (a:C). Lower spectrum: nano-assembled film of HAC (a-C:H).

of HD44179 and other objects (Peeters *et al.* 2002). Emission at this wavelength is characteristic of non-aromatic C=C groups (Duley 2000), although it could also arise from an overtone and/or combination band involving ring vibrations (Allamandola *et al.* 1989).

Spectra of several CNPs samples in the region of the 3.3 μm aromatic CH stretching band are shown in Figure 5. All have a well-defined spectral feature that can be associated with the CH stretching band in PAH molecules. As expected, since these samples are recorded at 300 K, the wavelength of this feature is blue-shifted from the type A and type B interstellar emission bands. The FWHM of the laboratory feature in Figure 5a is virtually identical to that of the type A emission band (2248 37 cm^{-1}, $\sim$0.04 μm), while that in Figure 5b is somewhat narrower ($\sim$23 cm^{-1}, $\sim$0.025 μm). The wavelength of the aromatic CH band in our samples ranges between 3.27 μm (3057 cm^{-1}) in Figure 5a to 3.255 μm (3072 cm^{-1}) in Figure 5d. This variation arises because of detailed changes in the molecular composition of individual samples and reflects the fact that CNPs, as prepared in our experiments, are not identical. Nanoscale inhomogeneities in these samples can arise from variations in the size and substitution of PAH components, as well as from inclusion of different amounts of non-aromatic hydrocarbons.

A direct comparison between the profiles of the laboratory and interstellar aromatic CH bands (Hu & Duley 2008) yields a good fit to both the type A and type B spectra after shifts of -18 and -24 cm^{-1} ($+$ 0.019 and $+$ 0.026 μm), respectively, in the laboratory spectra. These shifts are consistent with the higher effective excitation temperature, T_{ism}, of emitters in interstellar sources (Joblin *et al.* 1995). From the measurements of Joblin

388

Walt W. Duley

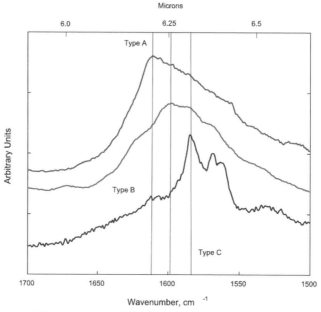

Figure 4. Spectra of three samples exhibiting a range of 6.2 μm profiles. Type A, B and C designations are from Peeters *et al.* (2002).

et al. (1995), the aromatic CH band is expected to shift by $\sim -0.033(T_{\rm ism} - T_{\rm lab})$ cm^{-1}. Then with $T_{\rm lab} = 300$ K, shifts of -18 and -24 cm^{-1} as required to fit the type A and type B profiles with the laboratory spectrum, correspond to $T_{\rm ism} \sim 850$ and 1025 K, respectively. These values are in agreement with estimated excitation temperatures for the 3.29 μm feature in interstellar sources.

The laboratory experiments of Joblin *et al.* (1995) predict that the width of the 3.29 μm feature should also increase with temperature, but this is not consistent with the fits shown Hu *et al.* (2008). This discrepancy likely occurs because the profile of the Raman line is broadened by delocalization (on a timescale of 1 psec) of the excitation centered on the CH stretching band, rather than as a result of thermal heating. Thus, the width of the Raman feature more closely reflects the processes occurring in the excitation of the interstellar emitters, resulting in a spectral width that differs from that of thermal broadening. In this context, we have previously found that line widths as small as several cm^{-1} are observed in SERS spectra of CNPs at longer wavelengths (Hu & Duley 2007, 2008). Narrowing of spectral lines is expected at long wavelengths because of a reduction in the rate of energy de-localization at lower vibrational energies.

Our laboratory data is in agreement with a model in which the type A band is the emission counterpart of the DISM absorption feature at 3.275 μm (Hu & Duley 2008). We reach this conclusion based on the following information: i) SERS spectra of CNPs at 300K show an aromatic CH band at 3.27 μm (3057 cm^{-1}) only -0.0043 μm ($+4$ cm^{-1}) away from the DISM absorption feature, ii) the profile of this feature closely matches that of the type A emission peak, iii) thermal emission spectra of our samples show that the CH peak occurs at 3.294 $\pm$ 0.002 μm (3036 $\pm$ 2 cm^{-1}) at 800 K (Grishko & Duley 2000). CNP samples that exhibit type A bands in our spectra also contain an aliphatic component in the form of chains bonded to PAH structures (Hu *et al.* 2007b), suggesting that PAHs in the DISM likely have aliphatic side-chains. A model in which CH groups on PAHs are replaced by aliphatic side-chains in diffuse clouds is consistent

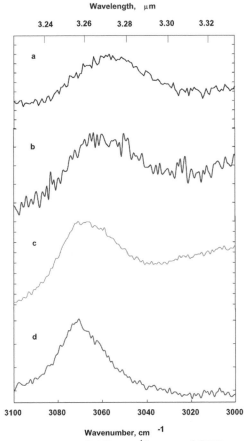

Figure 5. SERS spectra in the 3.29 μm (3040 cm^{-1}) range of CNP samples at 300 K. Individual peaks correspond to excitation of the CH stretching mode in aromatic (PAH) molecular groups. Other data suggests that the average size of the PAH structures in these samples decreases from spectrum (a) to spectrum (d). PAHs in sample (d) may contain as few as one or two C6 rings.

with the observation that the 3.4 μm absorption band from CH$_2$ and CH$_3$ groups is much stronger than the aromatic CH band in DISM material (Dartois *et al.* 2007) and that the 3.29 μm CH emission feature is very weak under these conditions (Rouan *et al.* 1999).

It is notable that a broad absorption feature, likely attributable to another type of PAH, is observed at 3.255 μm (3072 cm^{-1}) in spectra of molecular clouds (Sellgren 2001). We see a feature at this wavelength in many of our CNP samples containing small rings (Figure 5c-d) and associate it with benzene, naphthalene and anthracene molecules having aliphatic side-chains. It is possible that small PAH molecules of this type are only present in these heavily obscured objects where they are shielded from UV radiation.

4. Conclusions

Laboratory spectra of CNPs reproduce many of the infrared emission features, including their spectral widths. The agreement between the specral linewidths observed in SERS spectra and that seen in interstellar emission occurs because the Raman process results in similar internal vibrational energy excitation. Spectral shifts required to fit the emission peak of the 3.29 μm band is consistent with vibrational heating to $\sim$ 800–1,000 K. This suggests that the DISM absorption band at 3.275 μm arises from the same chemical species as that which produces type A emission in many astronomical sources.

These emitters are indicated to be relatively small (2-7 ring) PAH molecules with one or more attached aliphatic side-chains.

This research was supported by a grant from the NSERC of Canada. The author thanks Anming Hu for providing data and Q. B. Lu for the use of laser facilities.

References

Allamandola, L. J., Tielens, A. G. G. M., & Barker, J. R. 1989, *ApJS*, 71, 733

Dartois, E. 2007, *A&A*, 463, 635

Dieringer, J. A. 2006, *Faraday Disc.*, 132, 9

Duley, W. W. 2000, *ApJ*, 528, 841

Duley, W. W. & Williams, D. A. 1981, *MNRAS*, 196, 269

Duley, W. W. & Williams, D. A. 1983, *MNRAS*, 205, 67P

Duley, W. W., *et al.* 2005, *ApJ*, 626, 923

Etchegoin, P., *et al.* 2003, *Chem. Phys. Lett.*, 367, 223

Ferrari, A. C. & Robertson, J. 2000, *Phys. Rev. B*, 61, 14095

Geballe, T. R. 1997, in: Y. J. Pendleton & A. G. G. M. Tielens (eds.), *From Stardust to Plan- etisimals*, ASP Conf. Ser. 122, (San Francisco:ASP), p. 109

Grishko, V. & Duley, W. W. 2000, *ApJ (Letters)*, 543, L85

Hu, A., Alkhesho, I., Duley, W. W., & Zhou, H. 2006, *J. Appl. Phys.*, 100, 084319

Hu, A., Alkhesho, I., Zhou, H., & Duley, W. W. 2007a, *Diamond and Related Materials*, 16, 149

Hu, A. & Duley, W. W. 2007, *ApJ*, 672, L81

Hu, A. & Duley, W. W. 2008, *ApJ*, in press

Hu, A., Lu, Q-B., Duley, W. W., & Rybachuk, M. 2007b, *J. Chem. Phys.*, 126, 154705

Hudgins, D. M. & Allamandola, L. J. 1999, 513, L69

Hudgins, D. M. & Allamandola, L. J. 2004, in: A. N. Witt, G. C. Clayton, & B. T. Draine (eds.), ASP Conf. Ser. 309, *Astrophysics of Dust*, (San Francisco:ASP), p. 665

Joblin, C., Boissel, P., Leger, A., d'Hendecourt, L., & Defourneau, D. 1995, *A&A*, 299, 835,

Leger, A. & Puget, J. L. 1984, *A&A (Letters)*, 137, L5

Mapelli, C., Castiglione, C., Zerbi, G. & Mullen, K. 1999, *Phys. Rev. B*, 60, 12710

Moskovits, M. 1985, *Rev. Mod. Phys.*, 57, 78

Peeters, E., *et al.* 2002, *A&A*, 390, 1089

Peeters, E., Mattioda, A. L., Hudgins, D. M. & Allamandola, L. J. 2004, *ApJ (Letters)*, 617, L65

Rouan, D., Le Coupanec, P., Lacombe, F., Tiphene, D., Gallais, P., Leger, A., & Boulanger, F. 1999, in *The Universe as seen by ISO*, ESA, SP-427, p. 743

Scott, A., Duley, W. W., & Jahani, H. R. 1997, *ApJ (Letters)*, 490, L175

Sellgren, K. 2001, *Spectrochimica Acta*, A57, 627

Sloan, G. C., *et al.* 2005, *ApJ*, 632, 956

Tokunaga, A. T. 1996, in: H. Onaka, T. Matsumoto & T. L. Roelig (eds.), *Diffuse Infrared Radiation and the IRTS*, ASP Conf. Ser. 124, (San Francisco:ASP), p. 149

van Diedenhoven *et al.*, 2004, *ApJ*, 611, 928

Werner, M. W., *et al.* 2004, *ApJS*, 154, 309

Discussion

MULAS: You compared your spectrum of the 3.3 μm band with the interstellar one, applied the shift due to temperature, and showed that you still have a small residual. I'd note that in the interstellar medium you will also have overtones, and they will be exactly where you have the residual.

DULEY: Yes, that's the possibility.

MULAS: Also, in one of your slides you showed the absorption of the 3.3 μm aromatic band in a molecular cloud. Could an alternative explanation be simply that, in that case, you have a molecule that is at very low temperature, so you see the reverse of this shift?

DULEY: It's too much of a shift. It has to be a different species.

HENNING: These are very nice data in the infrared. Do you have any absorption spectra for the optical or UV to constrain the absorption spectrum?

DULEY: I have a new student who's going to do a lot of work in these other regions.

SMITH: Have you taken a look at the 3.4 μm feature?

DULEY: Yes, we have a lot of data on that, but we haven't quite sorted out what is going on. There are a few surprises there that I think will be quite interesting. We've been working our way through the spectrum, and taking as it comes ...

SARRE: One of the key observational aspects is the strength of bands in the 5 to 6 μm region. Can the laboratory data explain why the C-C modes are strong in some regions, and not in others?

DULEY: Well, for the C-C modes it depends on the wavelength. They can shift, and if they are not rings, they can really shift quite a bit. Keep in mind that you can have C-C modes but not have C-H modes if you don't have a high degree of hydrogenation. If you have a large particle that just has a lot of rings, then the C-C modes can be quite strong, and the other bands that are associated with hydrogen are really reduced. You can vary that ratio by changing the level of hydrogenation in the molecule.

SARRE: Is that really seen in the lab?

DULEY: Yes, we do see variations in composition. The composition depends on the gas in which samples are prepared.

ZIURYS: Did you ever do experiments where you add a little bit of oxygen or nitrogen to produce heteroatom structure?

DULEY: Yes, but it's not actually that easy to get these heteroatoms in there, particularly the nitrogen. We've tried, and we've got some data, but it's not something I am comfortable with. They don't seem to like to get into these molecules under our experimental conditions. There may be a channel that produces this in some special way, but so far we haven't really found it.

ZIURYS: How do you ablate your graphite? Do you use a graphite rod?

DULEY: We use highly oriented pyrolytic graphite. It's basically layers of graphene.

ZIURYS: Did you ever think of drilling a hole though it and running in oxygen and nitrogen?

DULEY: Yes, but it doesn't work well. It will oxidize around where you are heating, but it doesn't get into the product very easily.

Walt Duley (left) and Alan Tokunaga at coffee.

Organic Matter in Space
Proceedings IAU Symposium No. 251, 2008
S. Kwok & S. Sandford, eds.

© 2008 International Astronomical Union
doi:10.1017/S1743921308022023

Laboratory analogues of hydrocarbonated interstellar nanograins

T. Pino[1], A. T. Cao[1], Y. Carpentier[1], E. Dartois[2], L. d'Hendecourt[2], and Ph. Bréchignac[1]

[1]Laboratoire de Photophysique Moléculaire, UPR 3361, Université Paris-Sud,
91405 Orsay Cedex, France
email: philippe.brechignac@u-psud.fr

[2]Institut d'Astrophysique Spatiale, UMR-8617, Université Paris-Sud,
91405 Orsay Cedex, France

Abstract. Carbonaceous extraterrestrial matter is observed in a wide variety of astrophysical environments. The spectroscopic signatures revealed a large variety of chemical structure illustrating the rich carbon chemistry that occurs in space. In order to produce laboratory analogues of carbonaceous cosmic dust, a new chemical reactor has been built in the Laboratoire de Photophysique Moléculaire. It is a low pressure flat burner providing flames of premixed hydrocarbon / oxygen gas mixtures, closely following the model system used by the combustion community. In such a device the flame is a one-dimensional chemical reactor that offers a broad range of combustion conditions and sampling which allows production of many and various byproducts. In the present work, we have studied: i) the infrared transmission spectra of thin film deposit samples whose nature ranges from strongly aromatic to strongly aliphatic materials; ii) the resonant two-photon photoionisation spectra of gas phase PAHs formed in the flame.

Keywords. Astrochemistry, methods: laboratory, ISM: dust, molecules

1. Introduction

The ubiquitous presence of complex organic matter in many and various astrophysical environments is a relatively recent observational fact. More specifically, carbon chains, large aromatic-like clusters or molecules (PAHs), hydrogenated amorphous carbon, and diamond are materials that are directly observed or thought to be components of the interstellar medium (ISM). Most importantly it is now widely accepted that interstellar "nanograins" having sizes intermediate between those of the PAHs easily accessible to laboratory studies and those of the "standard" interstellar grains (ca 50 nm) are ubiquitous "free-flyers" in ISM.

These nanograins await characterization of their intimate structures, therefore laboratory analogues of these particles have to be spectroscopically characterized in the gas phase. This is the goal of the new experimental set-up, recently built at the Laboratoire de Photophysique Moléculaire, devoted to the production and characterization of such carbonaceous nanoparticles using a hydrocarbon-rich, premixed, flat, low-pressure, hydrocarbon / oxygen flame. The mass distribution of products in the reactor is monitored by a high resolution Time-Of-Flight spectrometer, after sampling the flame through a quartz cone. The spectroscopic properties can be measured both in absorption in solid phase, using thin films deposits, and in the gas phase using laser techniques. The results obtained so far from such measurements have revealed a strong evolution from aromatic materials containing aliphatic substituents to large polymer-like soot particles.

these materials are not truly catalysts. The properties of these surfaces change during the course of reaction and become more efficient as the reaction proceeds to build up a macromolecular grain coating that would usually serve to shut down such activity (Johnson *et al.* 2007). Indeed amorphous iron silicate smokes that had accumulated a coating comprising 10% by mass carbon and 0.2% by mass nitrogen based on the total mass of the sample, remained an active and very efficient surface for production of nitrogen-bearing organic materials from a mixture of CO, N_2 and H_2. More recent work may provide a simple explanation for these observations: the carbonaceous grain coating is itself an efficient surface for the reduction of CO and N_2 by hydrogen to form a variety of organic materials.

2. Experimental Description

Llorca & Casanova (2000) demonstrated that FTT reactions occur under low pressures typical of the primitive solar nebula. Our experiments were designed to produce mixtures of solids and organics that could serve as analogs of primitive asteroidal material. Grains in protostellar nebulae are exposed to the ambient gas for hundreds or even tens of thousands of years at pressures ranging from 10^{-3} to 10^{-4} atm or less. We do not have such times available for laboratory experiments, although we can duplicate the total number of collisions a grain might experience with components of the ambient gas by running experiments for shorter times at higher pressures. In our laboratory, experiments last from about 3 days at temperatures of 873 K to more than a month at temperatures of 573 K. If an average experiment lasts a week (6.05×10^5 s) then we can simulate two centuries (6.3×10^9 s) of exposure to an ambient gas at 10^{-4} atm. by running experiments at ~1 atmosphere total pressure. Although these higher pressures could conceivably effect the products synthesized in our experiments, we believe that the effects of temperature are much more important.

The experiments were very simple (see Figure 1 and Hill & Nuth 2003). We load ~25 cm^3 of catalyst into a glass finger through which gas can circulate by means of a glass tube that extends to the bottom of the finger. The finger is heated via an external mantle to a controlled temperature. We evacuate the system to a pressure less than ~0.1 Torr, then fill the system with gas (75 Torr CO, 75 Torr N_2, 550 Torr H_2) to a total pressure of 700 Torr. We then begin to circulate gas using a bellows pump, begin heating the finger containing the catalyst and record our first infrared spectrum of the gas (only CO is detected in this spectrum) using an FTIR spectrometer. The gas fluidizes the catalyst. The finger is plugged at the top of the heater using glass wool to contain the grains while letting the gas circulate. As the experiment proceeds we use periodic FTIR spectra to follow the loss of CO and the formation of methane, water, and carbon dioxide, and monitor smaller spectral features due to ammonia and N-Methyl Methylene Imine. Once the CO has been reduced to about ten percent of its starting concentration we take a final infrared spectrum, turn off the heater, cool the system to room temperature, evacuate it to less than ~0.1 Torr, then refill the system with fresh gas and begin a second run. Note that we do not use a fresh batch of catalyst for this second run. By making ~15 runs with the same catalyst, we simulate ~3,000 years of exposure of grains to nebular gas and build up a substantial coating of macromolecular carbon, nitrogen, and hydrogen.

3. Results

Figure 2a shows loss of CO with time for runs at ~873 K. The CO decays more slowly in the first run than in subsequent runs. The generation of methane as a function of time and run number is shown in Figure 2b; again, the rate is slower in the first run, but gets

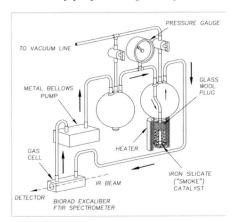

Figure 1. Simple experimental apparatus used to circulate reactive gas mixtures over potential catalysts at controlled temperatures and monitor the changes in the circulating gas vie infrared spectroscopy.

faster for subsequent runs. All gas phase products, with the exception of CO_2, follow the pattern set in Figure 2b by methane in experiments carried out at 573 K, 773 K and 873 K. All of our experimental runs followed this same pattern, with a much slower rate of change at lower temperatures, but a general increase in reaction rate after the catalyst was first exposed to the reactive gas. With more efficient catalysts and higher temperature experiments, the system quickly achieves the final "steady-state" reaction rate.

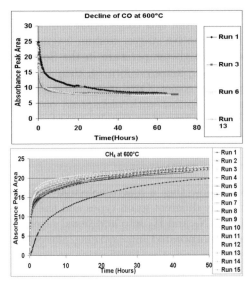

Figure 2. Top; changing abundances of CO as a function of time in four separate runs using the same catalyst. Bottom; changing abundances of methane in successive experimental runs at 873 K using the same catalyst.

For a typical, textbook catalyst, this result is counterintuitive. With each additional run, the catalyst forms slightly larger clumps (thus reducing surface area), the active metal atoms at the surface become more oxidized due to reaction with water generated

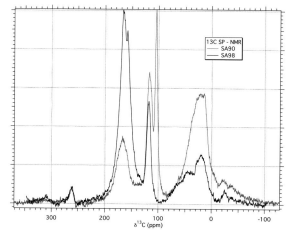

Figure 2. Single pulse ^{13}C NMR spectra of 2 tholins with different elemental compositions. Aliphatic, cyanides, and triazine and/or imines are identified as major compounds in these spectra.

insoluble fractions, and small molecular compounds appear to be of help in understanding the formation process and to identifying the fundamental building blocks of the tholins. Analysis of the soluble fraction appears consistent with the NMR results, as unsaturation is correlated with the presence of nitrogen. The key role of nitrogen in determining the unsaturation index is observed in carbon nitrides films studied in material science, as this atom favors clustering of sp^2 bondings. FT-ICR results also suggest a random accretion of specific molecular groups around well identified centers, consistent with the chemical selectivity pointed out by NMR and the disordered character of the solid evidenced by Raman spectroscopy and XRD.

3. Are tholins relevant analogs?

One of the key question about tholins is their relevancy for simulating extraterretrial organic solids. Because their chemical structure is far from being determined, and that of planetary organics cannot be fully assessed by spectroscopic techniques, no comparison can be achieved. Hence, the relevancy of tholins - as any other analogs like amorphous carbon, soots, etc. – should be defined as their ability to match astronomical data.

In the case of application to understanding Titan's atmosphere, some laboratory-produced tholins have optical constants in the visible range that are very similar to those of the planetary aerosols (Khare *et al.* 1984, Imanaka *et al.* 2004). The physical mechanisms controlling the visible absorption are electronic transitions, between π and π^* states forming valence and conduction bands (Daigo & Mutsukura 2004). There is however no simple or direct relationship between the electronic structure and the chemical

Table 1. Summary of chemical groups in tholins, as identified, lacking/minor or suspected species.

Identified	Minor or lacking	Suspected
Amines	Isocyanide	Hydrazoles
Cyanide	Unsaturated carbons	
CH$_2$/CH$_3$	Other azines	
Triazine, highly branched	Azoles	
Imines	Carbodiimide	

composition/structure. In other words, different compounds may exhibit similar absorption properties. This has been demonstrated by Tran *et al.* (2003), who show that various photolysis products exhibit very similar optical constants to those of the tholins of Khare *et al.* (1984), although they have very different chemical structures and compositions. The tholins are best considered as a class of synthetic hydrogenated carbon nitrides, which should be used in astrophysics like other analogs like amorphous carbon, hydrogenated amorphous carbons, soots, etc. (see de Bergh *et al.* 2008 for a review), but without any observational evidence they are "chemically pertinent".

Interestingly, the study of cosmomaterials available in the laboratory (e.g., cometary grains) does not suffer from this spectroscopic artifact. Their characterization takes advantage of a large array of micro-analytical techniques, which allow comparison of direct chemical and structural information. The presence of tholins or tholin-like compounds is adressed in the following section in the case of cometary dust.

4. N-rich organics in cometary dust

The presence of N-rich refractory organics in cometary grains and Insoluble Organic Matter (IOM) extracted from pristine carbonaceous chondrites has been evidenced by SIMS imagery (Aleon *et al.* 2003). N abundances up to 20 wt% have been recorded on tiny areas within these samples ($\sim \mu m$). HCN polymers have been proposed as analogs of such compounds, regarding their elemental composition and the presence of HCN in cometary ices. Studies of tholins as presented in the first section have also focused on HCN polymers. Both families of compounds exhibit structural and spectroscopic similarities, though they are not identical. This suggests that tholins may also be used as analogs, along with HCN polymers, for IOM and other N-rich refractory organics. Tholins are generally chemically and optically more homogeneous at the micrometric scale, they can be produced under reproducible conditions, and they are likely more stable with time than HCN polymers. Furthermore, depending upon experimental compositions, they can exhibit a much broader range of chemical variations, hence providing a large set of analogs and calibration standards.

There is to date no direct evidence of the presence of HCN polymers or similar carbon nitrides in cometary grains. The characterization of cometary grains or IOMs at the micrometric scale can be achieved by Raman and IR microspectroscopies. 244 nm UV Raman micro-spectroscopy, which was used for studying tholins, has been applied to 2 stratospheric IDPs and a series of IOMs extracted from various carbonaceous chondrites (Murchison, Cold Bokkeveld, Murray, Orgueil, Alais, Tagish Lake, Renazzo). The first-order carbon bands in spectra of IOMs look quite similar, with a narrow and intense G band and a weak D band. For all chondrites studied, no HCN polymers or similar compounds were detected. In the Alais chondrites, the cyanide function has been identified thanks to a very faint band at 2225 cm^{-1} (Figure 3).

The Raman spectra of IDPs were acquired on whole particles without prior demineralization. The most C-rich IDP (L2008X3) exhibited Raman spectra with a useful signal to noise ratio, which all contained the first-order carbon bands at each measurement points (Figures 3 and 4). The shape of these bands are different from either chondritic IOMs or HCN polymers and tholins. They do not appear as a broad feature pointing to an amorphous structure, but rather as two distinct G and D bands consistent with a polyaromatic structure, the former being broader than in IOMs and the second more intense. These measurements are consistent with systematic trends observed in 514 nm Raman spectra (Figure 5), acquired on series of IDPs and carbonaceous chondrites (Quirico *et al.* 2005, Quirico *et al.*, in preparation). There are several ways for interpreting such data and

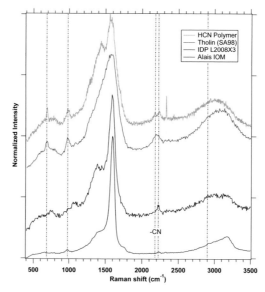

Figure 3. 244 nm Raman spectra of Alais (CI1) IOM and a HCN polymer.

complementary measurements are required to go further. Note that similar differences have been recently reported between carbonaceous chondrites and Antarctic micro meteorites (AMMs) (Dobrica *et al.* 2008).

Another significant difference with chondritic IOMs is the presence of spots exhibiting a strong -CN band (2225 cm^{-1}) (Figure 3). Such an observation is obviously consistent with the N-rich areas evidenced by SIMS imaging. These results demonstrate that 244 nm UV Raman spectroscopy is a powerful imaging technique for localizing N-rich organics, and systematic surveys on series of IDPs and AMMs are now in progress. Some N-rich areas are definitely not HCN polymers. More extensive surveys on chondritic IOMs, stratospheric IDPs, and AMMs, combined with SIMS imagery, are under way and these should provide new insights on the presence of tholin-like or HCN polymers.

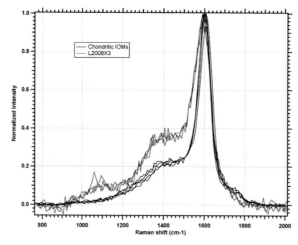

Figure 4. 244 nm Raman spectra of stratospheric IDP L2008X3 and chondritic IOMs. Note the broader G band and the more intense D band.

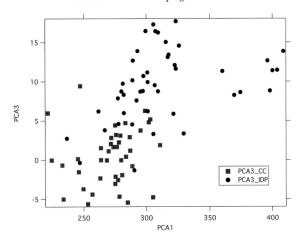

Figure 5. PCA analysis of 514 nm Raman spectra presented in Quirico *et al.* (2005). Carbonaceous chondrites and IDPs form distinct groups, suggesting structural differences in their polyaromatic networks.

References

Aleon, J., Robert, F., Chaussidon, M., & Marty, B. 2003, *Geochim. Cosmochim. Acta*, 67, 3773

Bernard, J-M, Quirico, E., Brissaud, O., Montagnac, G., Reynard, B., McMillan, P. F., Coll, P., Nguyen, M-J, Raulin, F., & Schmitt B. 2006, *Icarus*, 185, 301

Coll, P., Coscia, D., Gazeau, M-C, Guez, L., & Raulin, F. 1998, *Origin of life and evolution of the biosphere*, 28, 195–213

Cruikshank, D. P., Imanaka, & Dalle Ore, C. M., 2005, *Advances in Space Research*, 36, 178–183

Daigo, Y. & Mutsukura, N. 2004, *Diam. Rel. Mat.*, 13, 2170–2173

de Bergh, C., Schmitt, B., Moroz, L. V., Quirico, E & Cruikshank, D. P. 2008, in *The Solar System Beyond Neptune*, 483–506

Derenne, S., Quirico, E., Szopa, C., Cernogora, G., Schmitt, B., Lees, V., & McMillan P. F. 2008, *LPSC Meeting*, Abstract 1391

Dobrica, E., Engrand, C., Quirico, Montagnac, G., & Duprat J. 2008, *Meteor. and Plan. Sci.*, Meteoritical Society Meeting, Matsue, Japan, Abstract 50202

Ferrari, A. C., Rodil, S. E., & Robertson, J. 2003, *Phys. Rev. B*, 67, 155306

Imanaka, H., Khare, B. N., Elsila, J. E., Bakes, E. L. O., McKay, C. P., Cruikshank, D. P., Sugita, S., Matsui, T., & Zare, R. N. 2004, *Icarus*, 168, 344

Khare, B. N., Sagan, C., Thompson, W. R., Arakawa, E. T., Suits, F., Callcott, T. A., Williams, M. W., Shrader, S., Ogino, H., Willingham, M. W., & Nagy, B. 1984, *Adv. Space Res.*, 4, 59

Mutsukura, N. & Akita, K-I. 1999, *Thin Solid Films*, 349, 115–119

Quirico, E., Borg, J., Raynal, P-I, Montagnac, G., & d'Hendecourt, L. 2005, *Plan. Spac. Sci.*, 53, 1443–1448

Quirico, E., Montagnac, G., Lees, V., McMillan, P. F., Szopa, C., Cernogora, G., Rouzaud, J-N, Simon, P., Bernard, J-M, Coll, P., Fray, N., Minard, R. D., Raulin, F., Reynard, B., & Schmitt, B., *Icarus*, submitted

Sarker, N., Somogyi, A., Lunine, J., & Smith M. A. 2003, *Astrobiology*, 3, 719

Tran, B. N., Joseph, J. C., Ferris, J. P., Persans, P. D., & Chera, J. J. 2003, *Icarus*, 165, 379–390

Zhang, Z., Leinenweber, K., Bauer, M., Garvie, L. A., McMillan, P. F., & Wolf, G. H. 2004, *J. Am. Chem. Soc.*, 123, 7788–7796

Discussion

MATTHEWS: You have made a direct comparison between a sample of our HCN polymer and various tholins. What are your conclusions about the existence of HCN polymer in the tholins?

QUIRICO: They are different compounds, but there are some striking similarities. I refer back to the NMR results, which point to the presence of similar functional groups. The main similarity lies in a peak that we assign as imines and/or triazine functional groups, the presence of cyanide, and the lack of any unsaturated carbons (aromatics, etc.). Nevertheless, in both NMR and infrared data, many bands are narrower. Overall, I would say that though there are some similarities in the chemical composition, HCN polymer are chemically simpler than tholins.

ZINNER: I want to make a comment to remove some confusion you might have caused by referring to areas with high nitrogen concentrations as 'hot spots.' The 'hot spots' discussed yesterday in various isotopic talks were regions which have large isotopic anomalies – specifically for nitrogen these are regions with high ^{15}N excesses, but not an enhancement in the nitrogen concentration.

QUIRICO: The main point I want to make is that there are some areas in the stratospheric dust that show very localized nitrogen concentrations.

NITTLER: You did some component analysis and showed a plot that discriminates between the IDPs and the chondrites, but you didn't discuss what the actually components are. Is there any physical reality to the components the analysis gives you?

QUIRICO: They do not have any physical reality. This just provides a means to characterize the spectral variation, which is more sensitive than current fits using two bands.

CODY: UV Raman spectroscopy gives you enormous signal enhancement, but it can also cause some photochemistry. Are you concerned about photochemical alteration caused by exposure of the IDP organics or the meteoritic organics to your UV laser?

QUIRICO: The issue of photostability is a major issue in Raman spectroscopy. There are four major problems, whatever the wavelength excitation: heating (reversible); annealing; photooxidation and photolysis. We used considerable care to minimize those effects, by using a rotating sample holder, and performing time-resolved measurements at the same spot location. Though we cannot exclude the full lack of sample alteration, those effects were strongly minimized in our experiences.

Organic Matter in Space
Proceedings IAU Symposium No. 251, 2008
S. Kwok & S. Sandford, eds.

© 2008 International Astronomical Union
doi:10.1017/S1743921308022060

Quenched carbonaceous composite (QCC) as a carrier of the extended red emission and blue luminescence in the red rectangle

S. Wada[1], Y. Mizutani[1], T. Narisawa[2], and A.T. Tokunaga[3]

[1]Dept. of Applied Physics and Chemistry, Univ. of Electro-Communications,
Chofugaoka, Chofu, Tokyo 182-8585, Japan
email: wada@pio.jp

[2]Center for Instrumental Analysis, Univ. of Electro-Communications,
Chofugaoka, Chofu, Tokyo 182-8585, Japan

[3]Institute for Astronomy, Univ. of Hawaii,
2680 Woodlawn Dr., Honolulu, HI 96822
email: tokunaga@ifa.hawaii.edu

Abstract. Filmy-QCC is an organic material synthesized in the laboratory, and it exhibits red photoluminescence (PL). The peak wavelength of the PL ranges from 650 to 690 nm, depending on the mass distribution of polycyclic aromatic hydrocarbon (PAH) molecules, and the emission profile is a good match for that of the extended red emission in the Red Rectangle nebula. The quantum yield of the PL ranges from 0.009 to 0.04. When filmy-QCC is dissolved in cyclohexane, it exhibits blue PL in the wavelength range of 400–500 nm with a quantum yield of 0.12–0.16. The large width of the red PL and the large wavelength difference between the PL of the filmy-QCC as a solid film and in a solution indicate that there is a strong interaction between the components of filmy-QCC. The major components of filmy-QCC are PAHs up to 500 atomic mass units. Our laboratory data suggest that the blue luminescence observed in the Red Rectangle nebula is probably caused by small PAHs in a gaseous state, and the extended red emission is caused by larger PAHs in dust grains.

Keywords. Methods: laboratory, astrochemistry, (ISM:) planetary nebulae: individual (Red Rectangle), stars: individual (HD44179)

1. Introduction

Extended red emission (ERE), a broad red emission band at 540–950 nm, was first observed in the Red Rectangle (RR) nebula by Cohen *et al.* (1975). ERE of different widths and central wavelengths have been found in reflection nebula, planetary nebula, compact H II regions, and the interstellar medium (see Witt & Vijh 2004 for a review). The ERE arises from photoluminescence. Experiments have shown that various materials emit red luminescence. However, it is thought that a specific class of material gives rise to the ERE. Many candidates for the carrier have been proposed, for example, polycyclic aromatic hydrocarbons (PAHs, d'Hendecourt *et al.* 1986), quenched carbonaceous composite (QCC, Sakata *et al.* 1992), hydrogenated amorphous carbon (HAC, Furton & Witt 1992), carbon clusters (Seahra & Duley 1999), crystalline silicon nanoparticles (Ledoux *et al.* 2001), nanodiamonds (Chang *et al.* 2006), and doubly ionized PAH ions (divalent cations, Witt *et al.* 2006). Koike *et al.* (2002) have also suggested that the ERE is caused by thermoluminescence of silicates.

The RR nebula is illuminated by HD 44179, which is a binary star. The nebula shows many dust features from the ultraviolet to the infrared spectral region, including the infrared emission features (IEF, Sellgren 2001). Recently, blue luminescence (BL) was detected in the RR by Vijh *et al.* (2004), They suggested small neutral PAHs in the gas phase (three- and four-ring PAHs) could explain the BL.

One hypothesis that we explore here is that the ERE, BL, and IEF arise from related materials. We present in this paper additional experiments on a synthetic dust analog, Quenched Carbonaceous Composite (QCC), and discuss possible carriers of features in the RR.

2. Experiments and Results

Preparation of Filmy-QCC. The preparation method of the filmy-QCC was the same as described in previous papers (Sakata *et al.* 1992, Sakata *et al.* 1994). A schematic of the apparatus and a photograph of it is shown in Figure 1. A. Sakata and co-workers at the University of Electro-Communications designed and built the apparatus to produce QCC. Most of the apparatus was fabricated by themselves. The apparatus was originally made for the study of the formation of molecules observed in the interstellar medium. Polyynes and PAHs are the main products from the plasmic gas (Sakata 1980). However abundant solids were produced from the gas, and it was quickly discovered to have bands similar to that of the IEF.

Two types of material were formed (see Figure 1). One is an organic material (filmy-QCC) and the other is a carbonaceous material (dark-QCC) with onion-like structure (Wada *et al.* 1999). A detailed description of the nature of QCC materials is given by Wada & Tokunaga (2006).

Mass Spectroscopy. Filmy-QCC and dark-QCC were analyzed by a two-step time-of-flight mass spectrometer (TOF-MS) by S. Gillette and T. Mahajan (Stanford University). The samples were heated with a pulsed CO_2 laser, and then the evaporated gases were

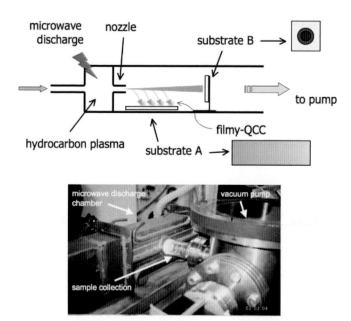

Figure 1. Top: The experimental setup for producing QCC samples. The hydrocarbon plasma is produced in a microwave discharge chamber. The plasmic gas is injected through a small orifice into a vacuum chamber where the QCC sample is collected. Filmy-QCC was collected on a quartz substrate A, located on the wall of the vacuum chamber. Dark-QCC was collected on substrate B, located in the plasmic beam. Bottom: Photograph of the QCC apparatus.

ionized with a YAG laser at a wavelength of 266 nm. $C_m H_n$ peaks of a given carbon number (m) and hydrogen number (n) were detected. Peaks with a different number of hydrogen atoms form a peak group. The central mass of the envelope decreases as the distance of the sampling location from the nozzle increases. The central mass is at 398 atomic mass units (amu) at 5 mm and 252 amu at 70 mm. Larger molecules are therefore formed in the deposit closer to the nozzle. Thus, the major components in the filmy-QCC are compact PAHs.

Two kinds of growth reaction are possible. One is the addition reactions of PAHs to each other, and the other is addition reactions of small radicals to PAH molecules. In the production of both dark-QCC and filmy-QCC, PAH growth is enhanced by high temperature and high density of active species. Schematic diagrams illustrating the structure of the PAH components in the filmy-QCC are presented in Figure 2. The diagrams are based on the mass spectroscopy and analysis of images obtained with a transmission electron microscope (TEM). Curved and parallel structure are often observed in the TEM images of the filmy-QCC (Goto *et al.* 2000). Therefore, we included five-membered carbon rings together with flat PAHs, dihydro-PAHs, and substituted PAHs, and it seems that the PAH molecules are stacked (oriented) locally by intermolecular forces.

Absorption Spectroscopy. Absorption spectra of the filmy-QCCs were measured by a U-3300 UV-VIS spectrophotometer (Hitachi Co). As shown in Figure 3, the absorbance of the filmy-QCC as a solid condensed at 20 and 27 mm from the nozzle has two strong peaks located around 210–230 and 310–320 nm. On the other hand, the absorbance of the samples collected at 5 and 13 mm are composed of two peaks at 220–230 nm and 370–380 nm. Other small absorbance peaks are found in the 400–500 nm region. Thus the filmy-QCC with higher mass PAHs has absorption features at longer wavelengths compared to the filmy-QCC with lighter mass PAHs.

PL of the Filmy-QCC as a Solid and Dissolved in Cyclohexane. We measured the PL spectra of the filmy-QCC with a method similar to that of Sakata *et al.* (1992), and this is shown in Figure 4. The filmy-QCC samples were easily oxidized in air under irradiation of UV. Therefore, we placed them in a vacuum cell and irradiated them with a 365 nm Hg lamp from outside of the cell. The filmy-QCC dissolved in cyclohexane was

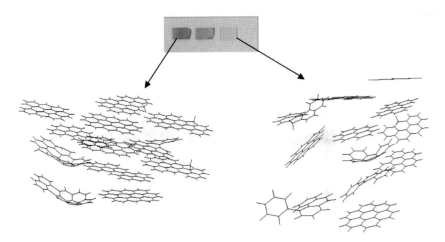

Figure 2. Simplified schematic diagram of the PAH components in the filmy-QCC based on the mass spectroscopy. The material is darker near the nozzle and has larger PAH components (right), compared to the material farther from the nozzle which is lighter in color and has smaller PAH components. Filmy-QCC is a complicated mixture of hydrogenated materials, and this figure does not show all of the components of filmy-QCC.

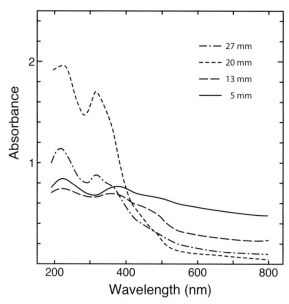

Figure 3. Absorbance spectra of the filmy-QCC, where the absorbance is the quantity $-\log_{10}$(transmitted light/incident light). The filmy-QCC was collected on a quartz glass substrate with a length of 40 mm. Spectra were measured using an aperture of 3 mm in diameter at a point along the substrate at 5 mm (solid line), 13 mm (long-dash line), 20 mm (short-dash line), and 27 mm (dot-dash line) from the edge adjacent to the nozzle of the apparatus. These distances correspond to the distance from the nozzle when the QCC was made.

illuminated with a 366 nm deuterium lamp, and the PL was measured with a Hitachi F-4500 fluorescence spectrometer. Dissolved oxygen gas was removed from the cyclohexane before we used it. The filmy-QCC shows PL at 400–500 nm. The PL spectra show a similar trend as with the absorption spectra in that the filmy-QCC with higher mass PAHs has a PL peak at longer wavelengths compared to the filmy-QCC with lighter mass PAHs.

Quantum Yield of the PL. A quantum yield measurement of the PL of filmy-QCC was carried out using films with known quantum yield. We made two standard films with vapor deposition in a vacuum chamber. These were N,N′-diphenyl-N,N′-bis(3-methylphenyl)-1,1′-biphenyl-4,4′- diamine (TPD, quantum yield = 0.35) and N,N′-diphenyl-N,N′-bis(1-naphthylphenyl)-1,1′-biphenyl-4,4′- diamine (NPD, quantum yield = 0.41) (Mattoussi *et al.* 1999). In the case of the filmy-QCC in solution, the quantum yield was obtained by comparison to a standard solution of 9,10-Diphenylanthracene.

The quantum yield of PL is the ratio of the number of luminescence photons to that of the number of photons absorbed, $n(\text{PL})/n(\text{absorbed})$. The quantum yields of the filmy-QCCs are in the range of 0.009–0.04. Quantum yields obtained by the TPD as a standard film are a little smaller than those obtained by the NPD standard. For these measurements, we set up the samples and standard films in the same way. The surface roughness of the films, stray light from the substrate and quartz container, and differences of refractive indices between samples and standard films all have an effect on the measurements. Therefore, we think that the quantum yields of the QCC-films have an accuracy of only one significant figure. Smith & Witt (2002) estimated the photoluminescence efficiency (quantum yield) in various types of astrophysical sources, and they found that it is in the range of 0.1% to 10%.

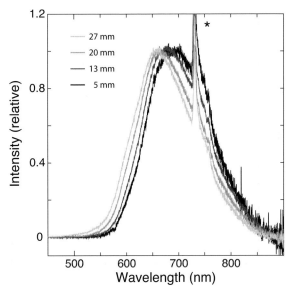

Figure 4. Photoluminescence spectra of the filmy-QCC excited by 365 nm line of the Hg lamp. Samples were collected at 5 mm, 13 mm, 20 mm, and 27 mm from the nozzle (shown from right to left). The spectra are normalized at the peak. An emission line from the Hg lamp is marked with the * symbol.

The quantum yields of the filmy-QCC in the cyclohexane solution are 0.12–0.16. As expected, this is higher than those of the filmy-QCC in the solid state since in the solid form energy is transferred to adjacent molecules. This increases the probability of energy loss through nonradiative processes, thus decreasing the PL yield.

3. Implications for the ERE and BL in the RR

Observational Characteristics of ERE and BL in the RR Nebula. The ERE is clearly observed on the wall of the biconical outflow structure of the RR nebula. In addition, the ladder structure of the RR was also clearly observed with a red filter (Cohen *et al.* 2004). If the red color of the ladder is caused by ERE, the carriers formed before the recent violent outflow event. Cohen *et al.* (2004) reported that the mass loss started about 14,000 yr ago and the outbursts are becoming more frequent.

The peak wavelength of the ERE was observed at the different locations in the RR nebula. Witt & Boroson (1990) found that the peak wavelength varies from ∼ 670 nm at 6″ south of the star to ∼ 645 nm at 10″ south. Ledoux *et al.* (2001), citing the work of Rouan *et al.* (1995), gave a peak wavelength of ∼ 755 nm at 2″ south of the star and ∼ 700 nm at 5″. The FWHM of the ERE peaks is broader as the wavelength of the peak increases. This suggests that the chemical composition, dust structure (arrangement of the molecules), dust size, and physical condition (charge) of the ERE carrier material changes with distance from the central star.

The BL spectrum shows some spectral structure and variability from location to location (Vijh *et al.* 2005), and it is slightly elongated in the east and west direction (Vijh *et al.* 2006). The intensity of the ERE is very strong at the wall of the bipolar outflow, but the BL is distributed around the central region, and the BL and the ERE are not spatially correlated (Vijh *et al.* 2006). Vijh *et al.* (2005) showed that there is a close spatial correlation of the 3.3 μm IEF with the BL.

Comparison of ERE and the PL of the Filmy-QCC. Although individual PAHs in the cyclohexane solution show blue-green PL, the mixture in filmy-QCC as a solid shows a red

PL at longer wavelengths. This suggests that there is strong interaction between excited PAHs with neighboring molecules in the solid filmy-QCC material. In our experiments, the PL peak of the filmy-QCCs ranged from 650 to 690 nm with a FWHM of 133–143 nm. From the mass analysis, the peak shift from 650 to 690 nm corresponds to the growth of PAH molecules in the filmy-QCC from 276 to 398 amu at the center mass of the distribution.

The PL peak wavelength of the filmy-QCC depends weakly on the size of molecules. Sakata *et al.* (1992) formed the filmy-QCCs on substrates at a given location with different temperatures. They showed that the PL peak of the sample formed on the higher temperature substrate occurs at longer wavelengths. Thus, the temperature of the substrate is an important factor for growth of PAH molecules. In their experiments, the filmy-QCCs show a peak ranging from 650 to 725 nm.

We show in Figure 5 the ERE spectrum 6″ south from the central binary system of the RR (Witt & Boroson 1990) and a PL of filmy-QCC condensed at 27 mm from the nozzle (central mass peak about 276 amu). The match of the central wavelength and the peak width is very good. We did not find variations of the PL band width of the filmy-QCC. All filmy-QCCs have a similar FWHM. Since the FWHM depends on the mass distribution of PAHs, we think that the right size distribution of PAHs is necessary for good matching to the width of the ERE band observed in the RR.

The PL spectra of filmy-QCC dissolved in cyclohexane overlaps with the spectra of BL. Since the BL peak occurs at a shorter wavelength than what we observe, the BL carriers are smaller PAHs than that of the filmy-QCC components. This supports the idea suggested by Vijh *et al.* (2005) that small PAH molecules in the gas phase are possible BL carriers.

Ionization State of the ERE Carrier. Since PAH molecules have a low ionization potential, we expect that they would be ionized in the biconical outflow cavities of the RR. In general, condensed PAHs possess lower ionization energy than the PAHs in the gaseous state because of the electronic polarization induced by the surrounding molecules. We obtained 5.3 eV for the work function of the filmy-QCC by UV photoelectron spectroscopy with He I. This corresponds to the energy of a photon with a wavelength of 234 nm. However, under irradiation with 172 nm excimer lamp, red emission was also observed

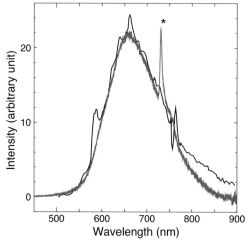

Figure 5. Comparison of the ERE spectrum 6″ south of the RR (black line) from Witt & Boroson (1990) and the PL of the filmy-QCC collected at 27 mm from the nozzle (grey line, see Figure 4). An emission line from the Hg lamp is marked with the * symbol.

clearly by excitation with a 365 nm Hg lamp. The shape of the PL feature does not change by irradiation with a 172 nm lamp. This suggests that neutral PAHs survive under 172 nm irradiation, although some PAHs are ionized. Therefore an important factor is the fraction of neutral PAH molecules in the dust grains that are in the radiation field surrounding the RR.

Decomposition of Dust Particles by UV Irradiation. Under irradiation by hard UV photons, ERE dust grains will gradually decompose because of (1) heating by the UV photons and (2) electrostatic disruption of dust grains due to a high degree of ionization as discussed by Waxman & Draine (2000). Because of the decomposition of the dust grains, isolated PAH molecules and ions are supplied in the outflow cavities. These PAH molecules and ions supplied by dust grains will lose most of their hydrogen atoms. The resulting products with a few hydrogen atoms are not exactly PAHs as defined in chemistry, although their skeletal structure is similar to PAHs. From this viewpoint, they can be considered to be fragments of soot. These products can emit infrared bands when they are heated stochastically, and they could be the IEF carriers. In this scenario, we suspect that the ERE carriers give rise to the IEF carriers through a process of decomposition and dehydrogenation.

4. Conclusions

We obtained the following experimental results about the properties of filmy-QCC:

(*a*) The major components of the filmy-QCC are PAHs. The sizes were found to be 200–500 amu.

(*b*) The filmy-QCC shows red PL with a peak wavelength ranging from 650 to 690 nm with a FWHM of 133–143 nm. Our experiments suggest that the PAHs are major emitters of the PL. The PL quantum yield of our filmy-QCC samples ranges from 0.009 to 0.04.

(*c*) A solution of the filmy-QCC dissolved in cyclohexane shows PL at 400–500 nm with a quantum yield of 0.12–0.16. The large difference in wavelength of the PL peak between the coagulated PAH molecules in filmy-QCC and the dispersed PAH molecules in a solution indicates strong intermolecular interaction of the PAHs with neighboring molecules in the filmy-QCC.

(*d*) We suggest that ERE emitters are PAHs in dust grains similar composition to the filmy-QCC and BL emitters are small gaseous PAH molecules those do not condense on dust grains.

More details will be presented in a forthcoming paper by Wada, Mizutani, Narisawa, & Tokunaga (in preparation).

Acknowledgements

We thank S. Gillette and T. Mahajan for providing the data shown in Figure 2. S. Wada acknowledges the support of Grant-in-Aid for Scientific Research(C) 17540215. This research has made use of NASA's Astrophysics Data System.

References

Chang, H.-C., Chen, K., & Kwok, S. 2006, *ApJ* (Letter), 639, L63
Cohen, M., *et al.* 1975, *ApJ*, 196, 179
Cohen, M., Van Winckel, H., Bond, H. E., & Gull, T. R. 2004, *AJ*, 127, 2362
Furton, D. G. & Witt, A. N. 1992, *ApJ*, 386, 587
d'Hendecourt, L. B., Léger, A., Olofsson, G., & Schmidt, W. 1986, *A&A*, 170, 91
Goto, M., Maihara, T., Terada, H., Kaito, C., Kimura, S., & Wada, S. 2000, *A&AS*, 141, 149
Koike, K., Nakagawa, M., Koike, C., Okada, M., & Chihara, H. 2002, *A&A*, 390, 1133

Ledoux, G., Guillois, O., Huisken, F., Kohn, B., Porterat, D., & Reynaud, C. 2001, *A&A*, 377, 707

Mattoussi, H., Murata, H., Merritt, C. D., Iizumi, Y., Kido, J., & Kafafi, Z. H. 1999, *J. Appl. Phys.*, 86, 2642

Rouan, D., Lecoupanec, P., & Léger, A. 1995, in: C. S. Jeffery (ed.), *Proc. 1st Franco- British meeting on the Physics and Chemistry of the Interstellar Medium, Newsletter on Analysis of Astronomical Spectra, no. 22*, p. 37

Sakata, A. 1980, in: B.H. Andrew (ed.), *Interstellar Molecules, IAU Symp 87*, (Dredrecht: Reidel), p. 325

Sakata, A., Wada, S., Narisawa, T., Asano, Y., Iijima, Y., Onaka, T., & Tokunaga, A. T. 1992, *ApJ* (Letter), 393, L83

Sakata, A., Wada, S., Tokunaga, A. T., Narisawa, T., Nakagawa, H., & Ono, H. 1994, *ApJ*, 430, 311

Seahra, S. S. & Duley, W. W. 1999, *ApJ*, 520, 719

Sellgren, K. 2001, *Spectrochimica Acta*, 57, 627

Smith, T. L. & Witt, A. N. 2002, *ApJ*, 565, 304

Vijh, U. P., Witt, A. N., & Gordon, K. D. 2004, *ApJ* (Letter), 606, L65

Vijh, U. P., Witt, A. N., & Gordon, K. D. 2005, *ApJ*, 619, 368

Vijh, U. P., Witt, A. N., York, D. G., Dwarkadas, V. V., Woodgate, B. E., & Palunas, P. 2006, *ApJ*, 653, 1336

Wada, S., Kaito, C., Kimura, S., Ono, H., & Tokunaga, A. T. 1999, *A&A*, 345, 259

Wada, S., & Tokunaga, A. T. 2006, in: F. J. M. Rietmeijer (ed.), *Natural Fulerenes and Related Structures of Elemental Carbon*, (Dordrecht: Springer), p. 31

Waxman, E., & Draine, B. T. 2000, *ApJ*, 537, 796

Witt, A. N., & Boroson, T. A. 1990, *ApJ*, 355, 182

Witt, A. N., Gordon, K. D., Vijh, U. P., Sell, P. H., Smith, T. L., & Xie, R.-H. 2006, *ApJ*, 636, 303

Witt, A. N. & Vijh, U. P. 2004, in: A. N. Witt, G. C. Clayton, & B. T. Draine (eds.), *ASP Conf. Ser. 309, Astrophysics of Dust*, (San Francisco: ASP), p. 115

Discussion

SLOAN: I remember quite a while ago that Adolf Witt had imaged the extended red emission in the Red Rectangle to be in a big X shape along the edges of the conical cavities north and south of the central star. If that is the case, then it seems to me that the carriers of the extended red emission are not newly formed grains, but are grains that haven't been destroyed yet because they are at the edge of the destruction zone.

HENNING: I think the structure is still the same; the interpretation is another question of course.

MULAS: I have simple technical question and another one about the Red Rectangle. The technical question – You showed how the photoluminescence spectrum of your samples changes with properties of the sample and you also showed the geometry with which you collect the light. Since emitted light has to travel some distance inside your sample before getting out, it probably undergoes some self-absorption. Did you compensate your spectrum for that? As your samples become more absorbing they may change the spectrum just due to that. About the Red Rectangle - It looks like it could be a problem to explain the luminescence by very small PAHs because you would have those very few PAHs, because they are so small and are very highly excited by every single photon, emitting more at 3.3 μm than you actually see. That may be a bit of the problem with the observations.

Organic Matter in Space
Proceedings IAU Symposium No. 251, 2008
S. Kwok & S. Sandford, eds.

© 2008 International Astronomical Union
doi:10.1017/S1743921308022072

Laboratory analogs of carbonaceous matter: Soot and its precursors and by-products

Cornelia Jäger[1], Harald Mutschke[2], Isabel Llamas-Jansa[2] Thomas Henning[3], and Friedrich Huisken[1]

[1]Laboratory Astrophysics Group of the Max Planck Institute for Astronomy
at the Institute of Solid State Physics, Friedrich Schiller University Jena
Helmholtzweg 3, 07743 Jena, Germany
email: `Cornelia.Jaeger@uni-jena.de`

[2]Astrophysical Institute and University Observatory, FSU Jena
Schillergässchen 3, 07745 Jena, Germany

[3]Max Planck Institute for Astronomy
Königstuhl 17, D-69117 Heidelberg, Germany

Abstract. Carbonaceous materials have been prepared in the laboratory by laser-induced pyrolysis of a mixture of hydrocarbons under different conditions and laser ablation of graphite in reactive gas atmospheres. We have investigated the soluble and insoluble parts of the condensed carbon powders with several spectroscopic and chromatographic methods in order to obtain information on the composition of the condensate. The results of these experiments have demonstrated that, at temperatures lower than 1700 K, the pyrolysis by-products are mainly PAHs, whereas at higher temperatures fullerenes and polyyne-based compounds are formed. The experimental findings point to different soot formation mechanisms with variable intermediates and end products. It has been found that soot extracts can contain more than 65 different polycyclic aromatic hydrocarbons (PAHs). Eventually, the study of the condensation pathways of soot particles and their precursors and by-products will permit the prediction of the spectral properties of carbonaceous matter in space.

Keywords. Laboratory, dust, extinction, infrared, stars

1. Carbonaceous cosmic dust analogs

Plenty of different analogs for the carbonaceous cosmic dust have been produced and characterized in the laboratories. Their spectral properties have been measured and compared to observations in order to obtain information on the exact composition and processing of the cosmic dust in different astrophysical environments. Hydrogenated amorphous carbon (HAC) was first introduced as a possible cosmic dust analog by Duley & Williams (1981). Sakata *et al.* (1983) produced a carbonaceous material from a hydrocarbon plasma called quenched carbonaceous composite (QCC). A few years later, Papoular *et al.* (1989) was the first to use coals of different states of aromaticity to model the carriers of the aromatic IR bands (AIBs). Detailed studies of the spectral properties of HAC films have been continued by Duley (1994), Grishko & Duley (2002), and Duley *et al.* (2005). In the last decade, the list of laboratory dust analogs was complemented by nanosized carbon grains produced by gas-phase condensation techniques such as laser pyrolysis (Herlin *et al.* 1998, Schnaiter *et al.* 1999, Jäger *et al.* 2006, Llamas *et al.* 2007, Jäger *et al.* 2007), resistive heating, arc discharge between carbon electrodes, or laser ablation coupled with condensation of grains in quenching gas atmospheres (Mennella *et al.* 1996, Schnaiter *et al.* 1998, Jäger *et al.* 1999, Jäger *et al.* 2008).

New and interesting structures are onion-like carbon grains. Already some years ago, these grains have been proposed as the carrier of the interstellar UV bump (de Heer & Ugarte 1993), however, the attribution remained questionable due to the shift of the measured UV absorption band in water. Onion-like carbon grains were also found in the granular and the dark QCC component produced by Wada *et al.* (1999). They found that the as-produced onions show a rather broad UV band around 220 nm. Annealing experiments shifted the peak to longer wavelengths. Tomita *et al.* (2002) produced so-called defective carbon onions by annealing of nanodiamonds and found an absorption band at 3.9 μm^{-1} for onions dispersed in water. Under annealing at temperatures higher than 1900 K, polyhedral onion particles were formed, and a further absorption band appeared at 4.6 μm^{-1}. A theoretical model was developed by the authors to complement the experimental results. Defective onions of 5 nm with hollow cores of 0.7 nm in diameter were found to fit the interstellar UV bump very well. In contrast, Chhowalla *et al.* (2003) measured a narrow UV band at 4.55 μm^{-1} for onion samples with sizes between 3 and 50 nm which were annealed at 873 K for 1h in air. The results of the UV studies are rather controversial and further efforts are necessary to understand the UV absorption properties of onion-like carbon grains.

Kerogen, a material also found in primitive meteorites and interplanetary dust particles, has also been considered as a possible cosmic dust analog. On Earth, the name kerogen describes a family of polymer-like organic materials which are formed in sedimentation and annealing processes of hydrocarbons. Kerogen occurs in different states of aromaticity. Pendleton & Allamandola (2002) have shown that the 3.4 μm IR band of kerogen material extracted from the Murchison meteorite indicates a striking similarity to the 3.4 μm IR profile of the diffuse interstellar medium (DISM), but there is no coincidence in the range between 5 and 10 μm. Papoular (2001) suggested that the 3.4 μm as well as the accompanying MIR bands in protoplanetary and planetary nebulae and in the DISM could be carried by kerogen-like dust that is formed in circumstellar envelopes of evolved stars.

Generally, the different proposed dust analogs represent related carbonaceous materials composed of aromatic and saturated aliphatic structural units. Functional groups such as $-CH_x$, $=C-H$, $-C=O$, $-C=C-$, $\equiv C-H$, $-C\equiv C-$, $-C-O-C-$, and $-C-OH$, either incorporated into the structure of the structural units (SUs) or bound to the edges of these units, give rise to the appearance of observable IR bands. Therefore, in the laboratory, for analogs consisting of C and H and typical functional groups, similar IR bands can be observed. Differences in band positions result from differences in the chemical neighborhood of these functional groups which affect slightly the bond distances or electronic densities of the bonds. Varying band ratios are caused by varying abundances of these groups in the structure.

One of the key parameters for a complete understanding of the interstellar carbon dust component and its processing in the interstellar radiation field is the formation process of carbon grains. However, the formation pathway of carbonaceous matter in astrophysical environments, as well as in terrestrial gas-phase condensation reactions, is not yet understood. Knowledge of the detailed condensation conditions of carbon dust in astrophysical environments may help to predict the consequential formation pathways and the structural and morphological properties of the condensing carbonaceous material.

Carbonaceous materials form in large quantities in circumstellar envelopes around carbon stars. It is assumed that carbon soot and polycyclic aromatic hydrocarbons (PAHs) are simultaneously formed as nano- and subnanometer-sized particles via gas-phase condensation. Several authors (Frenklach & Feigelson 1989, Cherchneff *et al.* 1992, Allain *et al.* 1997) have modeled the formation of PAHs in AGB stars. Whereas Frenklach &

Feigelson (1989) determined a small temperature range for the formation of PAHs in circumstellar environments, Cherchneff & Cau (1999) reconsidered the modeling of the PAH and carbon dust formation in carbon-rich AGB stars based on a developed physico-chemical model which describes the periodically shocked gas in the circumstellar shells close to the photosphere of the stars. The authors found that C_6H_6 formation begins at 1.4 R_*, and at a radius of 1.7 R_* the conversion of single rings to PAHs at a temperature of around 1700 K starts. This temperature is much higher than the temperature window calculated in previous studies.

2. Low- and high-temperature condensation of carbonaceous matter in the laboratory

In our laboratory, we performed a series of gas-phase condensation experiments comprising laser ablation of graphite in quenching gas atmospheres and laser pyrolysis of hydrocarbons. We have performed our experiments in two different temperature regimes to achieve high-temperature (HT) and low-temperature (LT) condensation. The study of the formation pathways has been performed by analytical characterization of the condensates, including soot and its by-products. Laser ablation of graphite and subsequent condensation of carbonaceous matter has been carried out in quenching gas atmospheres of He or He/H_2 mixtures at low pressures. A more detailed description of the method can be found in Jäger *et al.* (2008). The second harmonic of a pulsed Nd:YAG laser was employed to evaporate carbon from a graphite target. The laser power densities varied between 2×10^8 and 9×10^9 W cm^{-2}. The particles were extracted from the condensation zone by using a molecular beam technique.

Pulsed laser ablation of graphite in a quenching gas atmosphere is a complex non-equilibrium process, and sub-processes such as sublimation of the target material, ionization of the target species and the surrounding gas atoms and molecules, and deceleration of gas and condensates are rather difficult to describe. It is generally assumed that the temperature of the electrons, ions, and neutrals are different in an ablation plume. Therefore, we can only provide a lower limit of the temperature in the condensation zone. Our temperature estimation is based on different approaches. First we know that the sublimation of graphite at low pressures needs temperatures between 3600 and 4000 K. From the analysis of the HRTEM images of the condensates we can infer that no graphite was simply fractionalized or larger graphene layers ablated. Finally, the vibrational temperature of the laser-induced plasma generated by laser evaporation of a graphite target in a 10 mbar He atmosphere was found to range between 4000 and 6000 K for power densities between 0.5–2×10^9 W cm^{-2} (Iida & Yeung 1994).

In addition, we have investigated different carbon condensates prepared by laser-induced pyrolysis (LIP) of gas-phase hydrocarbon precursors (ethylene, acetylene, and benzene) using either a pulsed or a continuous-wave (cw) CO_2 laser. A detailed description of the experimental setup is given elsewhere (Llamas *et al.* 2007, Jäger *et al.* 2007). In the LIP experiments performed with the pulsed CO_2 laser, SF_6 was applied as a sensitizer. The laser radiation induces the dissociation of the reaction gas and the subsequent condensation of carbon nanoparticles which can either be extracted from the flow reactor by a particle beam extraction technique or by collection of the condensate in a filter. In the laser pyrolysis experiments with the cw laser, the temperatures in the condensation zone were determined with a pyrometer which provided a direct measurement of this important parameter. The temperatures achieved in pyrolysis experiments using pulsed CO_2 lasers have already been investigated by several authors. At similar pyrolysis conditions, but slightly lower power densities, temperatures around 3000 K were

Table 1. Experimental conditions for the HT and LT gas-phase condensation experiments.

Experiment	Precursor	Buffer gas	Laser	Laser power density (W cm^{-2})	Temperature (K)	Condensate
LA1	Graphite	He/H$_2$	pulsed	2×10^8–9×10^9	$\geqslant 4000$	fullerene-like soot and fullerenes
LP1	C$_2$H$_4$, C$_2$H$_2$ C$_6$H$_6$	He/Ar	pulsed	1×10^7–1×10^9	$\geqslant 3500$	fullerene-like soot and fullerenes
LP2	C$_2$H$_4$, C$_2$H$_2$	Ar	cw	5200		soot and 14 wt% PAHs
LP3	C$_2$H$_4$, C$_6$H$_6$	Ar	cw	5200	~ 1500	soot and 33 wt% PAHs
LP4	C$_2$H$_4$, C$_6$H$_6$	Ar	cw	6400		soot and 17 wt% PAHs
LP5	C$_2$H$_4$	Ar	cw	850	~ 1000	100 wt% PAHs

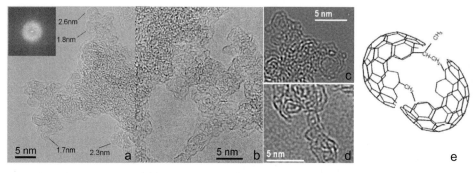

Figure 1. HRTEM images of fullerene-like carbon nano particles produced in a HT condensation process (LA1). Images a and b show soot grains with low and high hydrogen content, respectively. Images c and d present elongated fullerene particles and typical fullerene fragments. The structure e illustrates possible links between fullerene fragments in the soot condensates.

measured for comparable pressures of the He buffer gas (Kojima & Naito 1981, Doubenskaia *et al.* 2006).

The preparation conditions for all gas-phase condensation experiments are compared in Table 1. The first two rows contain experiments performed with pulsed lasers characterized by high power densities and, consequently, high temperatures of more than 3500 K in the condensation zone. The condensation of particles in laser pyrolysis experiments employing a cw laser proceeds at much lower temperature (less than 1700 K). We call these two different types of condensation HT and LT condensations, respectively.

3. Soot formation and structural properties of the condensates

HRTEM images of the carbonaceous matter produced in a HT gas-phase condensation process are shown in Figure 1.

The HRTEM micrographs reveal that very small fullerene-like particles are produced. These particles are composed of small, strongly bent graphene layers with varying lengths (L$_a$) and distances between these layers. The level of disorder depends on the employed condensation conditions which can be well observed in images a and b of Figure 1. Soot condensates produced only in helium atmospheres and at high pressures show more ordered and frequently closed fullerene cages. Grains containing higher contents of hydrogen are less ordered and do not show completely closed cages. In these grains, the cage fragments stick together by van der Waals forces or are linked by aliphatic –CH$_x$ groups (see sketch e in Figure 1). The small size of the grains points to a strong supersaturation of

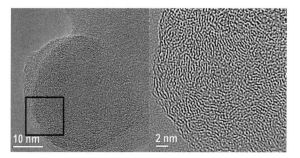

Figure 2. HRTEM images of typical large carbon nano particles produced in LT processes (LP2-4). The right micrograph displays a magnified section of the particle in the left image clearly showing the arrangement of the graphene layers in the particles.

carbon vapor in the condensation zone resulting in a high number of nucleation seeds. The further particle growth is exclusively due to coagulation.

A soot formation process, coupled with the formation of fullerenes, starting with polyyne chains, with subsequent spirocyclization and creation of saucer-shaped fullerene fragments has already been proposed by Kroto & McKay (1988). The generation of fullerenes and fullerene snatches in the carbonaceous condensate could be verified by using electron microscopy. The critical point in the creation of bent graphene layers or disturbed fullerene-like structures is the formation of these cage fragments from the chains. Quantum chemical molecular dynamics simulations have shed some light on the formation of fullerenes from C_2 fragments. High initial carbon densities are essential for such processes which can be found for example in carbon arc or laser ablation processes (Zheng *et al.* 2005). These authors found that the C_2 molecules quickly combine to long and branched carbon chains and macromolecules for temperatures above 2000 K. From the large chains, small cyclic structures with long carbon chains attached are formed (nucleation). The second step is a further ring condensation growth for example between two linear chains attached to a nucleus. Consequently, fullerene fragments of bowl shape with side chains are formed in this step (Irle *et al.* 2003).

In contrast, the condensed grains in LT processes are much larger and show more ordered and well developed planar graphene layers inside the particles (see Figure 2). Additionally, a mixture of PAHs was formed as by-products. This soluble component was analyzed by extraction of the soot in toluene, and the constituents of the extract were identified by chromatographic methods as a mixture of PAHs and partly hydrogenated PAHs. About 65 different PAHs have been analyzed, and 3-5 ring systems are by far most abundant. The PAHs are partly hydrogenated and contain saturated aliphatic-CH_2 and CH_3 groups. PAHs with masses up to 3000 Da were detected by application of matrix-assisted laser desorption/ionization mass spectrometry (MALDI TOF), but only in very low amounts. For example, a symmetric PAH molecule, $C_{222}H_{42}$, comprising 91 condensed rings and having a diameter of about 3 nm is a molecule of comparable mass. The large size and high internal order of the condensed soot grains points to a low supersaturation of the carbon vapor and the formation of a smaller number of stable nuclei compared to the HT condensation process. In the LT condensation, the further particle growth is dominated by condensation of intermediates on the surface of the seeds which means that PAHs continuously accumulate on the surface of the grains. Since larger PAHs have a lower volatility compared to the smaller ones, a preferred accumulation of large molecules during the surface growth process can be observed. The accumulation of the large molecules on the surfaces of the seeds can be confirmed by the results of the

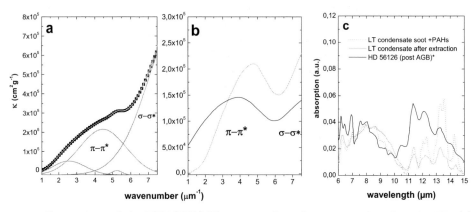

Figure 3. Comparison of the FUV/UV/VIS spectra of condensates produced in the HT (a) and LT (b) condensation process. A comparison between the IR spectral properties of a LT condensate with the observed spectrum of a post AGB star adapted from Hony *et al.* (2003) is shown in image (c).

HRTEM studies. A detailed analysis of the micrographs revealed a mean length L_a of the graphene sheets inside the particles of around 1.8 nm corresponding to PAHs with masses of around 1000 Da. The longest graphene layers have extensions of about 3 nm which corresponds to the highest-mass PAHs detected with MALDI TOF measurements.

4. Spectral properties of the HT and LT condensates

The spectral properties of the HT and LT condensates in the UV range differ considerably. Although the HT soot grains consist of about 50% sp^2 hybridized carbon atoms the UV spectra do not show distinct UV absorption bands (see Figure 3a). A deconvolution of the UV/VIS absorption profile by employing four Gaussians reveals weak and broad bands due to $\pi - \pi^*$ transitions between 4.45 and 3.84 μm^{-1}, for the samples with the highest and the lowest hydrogen content, respectively. The weak Gaussian at approximately 2.5 μm^{-1} is interpreted as a plasmon peak due to some larger, but strongly bent, graphene layers. The width of the $(\pi - \pi^*)$ main band and the appearance of the plasmon band accounts for a strong disorder in the carbon structures due to a broad distribution of curvatures and lengths of graphene layers in the small fullerene-like carbon grains (Llamas *et al.* 2007, Jäger *et al.* 2008). The very weak band at around 5.2 μm^{-1} (190 nm) can be attributed to the absorption of –C=O groups which could be identified in small amounts in the IR spectra of nearly all carbon soot samples. In contrast, the UV spectra of the LT condensates (Figure 3b) do show distinct UV absorption bands caused by $(\pi - \pi^*)$ transitions ranging between 3.6 and 4.9 μm. The exact position of this band depends on the internal structure of the soot grains and/or the content and composition of the soluble PAH component which also contributes to the spectra.

The IR spectral properties of the HT condensates are characterized by strong saturated and aliphatic $-CH_x$ absorptions at 3.4 μm. No aromatic =C–H has been observed in this spectral region. This supports the assumption that saturated aliphatic $-CH_x$ groups are responsible for the links between the fullerene fragments. A weak feature at 5.8 μm points to the incorporation of a small amount of –C=O groups inside the carbon structure. The presence of aromatic –C=C– groups can be identified at IR bands between 6.2 and 6.25 μm. Out-of-plane bending vibrations of aromatic =C–H groups can be observed in the range between 11 and 14 μm but their intensity is much lower compared to the LT condensates. Additionally, signatures for the presence of –C≡C– triple bonds can be seen in the spectra at 3.03 and 4.7 μm in in-situ measurements which points to the formation of polyynes as intermediates. A more detailed description of the IR spectral properties of soot particles produced in HT condensations can be found in Jäger *et al.* (2008).

As a consequence of the composition characteristic of the LT condensates which represent a mixture of soot and PAHs, aromatic IR bands (AIBs) such as 3.3, 6.2, 8.6, 11.3, 12.3, and 13.3 μm as well as aliphatic IR bands at 3.4, 6.8, and 7.25 μm can be identified in the IR spectra of such condensates. As shown in Figure 3c, the AIBs and aliphatic IR bands are superimposed on two broad plateau features around 8 and 12 μm. The spectral characteristics of LT condensates strongly resemble the observed IR spectra of post AGB stars and protoplanetary nebulae (Kwok *et al.* 2001, Hony *et al.* 2003, Hrivnak *et al.* 2007). The comparison shown in Figure 3c clearly reveals that nearly all of the observed IR bands are also present in the LT condensate produced in the laboratory. Differences in band ratios, apparent for the =C-H out-of-plane vibrational bands in the range between 11 and 14 μm for the condensate, containing soot and PAHs, result from the presence of small PAHs in the condensed material. In particular, the bands at 13.3, 13.5, and 14.3 μm are caused by 4-5 adjacent H atoms bound to the aromatic ring and argue for smaller PAHs or for special PAHs containing phenyl rings bound to larger PAH molecules. The extracted soot particles which have consumed all the large PAHs during the growth process show a much better coincidence with the observed bands. The comparison reveals that the LT condensate is a promising dust analog for carbonaceous materials produced in carbon-rich AGB stars.

5. Conclusions

The results of our laboratory experiments have demonstrated that there are two different soot formation mechanisms. In HT condensations at temperatures higher than 3500 K, very small fullerene-like particles and fullerenes, as by-products, are generated. The formation pathway of the soot is characterized by the formation of fullerene fragments from polyyne chains. LT condensates produced at temperatures lower than 1700 K consist of larger soot particles with long and rather plane graphene layers and mixtures of PAHs. Here, the formation process starts with a combination of small molecules resulting in the formation of aromatic benzene rings, the growth of larger and plane PAHs by subsequent C_2 addition to the aromatic rings, and the final growth of grains by the condensation of large PAHs on the surfaces of the nuclei. Low-temperature condensation is a very likely formation process of soot and PAHs in AGB stars. Condensation temperatures in our laboratory studies were found to be very similar to the temperature range for carbon dust condensation in carbon-rich AGB stars predicted by Cherchneff & Cau (1999).

Acknowledgements

This work has been supported by the Deutsche Forschungsgemeinschaft. We are grateful to Dr. I. Voicu for supplying some of the laser pyrolysis samples.

References

Allain, T., Sedlmayer, E., & Leach, S. 1997, *A&A*, 323, 163
Cherchneff, I., Barker, J. R., & Tielens, A. G. G. M. 1992, *ApJ*, 413, 445
Cherchneff, I. & Cau, P. 1999, in: T. Le Betre, A. Lèbre & C. Waelkens (eds.), *Asymptotic Giant Branch Stars*, Proc. IAU Symposium No. 191 (San Francisco:ASP), p. 251
Chhowalla, M., Wang, H., Sano, N., Teo, K. B. K., Lee, S. B., & Amaratunga, G. A. J. 2003, *Phys. Rev. Lett.*, 90, 155504
de Heer, W. A. & Ugarte, D. 1993, *Chem. Phys. Lett.*, 207, 480
Doubenskaia, M., Bertrand, Ph., & Smurov, I. 2006, *Surface & Coatings Technology*, 201, 1955

Duley, W. W. 1994, *ApJ* (Letters), 430, L133

Duley, W. W., Lazarev, S., & Scott, A. 2005, *ApJ*, 620, L135

Duley, W. W. & Williams, D. 1981, *MNRAS*, 196, 269

Frenklach, M. & Feigelson, E. D. 1989, *ApJ*, 341, 372

Grishko, V. I. & Duley, W. W. 2002, *ApJ*, 568, 448

Herlin, N., Bohn, I., Reynaud, C., Cauchetier, M., Galvez, A., & Rouzaud, J.-N. 1998 *A&A* 330, 1127

Hony, S., Tielens, A. G. G. M., Waters, L. B. F. M., & de Koter, A. 2003, *A&A*, 402, 211

Hrivnak, B. J., Geballe, T. R., & Kwok, S. 2007, *ApJ*, 662, 1059

Iida, Y. & Yeung, E. 1994, *Appl. Spectr.*, 48, 945

Irle, S., Zheng, G., Elstner, M., & Morokuma, K. 2003, *Nano Letters* 3, 1657

Jäger, C., Henning, Th., Schlögl, R., & Spillecke, O. 1999, *J. Non-Cryst. Solids* 258, 161

Jäger, C., Huisken, F., Mutschke, H., Henning, Th., Poppitz, W., & Voicu, I. 2007, *Carbon*, 45, 2981

Jäger, C., Krasnokutski, S., Staicu, A., Huisken, F., Mutschke, H., Henning, Th., Poppitz, W., & Voicu, I. 2006, *ApJS*, 166, 557

Jäger, C., Mutschke, H., Henning, Th., & Huisken, F. 2008, *ApJ*, submitted

Kojima, H. & Naito, K. 1981, *Ind. Eng. Chem. Prod. Res. Dev.*, 20, 396

Kroto, H. W. & McKay, K. 1988, *Nature*, 331, 328

Kwok, S., Volk, K., & Bernath, P. 2001, *ApJ* (Letters), 55, L87

Llamas-Jansa, I., Jäger, C., Mutschke, H., & Henning, Th. 2007, *Carbon*, 45, 1542

Mennella, V., Colangeli, L., Palumbo, P., Rotundi, A., Schutte, W., & Bussoletti, E. 1996, *ApJ*, 464, L191

Papoular, R. 2001, *A&A*, 378, 597

Papoular, R., Conrad, J., Giuliano, M., Kister, J., & Mille, G. 1989, *A&A*, 217, 204

Pendleton, Y. J. & Allamandola, L. J. 2002, *ApJS*, 138, 75

Sakata, A., Wada, S., Okutsu, Y., Shintani, H., & Nakada, Y. 1983, *Nature*, 301, 493

Schnaiter, M., Henning, Th., Mutschke, H., Kohn, B., Ehbrecht, M., & Huisken, F. 1999, *ApJ*, 519, 687

Schnaiter, M., Mutschke, H., Dorschner, J., Henning, Th., & Salama, F. 1998, *ApJ*, 498, 486

Tomita, S., Fujii, M., & Hayashi, S. 2002, *Phys. Rev. B*, 66, 245424

Wada, S., Kaito, Ch., Kimura, S., Ono, H., & Tokunaga, A. T. 1999 *A&A*, 345, 259

Zheng, G., Irle, S., & Morokuma, K. 2005, *J. Chem. Phys.*, 122, 014708

Discussion

SALAMA: Do you observe a continuous PAH size distribution, or do you instead observe many smaller PAHs?

JÄGER: PAHs with masses up to 3000 Da were detected in the MALDI-TOF mass spectra, but the smaller PAHs are by far the most abundant species.

SALAMA: Do you have any explanation for that?

JÄGER: My idea is that the large PAHs are consumed in the particle growth process. In low-temperature condensations, a small number of stable seed grains are formed simultaneously with a large amount of PAHs. The particle growth in the condensation under such conditions is dominated by a surface growth process which can be understood as a continuous accumulation of PAHs on the surfaces of the seed particles. The low-mass PAHs are much more volatile and they stay in the gas phase. The less volatile species accumulate on the surfaces of the nucleation seeds or small condensed particles and they built up the particles layer by layer. The largest graphene layers in the condensed particles were found to be around 3 nm corresponding to PAHs of approximately 3000 Da. I think this formation process explains why we do not see very abundant large PAHs.

Organic Matter in Space
Proceedings IAU Symposium No. 251, 2008
S. Kwok & S. Sandford, eds.

© 2008 International Astronomical Union
doi:10.1017/S1743921308022084

In situ observation
of structural alteration process
# of filmy quenched carbonaceous composite

Akihito Kumamoto[1,*], Yuki Kimura[1], Chihiro Kaito[1], and Setsuko Wada[2]

[1]Laboratory for Nano-structure Science, Ritsumeikan University, Japan
*email: rp008011@se.ritsumei.ac.jp
[2]Department of Applied Physics & Chemistry, Univ. of Electro-Communications, Japan

Abstract. The thermal alteration process of filmy quenched carbonaceous composite (filmy QCC) has been studied in situ by high-resolution transmission electron microscopy (HRTEM). HRTEM images of the as-prepared filmy QCC showed the typical amorphous carbon film structure. By heating above 300 °C, the structural alteration takes place. Curled graphene structure started to appear at 300 °C. Distorted onion-like structure similar to dark QCC appeared above 500 to 700 °C. The distorted onion-like structure that appears at 700 °C after heating for 30 minutes also appeared by heating at 450 °C for 2 hours.

Keywords. Methods: laboratory, dust, plasmas

1. Introduction

Many types of carbon and other carbonaceous materials have been proposed to explain the 217.5 nm feature. Sakata *et al.* (1983) had proposed and analyzed a quenched carbonaceous composite (QCC) material which is condensed from a hydrocarbon plasma. A brown-black carbonaceous materials named 'dark-QCC' shows a 217 nm absorption maximum (Sakata *et al.* 1983). A yellow-brown filmy material named 'filmy QCC' is collected on a wall surrounding the plasma beam. By thermal treatment at 500-700 °C, the filmy QCC is carbonized and shows a 217 nm absorption maximum (Sakata *et al.* 1994). The dark QCC is a coagulation of carbonaceous onion-like particles as elucidated by high resolution transmission electron microscopy (HRTEM) (Wada *et al.* 1999). In the present paper, the structural alteration of filmy QCC has been examined by in situ HRTEM observation in the temperature range of 300-700 °C.

2. Heating of filmy QCC

The filmy QCC was deposited onto a KCl cleavage crystal surface. The KCl was dissolved in water, and the isolated sample was collected onto carbon holey film supported by a standard electron microscopic grid. The samples were heated in an electron microscope using a special specimen holder in 6×10^{-6} Pa. Heating was carried out by controlled the rate of temperature rise to 5 °C per minute and maintained for 30 minutes at 100 °C intervals. Figure 1 shows the schematic presentation of the furnace specimen holder.

In a previous similar study, the structural alteration was different between the periphery and the central part of the film, i. e., turbostratic graphite structure and onion-like

434   Akihito Kumamoto *et al.*

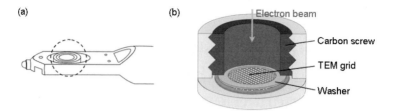

Figure 1. Schematic images of a heating head of the TEM holder (a). Broken line circle of the detail is shown as (b).

structure were seen, respectively (Kaito *et al.* 2003). We found in the present study that the periphery structural alteration was only seen in the strong electron irradiated region.

Figure 2 shows the filmy QCC of the regions which were heavily exposed to electron radiation. The difference between the periphery and inner parts are small. Upon heating to 300 °C, curled graphene was seen as indicated by arrows. By heating at 400 °C, the void contrast of the onion-like structure started to appear as indicated by arrows. By heating at 500 and 700 °C, onion-like spherule growth became apparent (Figure 2c and 2d). Figure 2e shows the results after heat treatment at 700 °C. The onion-like spherules are seen throughout the film. The basic structure is like that of dark QCC.

When the heating experiment was done at 450 °C for 2 hours, similar structural alteration, with the same onion-like spherules, was obtained.

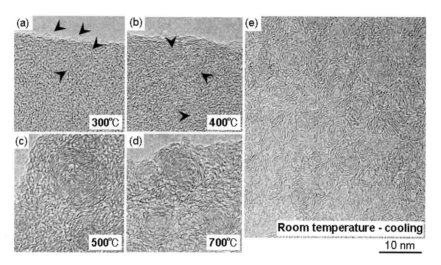

Figure 2. In situ images of heated filmy QCC at 300 °C, 400 °C, 500 °C, and 700 °C, and at room temperature after cooling.

References

Kaito, C., Kido, O., Wada, S., Kimura, Y., Suzuki, H., Sato, T., Kamitsuji K., & Kurumada, M. 2003, *Grain Formation Workshop*, vol. XXIII, pp.65
Sakata, A., Wada, S., Okutsu, Y., Shintani, H., & Nakada, Y. 1983, *Nature*, 301, 493
Sakata, A., Wada, S., Tokunaga, A. T., Narisawa, T., Nakagawa, H., & Ono, H. 1994, *ApJ*, 430, 311
Wada, S., Kaito, C., Kimura, S., Ono, H., & Tokunaga, A. T. 1999, *A&A*, 345, 259

Organic Matter in Space
Proceedings IAU Symposium No. 251, 2008
S. Kwok & S. Sandford, eds.

Ion irradiation effects on sooting flames by-products

R. Brunetto[1,2,3], **T. Pino**[2], **E. Dartois**[1], **A.T. Cao**[2], **L. d'Hendecourt**[1], **G. Strazzulla**[3], and **Ph. Bréchignac**[2]

[1]Institut d'Astrophysique Spatiale, UMR-8617, Université Paris-Sud,
91405 Orsay Cedex, France
email: rosario.brunetto@ias.u-psud.fr

[2]Laboratoire de Photophysique Moléculaire, UPR 3361, Université Paris-Sud,
91405 Orsay Cedex, France

[3]INAF – Osservatorio Astrofisico, via S. Sofia 78, I-95123 Catania, Italy

Abstract. Carbonaceous extraterrestrial matter is observed in a wide variety of astrophysical environments. Spectroscopic signatures reveal a large variety of chemical structure illustrating the rich carbon chemistry that occurs in space. In order to produce laboratory analogues of the carbonaceous cosmic dust, a new chemical reactor has been built in the Laboratoire de Photophysique Moléculaire. It is a low pressure flat burner providing flames of premixed hydrocarbon/ oxygen gas mixtures, closely following the model system used by the combustion community. In such a device the flame is a one-dimensional chemical reactor offering a broad range of combustion conditions and sampling which allows production of many and various by-products. In the present work, we have studied the effect of ion irradiation (200-400 keV), at the Laboratorio di Astrofisica Sperimentale in Catania, on several samples, ranging from strongly aromatic to strongly aliphatic materials. Infrared and Raman spectra were monitored to follow the evolution of the films under study, and characterize the irradiation process-induced modifications.

Keywords. Astrochemistry, methods: laboratory, ISM: dust, cosmic rays

1. Introduction

Carbonaceous extraterrestrial matter is observed in several astrophysical environments. Spectroscopic signatures observed by telescopes and satellites reveal a large variety of chemical structures, illustrating the rich carbon chemistry that occurs in space.

At the Laboratoire de Photophysique Moléculaire a new chemical reactor allows us to produce cosmic dust analogues (Cao *et al.* 2007). The chosen reactor is a low pressure flat burner providing sooting premixed flame (hydrocarbon / oxygen gas mixtures), a model system for the combustion community. The flame offers a broad range of combustion conditions and sampling, to synthesize a large variety of soot under controlled properties, ranging from hydrogenated amorphous carbon to strongly aromatic material.

It is known that irradiation processes play a important role in space in the life-cycle of carbonaceous dust. Consequently, in this work, we study the effects of ion irradiation (H^+, He^+, and Ar^{++}, energy of 200-400 keV), at the Laboratorio di Astrofisica Sperimentale in Catania, on several soot films, ranging from strongly aromatic to strongly aliphatic materials. The explored fluence range is $10^{14} - 10^{16}$ ions cm^{-2}.

Samples are monitored using infrared and Raman spectroscopy, to characterize the process before, during, and after irradiation. The experimental approach and setup are similar to the ones used previously in the Catania laboratory to characterize amorphization experiments of carbon-rich materials (Baratta *et al.* 2004, Brunetto *et al.* 2004).

2. Results

In Figure 1 we display two examples of Raman spectra of a soot sample (mix of aromatic and aliphatic structures) irradiated with 400 keV Ar^{++} ions and 200 keV He^+ ions.

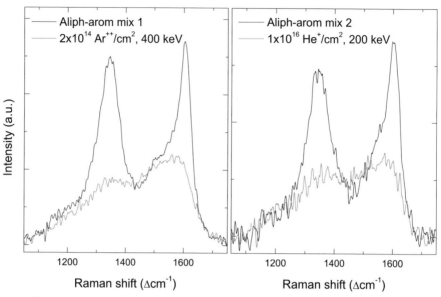

Figure 1. Raman spectra of soot sample (mix of aromatic and aliphatic structures) irradiated with 400 keV Ar^{++} ions (left) and 200 keV He^+ ions (right).

We measure the variations of relative peak sizes, positions, and widths of the so-called D (disorder) and G (graphitic) lines. These reflect the degree of disorder of the material (broader lines corresponding to more disordered materials). At the given energy, we observe an amorphization effect due to Ar irradiation $\sim$50 times more efficient than He.

We compare our results with those obtained in the analysis of extraterrestrial organic matter (Sandford *et al.* 2006) and other irradiation experiments of carbonaceous materials of astrophysical relevance (Baratta *et al.* 2004). We find that Raman spectroscopy confirms previous amorphization experiments on carbonaceous materials, closely following the evolutionary path of meteorites, interplanetary dust particles, and cometary grains collected by the Stardust mission.

IR spectroscopy shows the destruction of aromatic bonds (when present) partly converted to aliphatic C-H bonds, which are in turn destroyed at higher dose. The combined technique of soot production and modification by ion irradiation, enriches the possibilities to synthesize cosmic dust laboratory analogues with tailored properties. Astrophysical implications of the studied irradiation processes for the lifecycle of carbonaceous dust are under investigation.

References

Baratta, G. A., Mennella, V., Brucato, J. R., Colangeli, L., Leto, G., Palumbo, M. E., & Strazzulla, G. 2004, *J. Raman Spectrosc.*, 35, 487

Brunetto, R., Baratta, G. A., & Strazzulla, G. 2004, *J. Appl. Phys.*, 96, 380

Cao, A. T., Pino, T., Carpentier, Y., Dartois, E., Vasquez, R., Bréchignac, P., & D'Hendecourt, L. 2007, in: J. L. Lemaire, F. Combes (eds.), *Molecules in Space and Laboratory*, Paris, France, May 14-18, 2007. (S. Diana.), p. 81

Sandford, S. A. & 54 colleagues 2006, *Science*, 314, 1720

Organic Matter in Space
Proceedings IAU Symposium No. 251, 2008
S. Kwok & S. Sandford, eds.

Reactions of aromatics in space and connections to the carbon chemistry of Solar System materials

Max Bernstein

Astrophysics Branch, NASA/Ames Research Center
email: Max.Bernstein@nasa.gov

Abstract. Polycyclic aromatic hydrocarbons (PAHs) and related aromatic materials are thought to be the most abudant class of organic carbon in the universe, being present in virtually all phases of the ISM, and abundant in carbonaceous meteorites and asteroid and comet dust. The basic PAH skeleton is proposed to have formed in outflows of carbon rich stars, and isotopic measurements of extraterrestrial graphitic carbon is consistent with this notion. However, functionalized aromatics bearing oxygen atoms, aliphatic domains, and deuterium enrichments have been extracted from meteorites and more recently been measured in IDPs and Stardust retuned comet samples. Exposure of remnant circumstellar PAHs to energetic processing at low temperature in the presense of H_2O is the most parsimonious explanation for these observations.

We will present laboratory infrared spectra of various aromatic species and PAH cations in solid H_2O under conditions relevant for comparsion to absorptions attributed to PAHs observed towards objects embedded in dense clouds. In addition, we shall describe the reactions of PAHs under these conditions in the lab when they are exposed to energetic processing. Finally, we will propose a mechanism, and make specific predictions regarding the structures and distribution of deuterium that should be observed in extraterrestrial samples if low temperature ice radiation chemistry is playing a role in the formation of the molecules seen in Solar System materials.

Discussion

CODY: Did you ever see evidence in your quinoline experiments of nitrogen elimination? It's intriguing to think that if you are hydrogenating right next to the nitrogen, you might actually have some products where you've lost nitrogen.

BERNSTEIN: It's possible that it might have happened and that we did not see it. We tried to keep the doses extremely low to keep the chemistry fairly simple and to avoid blowing these things apart. I think quinoline (a naphthalene with a single nitrogen in one of its rings) is fairly stable. Preliminary indications are that compounds that have more than one nitrogen in them break apart much more easily. I think we were doing these experiments under conditions where we wouldn't have seen much nitrogen elimination.

KHARE: Your experiments are done at a very low temperature. At that temperature the products that you make won't fly away from the window. The only things that might escape will be things like N_2, CH_4, and CO.

BERSTEIN: Yes. In fact, not even them. We generally do these experiments on a salt window that is cooled to ~ 10 K. At this temperature the only thing that escapes from the ice is H_2, which you generate during the course of photolysis. If you do these experiments at temperatures above which some of the ice components, say methane, can sublime, then it will be harder to add that methyl group. The reaction efficiencies depend to some extent on the residence time of the associated molecules in the ice. If you are interested in reaction between a PAH and H_2O, it doesn't matter so much. If you are specifically interested in the reaction between a PAH and methane, then it would be more temperature dependent. However, the reactions would not change at all if you were to do them at lower temperatures.

SANDFORD: I am going to disagree on one point with both of you. When you photolyze these ices at these low temperatures, the photons break some of the molecules apart to produce ions and radicals. Very few of these are actually on the surface of the ice. The reason they don't escape is because they are trapped in the bulk of the solid H_2O-rich ice, not because the temperature is below the sublimation temperature of all the photoproducts. A lot of the group addition happens when the ice is warmed and its components become mobile as the H_2O sublimes away. So, in fact, at any temperature below the sublimation point of H_2O, you will get many of the very same products because you are just using the H_2O to store ions and radicals until they react during the warm up.

BERNSTEIN: That is true. If you were to vapor deposit a water:methane mixture at a low enough temperature that the methane is trapped, it would still be available for reactions even if the ice is later warmed.

KOBAYASHI: I understand that there are many aromatic compounds in space. Do you have any idea why aromatic amino acids are rarely found in meteorites or seen in other environments in space?

BERNSTEIN: My intuition, based on work we published in Nature in 2002 on the formation of amino acids, suggests that it's just easier to make simpler molecules. Using fairly simple starting materials, the kinds of molecules that are common in space, one certainly makes more glycine than anything else. The production of alanine or serine, where you are just adding one more carbon atom, shows considerably diminished yields. I imagine that we could make a naphthyl amino acid, but I think the more complicated the product is, the harder it is to make. The full answer is unlikely to be simple, however, since there are obviously multiple mechanisms leading to the amino acids that are seen in meteorites.

HENNING: You start with the assumption that PAHs in the diffuse interstellar medium have no side groups. Why do you make this assumption?

BERNSTEIN: I make that assumption for two reasons. First, we've heard the basic idea that in C-rich stellar outflows all the oxygen get used up to make carbon monoxide and any remaining carbon forms some sort of sooty material that contains just carbon and hydrogen. In this simple picture there's no O available for O-containing side groups. I am also presuming that side groups are going to be less stable against loss by photolysis in the diffuse ISM. It's possible that some functional groups could be present, but I'm simply starting with the worst-case scenario. I want to answer the question, why is there this richness in the chemistry of the aromatic compounds in meteorites? If I start with the simplest case, an aromatic molecule that has nothing on it, then it's not activated in any way. It's the toughest thing to drive in the chemistry. If I can get the process to work in that worst case scenario, then it should be even easier if you tell me that I get to start with methyl groups or bridging CH_2 groups.

HENNING: But then the process will be different, won't it?

BERNSTEIN: Yes, possibly, and then you get to make specific predictions. If we could get to the point where we could examine the hot spots in Stardust samples or in IDPs and say this or that functional group carries an isotopic enrichment, we could answer the question of whether they are circumstellar or whether they were introduced at another point. If there are clear correlations between certain functional groups and isotopic anomalies we could learn a great deal.

From left to right: Max Bernstein, Scott Sandford, and William Irvine (photo by Dale Cruikshank).

Cliff Matthews singing during his after-dinner speech at the banquet.

Organic Matter in Space
Proceedings IAU Symposium No. 251, 2008
S. Kwok & S. Sandford, eds.

Optical and chemical properties of tholins

Bishun N. Khare[1], Christopher P. McKay[2], Dale P. Cruikshank[2], Yasuhito Sekine[3], Patrick Wilhite[4], and Tomoko Ishihara[1]

[1] Carl Sagan Center, SETI Institute, Space Science Division, NASA Ames Research Center, USA
email: bkhare@mail.arc.nasa.gov

[2] Space Science Division, NASA Ames Research Center

[3] Department of Earth and Planetary Science, University of Tokyo, 7-3-1 Hongo, Bunkyo, Tokyo 113-0033, Japan

[4] Center for Nanostructures, School of Engineering, Santa Clara University, Santa Clara California 95053-0569

Abstract. For over three decades tholins have been synthesized from mixtures of the cosmically abundant gases CH_4, C_2H_6, NH_3, H_2O, HCHO, N_2, and H_2, previously in the Laboratory for Planetary Studies at Cornell University and in recent years at NASA Ames Research Center. The tholin synthesized by UV light or spark discharge on sequential and non-sequential pyrolysis GC-MS revealed hundreds of compounds, and on hydrolysis produced a large number of amino acids including racemic protein amino acids. Optical constants have been measured of many of the tholins, tholins produced from a condensed mixture of water and ethane at 77 K, poly HCN, and Titan tholin produced on electrical discharge through a mixture of 90% N_2 and 10% CH_4. Its optical constants were measured from soft x-rays to microwave for the first time.

Here we report the absorption properties of Titan tholin that is produced in the temperature range 135 to 178 K where tholins are produced by magnetospheric charged particles, then pass through lower temperature at 70 K and finally to the ground at 95 K. While descending to the ground, it gets coated and processed on the way by other sources of energy such as long UV and cosmic rays. It is therefore expected that the stable products of CH_4 photolysis react with Titan tholin to recycle the CH_4 supply in Titan's atmosphere. Furthermore, the reactions of gaseous C_2H_6 with the reactive materials on the surface of the tholin could incorporate atmospheric C_2H_6 into the tholin and therefore might reduce the deposition rate of C_2H_6 onto the ground of Titan.

Discussion

LORENZ: You suggest that the carbon in tholin might be recycled into the atmosphere as methane but I don't understand how you can recycle methane if the hydrogen escapes to space. The hydrogen budget doesn't close. Can you explain how you get around that?

KHARE: Well there is some evidence that hydrogen is absorbed as well, but I am still trying to confirm this using deuterium.

LORENZ: But on the planetary scale hydrogen is escaping from Titan.

KHARE: There is a possibility that some hydrogen may be absorbed on the tholin. If tholin could behave like a chromatographic column, it could absorb species and release them as a gas when it absorbs energy. In Titan's case, you have all kinds of energy falling on the tholin at various heights. It is very important to consider the concept that the real Titan tholin is likely coated with materials as it descends to the ground. The tholin has absorption characteristics and maybe catalytic properties too. Our job now is to create tholin under Titan temperatures and pressures and then examine its properties without exposing it to air.

LORENZ: I still maintain that, whatever exchanges and reactions occur, on the planetary scale you cannot solve the methane survival problem with tholin.

KHARE: Our current research tells us that tholin can be at least a partial contributor to the available methane for astronomical periods of time with the same mixing ratio.

Bishun Khare and Delphine Nna-Mvondo (photo by Dale Cruikshank)

Organic Matter in Space
Proceedings IAU Symposium No. 251, 2008
S. Kwok & S. Sandford, eds.

Photochemistry of interstellar/circumstellar ices as a contributor to the complex organics in meteorites

Michel Nuevo and Scott A. Sandford

NASA Ames Research Center, Space Science Division
Mail Stop 245-6, Moffett Field, CA 94035, USA
email: `michel.nuevo-1@nasa.gov`, `scott.a.sandford@nasa.gov`

Abstract. The UV irradiation of interstellar/circumstellar ice analogs is known to lead to the formation of organic compounds such as amino acids and maybe nucleobases. In this work, the mechanisms of formation and distribution of amino acids, chosen as tracers for the organic compounds formed in such experiments, are studied and compared with meteoritic data.

Keywords. Astrochemistry, methods: laboratory, ISM: molecules, ultraviolet: ISM, meteorites

1. Introduction

Organic molecules have been shown to be formed in interstellar/circumstellar environments consisting of gas and dust particles covered with a thin ice layer containing mainly H_2O, and then CO, CO_2, CH_3OH, NH_3, and CH_4 (Gibb *et al.* 2004), and subjected to UV-photon and cosmic-ray radiations (Mathis *et al.* 1983). Those organics are then incorporated into protostellar nebulae, such as the one in which our Solar System formed, as well as into comets and interplanetary dust particles, that can be collected in Earth's atmosphere (Matrajt *et al.* 2005) and onboard spacecraft (Sandford *et al.* 2006). Their recent analysis shows the presence of a non-negligible organic fraction. Some compounds, such as amino acids and nucleobases, are of particular interest for prebiotic chemistry. Amino acids have been identified in organic residues formed from the UV irradiation of interstellar ice analog mixtures (Bernstein *et al.* 2002, Muñoz Caro *et al.* 2002, Nuevo *et al.* 2008), and have been detected in carbonaceous chondrites such as Murchison (Cronin & Pizzarello 1997). However, their detection in the interstellar medium (ISM) is still debated (Kuan *et al.* 2003). Nucleobases have also been found in meteorites (van der Velden & Schwartz 1977, Stoks & Schwartz 1979), but have so far never been observed in the ISM (Charnley *et al.* 2005).

2. Experimental setup and protocol

In a vacuum chamber (a few 10^{-8} torr), gases are deposited on a substrate cooled down to 15–20 K. These gases are chosen to mimic the composition of the ISM. Under such physical conditions, the gases condense on the substrate in a thin ice layer, which is simultaneously irradiated by a UV H_2 lamp, emitting UV photons at 121.6 nm (Lyman α) and around 160 nm. After irradiation, the system is warmed to room temperature, where an organic residue remains on the substrate and can be analyzed. Such residues contain a large variety of compounds, including amino acids (Bernstein *et al.* 2002, Muñoz Caro *et al.* 2002, Nuevo *et al.* 2008). Liquid (LC) and gas (GC) chromatographies are the main techniques used to analyze residues. Such techniques separate residues into their individual compounds by differences of diffusion rates through a chemically-active column.

3. Results

Formation and distribution of amino acids. Analyses of these organic residues show that amino acids *always* form as long as the starting ice mixture contains the 4 elements C, H, N, and O (Nuevo *et al.* 2008). The irradiation time and the temperature (10–80 K) have no significant effect on their final distribution. In most of cases, glycine is found to be the most abundant amino acid formed, the abundance of the other compounds decreasing exponentially with their molecular mass, indicating that such compounds were formed via non-biological processes. However, the formation mechanisms of amino acids are not clearly understood. Laboratory studies show that there are in fact several competing mechanisms leading to their formation, and that their distribution in the laboratory is different from what is observed in meteorites (Elsila *et al.* 2007, Nuevo *et al.* 2008).

Comparison with meteoritic amino acids. The abundances of the 3 most abundant amino acids formed in laboratory organic residues, namely glycine, alanine, and serine, have been compared with their abundances in 1 g of the Murchison meteorite (Nuevo *et al.* 2008 and references therein). In the residues, alanine was found to be produced with abundances half that of glycine, whereas in Murchison these two amino acids have similar abundances. Serine is formed with lower relative abundances in Murchison than in residues. These discrepancies indicate that the formation mechanisms of amino acids are different in the laboratory and in meteorites, or that Murchison had subsequent alteration, such as the (photo-) chemical process or decomposition of high molecular mass amino acids into smaller compounds such as glycine and alanine. However, this comparison is biased since mainly only proteinaceous amino acids have been searched for in organic residues (Nuevo *et al.* 2008). The search for non-proteinaceous amino acids such as α-aminobutyric acid or isovaline in residues, which have been detected in Murchison, may allow a better understanding of the formation mechanisms of such compounds.

Nucleobases. Nucleobases are complex molecules consisting of one (pyrimidine-based) or two (purine-based) rings containing nitrogen atoms. They are probably key compounds for prebiotic chemistry. The formation and photo-stability of pyrimidine- and purine-based compounds are currently being studied at NASA Ames, where pyrimidine mixed with H_2O ice is UV irradiated at low temperature. The early results show that such mixtures lead to the formation of nucleobase-like compounds.

References

Bernstein, M., Dworkin, J., Sandford, S., Cooper, G., & Allamandola, L. 2002, *Nature*, 416, 401

Charnley, S. B., *et al.* 2005, *Adv. Sp. Res.*, 36, 137

Cronin, J. R. & Pizzarello, S. 1997, *Science*, 275, 951

Elsila, J. E., Dworkin, J. P., Bernstein, M. P., Martin, M. P., & Sandford, S. A. 2007, *ApJ*, 660, 911

Gibb, E. L., Whittet, D. C. B., Boogert, A. C. A., & Tielens, A. G. G. M. 2004, *ApJS*, 151, 35

Kuan, Y.-J., Charnley, S. B., Huang, H.-C., Tseng, W.-L., & Kisiel, Z. 2003, *ApJ*, 593, 848

Mathis, J. S., Mezger, P. G., & Panagia, N. 1983, *A&A*, 128, 212

Matrajt, G., Muñoz Caro, G. M., Dartois, E., d'Hendecourt, L., Deboffle, D., & Borg, J. 2005, *A&A*, 433, 979

Muñoz Caro, G. M., Meierhenrich, U. J., Schutte, W. A., Barbier, B., Arcones Segovia, A., Rosenbauer, H., Thiemann, W. H.-P., Brack, A., & Greenberg, J. M. 2002, *Nature*, 416, 403

Nuevo, M., Auger, G., Blanot, D., & d'Hendecourt, L. 2008, *Orig. Life Evol. Biosph.*, 38, 37

Sandford, S. A., *et al.* 2006, *Science*, 314, 1720

Stoks, P. & Schwartz, A. 1979, *Nature*, 282, 709

van der Velden, W. & Schwartz, A. 1977, *Geochim. Cosmochim. Acta*, 41, 961

Organic Matter in Space
Proceedings IAU Symposium No. 251, 2008
S. Kwok & S. Sandford, eds.

Radiation chemistry approach to the study of ice analogs

Maria Colin-Garcia, Alicia Negrón-Mendoza, and Sergio Ramos-Bernal

Instituto de Ciencias Nucleares, Universidad Nacional Autónoma de México,
Circuito Exterior, Cd. Universitaria, 04510 México, D. F., México
email: `negron@nucleares.unam.mx`

Abstract. The aim of this work is to study the chemistry of the irradiation of frozen solutions of HCN. This compound has been detected in comets and other icy bodies. The CN group might have made its first appearance in the early stages of chemical evolution. Therefore, is behavior under irradiation at low temperature is relevant for chemical evolution studies and icy bodies.

Keywords. Astrobiology, astrochemistry, molecular processes, laboratory, comets

1. Introduction

In the light of the present knowledge of ices in the Solar System, a radiation chemistry approach can be very useful to study the behavior of compounds connected not only to these bodies, but also to interstellar chemistry.

Comets are thought to be carriers of material that was initially available on Earth. The Deep Impact mission to Comet 9P/Tempel 1 showed the presence of H_2O, C_2H_6, C_2H_6, HCN, CO, CH_3OH, H_2CO, C_2H_2, and CH_4 (Mumma *et al.* 2005). To this end, the organic inventory is of central importance concerning the origin of life.

The comet is exposed to radiation for more than 4.6 billion years. The amount of energy deposited in a cometary nucleus comes from internal and external sources (Whipple 1977).

In simulation experiments it is possible to use radiation sources which are more common in radiation chemistry laboratories, with the same LET as that protons at cosmic energies. The gamma radiation of ^{60}Co has an average LET of 0.23 KeV, similar to that of billon electron-volt protons. This means that the average energy deposit along the radiation paths, and so the distribution of reactive species and their reactions, is similar (Draganic *et al.* 1984).

2. Experimental

Hydrogen cyanide was generated in a special setup from KCN and sulfuric acid. The concentrations ranged from 0.001 to 0.2 moles dm^{-3} and the pH was 6-9. The frozen (77 K) samples were prepared in an argon atmosphere. The irradiation was carried out using a ^{60}Co source. The sample radiation doses were from 3 kGy to 419 kGy. After irradiation, the samples were thawed at room temperature. The analysis was made according to the procedures followed by Colin *et al.* (2008).

3. Results and discussion

The inventory of products include several types of compounds such as gases and non volatile products that remain in the solution. Among the gases are carbon monoxide, carbon dioxide, methane, ammonia, and hydrogen. To analyze non-volatile products, the solution was lyophilized and the dry residue hydrolyzed and analyzed for small molecules, to reveal the formation of biologically important compounds. The amounts and aspect of the dry residue depend on the absorbed dose. The infrared spectra of the dry residue showed characteristic absorption bands that suggest that their constituents are mainly polyamides, urea, and substituted urea.

The decomposition of HCN formed about 76% of the oligomeric material. This product constituted the main component of the dry residue. The estimated molecular weight for such oligomers is up to 20000 daltons. The formation of the oligomers was observed in all the concentrations and doses studied. Oligomer concentration increased with the dose, and after the depletion of cyanide the build up ended. It was found that cyanide solutions of 10^{-3} moles dm^{-3} irradiated at 2.5 kGy, at room temperature, are enough to produce these types of compounds.

4. Final Remarks

The chemical action of ionizing radiation consists essentially in producing chemically reactive species (free radicals, radical-ions). In water-dominated, CN-containing mixtures, radicals from water ices react by addition of abstraction with CN molecules and produce oligomers. The data obtained suggest that they are formed by polyamides and esters. Upon hydrolysis, these oligomers released compounds of biological significance like amino acids, purines and carboxylic acids. These oligomeric materials present the same characteristics in the frozen and in the liquid multi-component systems. The yields in ice were one order of magnitude lower. This is mainly due to the rigid structure of the ice and the limited mobility of the radicals at low temperature. The results of this study underline the importance of radiation-induced reactions as an energy source in extraterrestrial scenarios like in comets or other icy objects.

5. Acknowledgements

This work was supported by DGAPA-UNAM Grant IN223406. The support of the Posgrado en Ciencias Biológicas, UNAM for one of us (MC) is acknowledged.

References

Colín-García, M., Negrón-Mendoza, A., & Ramos-Bernal, S. 2008. *Astrobiology*, in press
Draganič, I. G., Draganič, S. D., & Vujosevič, S. 1984, *Icarus*, 60, 464
Mumma, M. J., Di Santi, M. A., Magee-Sauer, K., Bonev, B. P., Villanueva, G. L., Kawakita, H., Dello Russo, N., Gibb, E. L., Blake, G. A., Lyke, J. E., Campbell, R. D., Aycock, J., Conrad, A., & Hill, G. M. 2005, *Science*, 310, 270
Whipple, P. L. 1977, in: A. H. Delsemme (ed.), *Comets, asteroids, meteorites*, University of Toledo, (Toledo) p. 25

Organic Matter in Space
Proceedings IAU Symposium No. 251, 2008
S. Kwok & S. Sandford, eds.

Irradiation of mixed ices as a laboratory cometary model

Maria Colin-Garcia[1], Alicia Negrón-Mendoza[1], Sergio Ramos-Bernal[1], and Elizabeth Chacon[2]

[1]Instituto de Ciencias Nucleares, Universidad Nacional Autónoma de México, C. U. Mexico, D. F. Mexico 04510

[2]Facultad de Ciencias de la Tierra, Universidad Autónoma de Nuevo León
email: negron@nucleares.unam.mx

Abstract. Icy bodies in space are being irradiated continuously by ionizing radiation. Therefore, the transformation of organic molecules trapped in extraterrestrial ices might have been possible. This work studied a bulk irradiation of a mixture of some constituents of cometary nuclei. The results show that the formation of different compounds, among them ammonia, carbon dioxide, amines, ureas, free amino acids, and oligomeric material, yields carboxylic acids, amino acids, and purines upon hydrolysis.

Keywords. Astrobiology, astrochemistry, molecular processes, laboratory, comets

1. Introduction

Comets are minor bodies in our Solar System as old as the Solar System itself. They were thought to provide unique information about the pristine Solar System. However, chemical evolution processes that occurred at early stages of the Solar System and after could have transformed the organics within them. Irradiation of those bodies has occurred since they were formed from interstellar grains; UV photons and cosmic ions had already interacted with those grains (Hudson & Moore 1999). Later, during their storage in the reservoirs mentioned, their surfaces were continuously irradiated by cosmic rays. All of this exposure alters the chemical and physical properties of those bodies (Hudson & Moore 1999).

Comet nuclei are composed of rock, dust, water ice, and frozen gases such as carbon monoxide, carbon dioxide, methane, and ammonia. Also there are organic molecules such as methanol, hydrogen cyanide, formaldehyde, ethanol, and ethane, as well as perhaps more complex molecules such as long-chain hydrocarbons and amino acids.

The aim of this work is to give another contribution to the understanding of the evolution of the organics in comets, studying the products obtained by the behavior of an (aqueous-dominant) simplified cometary model exposed to high doses of ionizing radiation. The experiment was performed using gamma rays of the same LET (linear energy transfer) of the most abundant protons in the cosmic rays (Draganic *et al.* 1984). The mixture consisted of $HCN/CH_3OH/CH_3CN/C_2H_5CN/HCOOH$, in a proportion in which they appeared in a dense interstellar cloud (1: 0.6: 0.2: 0.1: 0.05). The samples were at 298 K and 77 K.

2. Experimental

Irradiations were carried out using a gamma ray source of ^{60}Co (Gamma-beam 651-PT) at ICN, UNAM. Frozen and liquid solutions were exposed to different radiation doses from 3 to 419 kGy. After irradiation, the samples were melted and the following measurements were made directly on the irradiated samples: determination of urea, pH of the irradiated solution, amino acids, and the non-reacted fraction of the reactants (Colín-García *et al.* 2008). For non-volatile compounds the bulk of irradiated solution (300 ml) was lyophilized. After this procedure, the dry residue was weighted and analyzed by IR spectroscopy and HPLC. An aliquot of the dry residue (50 mg) was hydrolyzed with hydrochloric acid diluted in methanol, 1: 5. Then, samples were esterified and analyzed by gas chromatography (Colín-García *et al.* 2008).

3. Results and discussion

The radiolysis of the multi-component system is a composite process that can be considered roughly as the sum of the behavior of the individual constituents of the system. Cross-reactions may occur. Ionizing radiation induced an abundant formation of a dark-colored mixture of non-volatile compounds. The inventory of radiolytical products includes gases, amines, amino acids, carboxylic acids, and oligomeric material. There is formation and accumulation of radiolytic products. At increasing doses, the primary chemical constituents of the model system become depleted, while the radiolytic products, once formed, accumulate.

The amount and aspect of the dry residue depend on the dose received and the temperature of irradiation. In both sets of experiments the dry residue increased with dose.

The nature of the oligomers is complex, mainly consisting of polyesters and polyamides with urea fragments. Carboxylic acids were detected in the hydrolyzed material. The products detected at 77 K were malonic, succinic, glutaric, carboxysuccinic, citric, and tricarballylic acids. Free amino acids are produced in very small quantities. The production of these compounds is also dose and temperature dependant.

4. Final Remarks

The oligomer is produced by addition or abstraction reactions of the radicals from water ices radiolysis and CN-containing molecules. The radiochemical yields in ice were one or two orders of magnitude lower. This is due mainly to the rigid structure of the ice and the limited mobility of the radicals at low temperature. The results found in this study underline the importance of radiation-induced reactions as an energy source in extraterrestrial scenarios like comets or other icy objects.

Acknowledgements

This work was supported by DGAPA-UNAM Grant IN223406. The support of the Posgrado en Ciencias Biológicas, UNAM for one of us (MC) is acknowledged.

References

Colín-García, M., Negrón-Mendoza, A., & Ramos-Bernal, S. 2008, *Astrobiology*, in press
Draganič, I. G., Draganič, Z. D., & Vujosevič, S. 1984, *Icarus*, 60, 464
Hudson, R. L. & Moore, M. H. 1999, *Icarus*, 140, 450

Organic Matter in Space
Proceedings IAU Symposium No. 251, 2008
S. Kwok & S. Sandford, eds.

Photodesorption of ices – Releasing organic precursors into the gas phase

Karin I. Öberg[1], Ewine F. van Dishoeck[2], and Harold Linnartz[1]

[1]Raymond and Beverly Sackler Laboratory for Astrophysics
Leiden Observatory, Leiden University
P.O. Box 9513, NL–2300 RA Leiden, the Netherlands
email: oberg@strw.leidenuniv.nl, linnartz@strw.leidenuniv.nl

[2]Leiden Observatory, Leiden University
P.O. Box 9513, NL–2300 RA Leiden, the Netherlands
email: ewine@strw.leidenuniv.nl

Abstract. A long-standing problem in interstellar chemistry is how molecules can be maintained in the gas phase at the extremely low temperatures in space. Photodesorption has been suggested to explain the observed cold gas in cloud cores and disk mid-planes. We are studying the UV photodesorption of ices experimentally under ultra high vacuum and at astrochemically relevant temperatures ($15 - 27$ K) using a hydrogen discharge lamp (7-10.5 eV). The ice desorption during irradiation is monitored using reflection absorption infrared spectroscopy and the desorbed species using mass spectrometry. We find that both the UV photodesorption rates and mechanisms are highly molecule specific. CO photodesorbs without dissocation from the surface layer of the ice. N_2, which lacks dipole allowed electronic transitions in the range of the lamp, does not photodesorb. CO_2 desorbs through dissociation and subsequent recombination from the top few layers of the ice. At low temperatures ($15 - 18$ K) the derived photodesorption rates are $\sim 10^{-3}$ for CO and CO_2 and $< 2 \times 10^{-4}$ for N_2 ice per incident photon.

Keywords. Astrochemistry, molecular data, molecular processes, methods: laboratory

1. Introduction

CO and CO_2 ice are among the most abundant species in star forming regions, after H_2, and, e.g., Tielens & Charnley (1997) show that they form the building blocks of a complex organic chemistry once desorbed into the gas phase. A long-standing problem is how these molecules can be maintained in the gas phase at the low temperatures found in star forming regions, where all molecules other than H_2 should stick on dust grains on timescales shorter than the cloud and outer disk lifetimes. Yet lines of gaseous CO are detected towards dark clouds (e.g., Bergin *et al.* 2002). Similarly, Piétu et al. (2007) observed abundant gas-phase CO, which cannot be explained by thermal equilibrium chemistry, in the cold parts of protoplanetary disks. Very recently Sakai *et al.* (2008) detected cold HCO_2^+ (tracing CO_2), toward a protostar as well. Photodesorption of ices has been suggested to explain these kinds of observations and Öberg *et al.* (2007) showed that CO photodesorption is an efficient process with a rate of $3(\pm 1) \times 10^{-3}$ photon^{-1}. We also reported that N_2 does not photodesorb. In this study we extend the CO and N_2 photodesorption study to include different ice temperatures and morphologies. We also present the first results on the photodesorption rate of CO_2 and its dependences on ice thickness, UV flux, irradiation time and total UV dose.

2. Experimental and Data Analysis

In the photodesorption experiments, thin ices of 2 to 100 monolayers (ML) are grown at 15–27 K on a gold substrate under ultra-high vacuum conditions ($P \sim 10^{-10}$ mbar). The ice films are subsequently irradiated with UV light from a broadband hydrogen microwave discharge lamp, which covers 120–170 nm (7–10.5 eV). The setup allows simultaneous detection of molecules in the gas phase by quadropole mass spectrometry (QMS) and in the ice by reflection absorption infrared spectroscopy (RAIRS) using a Fourier transform infrared spectrometer. The ice UV destruction rate is determined by RAIR spectroscopy of the ice during irradiation as described by Öberg *et al.* (2007). In general we define the photodesorption rate as the destruction rate of the original ice minus the formation rate of other ice species. N_2 has no permanent dipole moment and is not detected with RAIRS. Instead the photodesorption upper limit is constrained using only QMS measurements.

3. Results and Discussion

In the extended investigation of CO ice photodesorption, we find that the desorption rate decreases with increasing temperature such that it is almost a factor of 3 lower at 27 K compared to at 15 K. When the ice is deposited at 27 K and then cooled down to 16 K before irradiation the rate is a factor of 3 lower as well, showing that the CO ice photodesorption depends on the ice structure (as modified by annealing) rather than temperature. An additional experiment with a CO:N_2 = 4:4 ML mixed ice indicates a slight N_2 co-desorption with CO at a rate of $\sim 3 \times 10^{-4}$ photon^{-1}. This N_2 co-desorption may provide the means of releasing N_2 ice into the gas phase non-thermally, despite its low UV absorption cross section.

In the new CO_2 ice experiments, the derived CO_2 photodesorption rate has no measurable dependence on lamp flux, irradiation time or total photon dose at 18 K. There is also no measurable isotope dependence between $^{13}C^{18}O_2$ and $^{13}C^{16}O_2$. The CO_2 photodesorption rate at 18 K is however dependent on ice thickness up to several monolayers; the rate increases from ~ 0.7 to 1.8×10^{-3} photon^{-1} when the ice is grown from 2 to 10 ML. Beyond 10 ML, the rate is thickness-independent, indicating that only the top 10 ML can photodesorb. This is in contrast with the surface layer desorption of CO ice. The difference can be understood from the different photodesorption mechanisms of CO and CO_2; CO photodesorption is predicted to occur on surfaces only (Takahashi & van Hemert, in prep.), while the dissociation of CO_2 prior to desorption produces energetic fragments, which can penetrate several ice layers. When including photodesorption of different ices in astrophysical models these different thickness dependences need to be taken into account to accurately predict the photodesorption efficiencies.

In general this study shows that at least two ices are efficiently photodesorbed and we recommend that photodesorption be included in models where non-thermal desorption is expected to affect the model outcome.

References

Bergin, E. A., Alves, J., Huard, T., & Lada, C. J. 2002, *ApJ* (Letters), 570, L101

Öberg, K. I., Fuchs, G. W., Awad, Z., et al. 2007, *ApJ* (Letters), 662, L23

Piétu, V., Dutrey, A., & Guilloteau, S. 2007, *A&A*, 467, 163

Sakai, N., Sakai, T., Aikawa, Y., & Yamamoto, S., 2008, *ApJ* (Letters), 675, L89

Tielens, A. G. G. M. & Charnley, S. B. 1997, *Origins of Life and Evolution of the Biosphere*, 27, 23

Organic Matter in Space
Proceedings IAU Symposium No. 251, 2008
S. Kwok & S. Sandford, eds.

Electron, proton and ion induced molecular synthesis and VUV spectroscopy of interstellar molecules in the ice phase

Bhalamurugan Sivaraman[1], Sohan Jheeta[1], Nigel Mason[1], Adam Hunniford[2], Tony Merrigan[2], Bob McCullough[2], Daniele Fulvio[3], Maria Elisabetta Palumbo[3], and Marla Moore[4]

[1]Department of Physics and Astronomy, The Open University, Walton Hall, Milton Keynes, MK7 6AA, UK
email: b.sivaraman@open.ac.uk; s.jheeta@open.ac.uk; n.j.mason@open.ac.uk

[2]International Research Centre for Experimental Physics, Queen's University in Belfast
email: c.a.hunniford@qub.ac.uk; tmerrigan01@qub.ac.uk; rw.mccullough@qub.ac.uk

[3]Catania Astrophysical Observatory, Catania University
email: dfu@oact.inaf.it; elisabetta.palumbo@oact.inaf.it

[4]Cosmic ice lab, NASA, Goddard Space Flight Center
email: Marla.H.Moore@nasa.gov

Abstract. Planets and their moons are constantly subjected to irradiation from both their respective planetary magnetospheres and the solar wind. Energetic particles (electrons, protons and ions) in such radiation may induce complex chemistry within the icy mantles of such bodies, producing many organic compounds. Such processes can be simulated in laboratory experiments. In this report we present recent results from experiments exploring both molecular synthesis and the morphology of such ices.

The morphology of any ice may be characterised by IR and Vacuum Ultra-Violet (VUV) spectroscopy. The latter is particularly useful for studying ices in which infrared inactive molecules like oxygen (O_2) are common. We have shown that oxygen forms dimers in typical planetary ices and that, in contrast to previous analysis, many of the chemical reactions within the ice involve such dimer (and larger cluster) chemistry. We also present the results of a series of experiments that explore electron, proton and ion irradiation on Solar System relevant ices such as carbon dioxide (CO_2) at different temperatures. Infrared spectra recorded before and after irradiation are used to identify and quantify molecules formed in such irradiation, e. g. ozone. These experiments show that the morphology of the ice plays a critical role in the chemistry.

Keywords. Astrochemistry, methods: laboratory, techniques: spectroscopic

---

1. Oxygen dimer

Molecular oxygen (O_2) has recently been found in interstellar space (Larsson *et al.* 2007). VUV spectroscopy is considered to be an effective tool to study the properties of this molecule in the ice phase. In our experiments molecular oxygen was deposited at a pressure of 1×10^{-7} mbar onto a cold CaF_2 window, kept at ~25 K, for different exposure times (15, 20, 25 and 32 second) and VUV spectra were recorded. The feature peaking at 180 nm was assigned to the $(O_2)_2$, dimer (Mason *et al.* 2006). The band was very weak when using thin O_2 ice layers but grew in intensity with increased deposition to form a thick O_2 ice (Figure 1). This suggests that more O_2 dimers are formed in thicker ice samples. Therefore, the proportion of dimers and monomers of O_2 vary with the ice thickness.

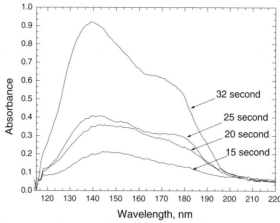

Figure 1. VUV spectra of oxygen ice recorded at different thickness and at $\sim$ 25 K.

2. Carbon dioxide

Laboratory experiments were carried out using CO_2 ice to study the processes of importance when magnetospheric ions impinge upon lunar surfaces. Both reactive (D^+, H^+) and non-reactive (He^+) ions, with energies ranging from 1.5−3 keV, were used to irradiate analogues of typical planetary ices. Irradiation was performed at two temperatures (30 and 80 K) to explore the effect of ice density and morphology.

In pure CO_2 ice, products include CO, O_3 and CO_3. In contrast to initial expectations, the largest concentration of O_3 was observed at the lower temperature (Figure 2). This is ascribed to O atoms being more localised in the lower temperature ice (i.e., within a matrix) whilst at higher temperatures they are free to migrate through the ice and more reaction pathways are opened, thereby hindering further O_3 production. At 80 K, a small fraction of O_3 can also be lost due to sublimation. The same results were observed in recent experiments (Moore *et al.* 2008) using 10 keV electrons and 0.8 MeV protons on CO_2 ices.

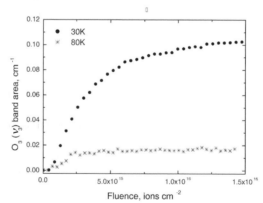

Figure 2. O_3 growth during 2.12 keV D^+ ion irradiation on pure CO_2 ice kept at 30 and 80 K.

References

Larsson, B., *et al.* 2007, *A&A*, 466, 999

Mason, N. J., *et al.* 2006, *Faraday discussions*, 133, 311

Moore, M., Hudson, R. L., Sivaraman, B., & Mason, N. J. 2008, *The formation and destruction of ozone in ices relevant to planetary and satellite surfaces*, Abstract submitted to COSPAR 2008

Organic Matter in Space
Proceedings IAU Symposium No. 251, 2008
S. Kwok & S. Sandford, eds.

© 2008 International Astronomical Union
doi:10.1017/S1743921308022175

Hydrogen cyanide polymers connect cosmochemistry and biochemistry

Clifford N. Matthews[1]† and Robert D. Minard[2]

[1]University of Illinois, Chicago, IL USA
email: cnmatthews@cs.com

[2]Penn State Astrobiology Research Center and Department of Chemistry, Penn State
University, University Park, PA 16802 USA

Abstract. To understand the origin of protein/nucleic acid based life as we know it on Earth, we must "follow" the nitrogen. Because of its unique hydrogen bonding characteristics, nitrogen is the key element in catalytic and/or informational proteins and nucleic acids essential to cell function and reproduction. We present evidence that HCN is the original source of prebiotic protein and nucleobase nitrogen. We also present chemically rational models supporting the radical hypothesis that the polymerization of HCN yields ab initio mundi prebiotic protein and polynucleobase macromolecules of sufficient size and complexity to allow the spontaneous generation of pre-RNA World biopolymers capable of catalysis and information transfer.

Keywords. Astrochemistry, astrobiology, comets: general, ISM: molecules, infrared: ISM, meteors, meteoroids, molecular processes, planets and satellites: individual (Earth, Titan)

The idea that the unidentified IR bands (UIBs) from diffuse atomic and molecular clouds are attributable to polycyclic aromatic hydrocarbons (PAHs) has gained considerable popularity since their discovery in 1973 (Bernstein *et al.* 1999). However, PAHs are not the only possible explanation for these strong IR emissions and they could arise from other forms of carbonaceous material containing sp^2 hybridized carbon (such as C=N) or from organic nanoparticles (Thaddeus 2006). Since unfunctionalized PAHs have no role in biochemistry today, their involvement in prebiotic chemistry is speculative (Platts 2004, Ehrenfreund *et al.* 2006).

In contrast to PAHs, the presence of HCN in comets, protostellar, planetary and lunar atmospheres is unequivocal. It is also well known that this HCN can polymerize spontaneously to HCN polymers under a wide variety of conditions. A much more credible link from cosmochemistry to biochemistry involves HCN polymers that may be part of organic nanoparticles, including PAHs, responsible for the UIBs. There is also evidence for the presence of these polymers in meteorites, on comets and Titan.

The polymerization of HCN is a spontaneous, exothermic reaction initiated by bases such as ammonia or free radicals from ionizing radiation and occurring over a wide range of temperature and pressure in both polar (H_2O) or non-polar (hydrocarbon) phases and on surfaces (Matthews *et al.* 2006). For example, a toluene solution of 1M HCN with 1 mol% of triethylamine as a base catalyst changes from yellow to orange to brown to black over a few days with almost all the HCN ultimately being converted to a solid black precipitate consisting of roughly 1 micron polymeric particles.

Incredibly, regardless of the conditions under which it is produced, when black HCN polymer is stirred with water, it partially dissolves to yield solutions containing small but

† Present Address: 64 Gothic St., Apt. 203, Northampton, MA 01060 USA

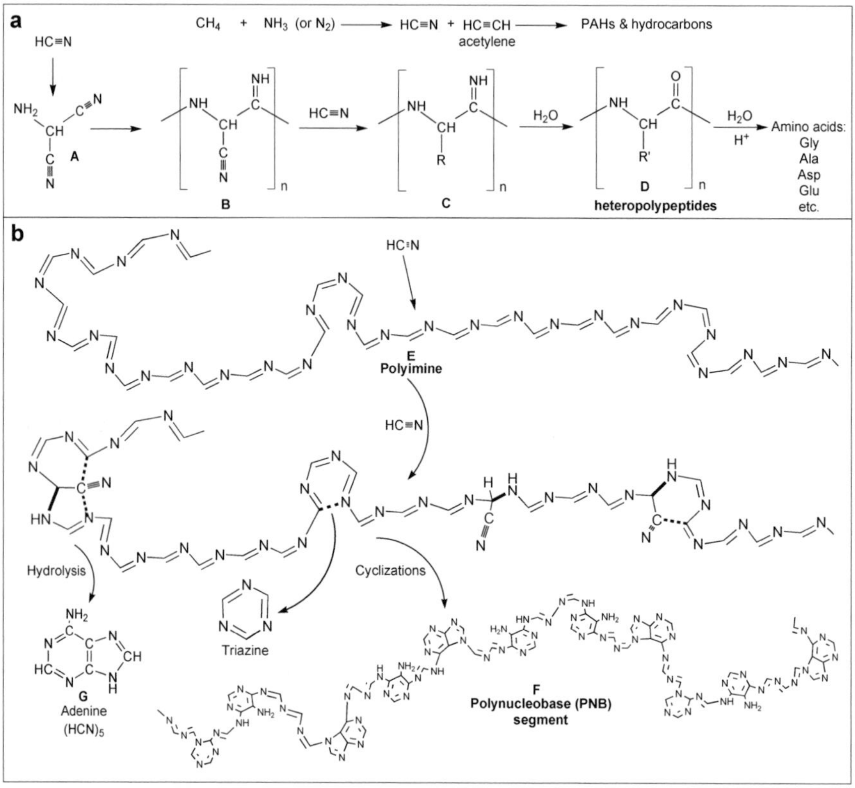

Figure 1. Pathways and structures proposed for HCN polymerization. A sample of HCN polymer may possess any or all of these structures including hybrids (multimers).

measurable quantities of a number of amino acids and nucleobases, the building blocks of both proteins and nucleic acids (Oró 1961, Oró & Kamat 1961). These solutions must also contain higher molecular weight macromolecules because on further hydrolysis by heating in acid or base (6N HCl 24 hr at 110° or pH 8 phosphate buffer, 140°, 3 days) greatly increased types and quantities of amino acids and nucleobases are detected. Glycine is the major amino acid produced (1 to 9%) with lesser amounts of alanine, aspartic acid, glutamic acid and several other amino acids (Matthews & Moser 1966, 1967). The nucleobases adenine, guanine, uracil and eight others were identified from the hydrolysis of solids produced in a dilute aqueous NH_4CN solution frozen at -78 °C for 27 years (Miyakawa *et al.* 2002). Without hydrolysis, the yields of amino acids or nucleobases were at least 10 times lower or were undetectable. Eight of twelve nucleobases were undetectable without hydrolysis.

There have been many studies showing how small molecules such as HCN, aldehydes, ammonia, or HCN tetramer can be converted into prebiotic monomers: amino acids via the Strecker synthesis or nucleobases via HCN tetramer chemistry (see below). There are two major problems with monomers: 1) they are water soluble and would be infinitely dilute in the primitive ocean; 2) condensing the amino acids or nucleobases into polymers with catalytic or informational capabilities would be virtually impossible in the watery ocean environment. Matthews & Moser (1967) showed that water is not required initially to form amino acids, as it would be if the Strecker route was involved. When methane and ammonia (or nitrogen) are sparked in the absence of water, hydrogen cyanide and acetylene are formed and both polymerize, the acetylene yielding primarily hydrocarbons and PAHs and the HCN yielding primarily HCN polymer (see Figure 1). Here, we propose models whereby HCN polymerizes directly to form macromolecules containing polypeptide and linked nucleobase segments.

First reported by Proust (1808), HCN polymer has still not been fully characterized because of the complex heterogeneous nature of its structure. We can, however, rationalize the hydrolysis products with models that are consistent with known mechanistic organic chemistry (Figure 1). In the 1960's, Matthews proposed a mechanism (Figure 1a). Polyaminomalononitrile, **B**, can be considered an addition polymer of the reactive trimer aminomalononitrile, **A**. Cumulative reactions of HCN on the highly activated nitrile groups of **B** then yield the heteropolyamidines **C** which are readily converted by water to heteropolypeptides, **D**, with release of ammonia and CO_2. **D** can be hydrolyzed to amino acids as shown. There are several chemical experiments consistent with this model including deuterium exchange studies (Matthews *et al.* 1977) and synthesis and modification of an analogue, poly-α-cyanoglycine (Minard *et al.* 1975).

Using solid state NMR, Mamajanov & Herzfeld (2008) have shown that black HCN polymer contains the extended polyimine structure **E** shown in Figure 1b. A mechanism has been proposed (Figure 1b) for the transformation of the polyimine structure into a polynucleobase (PNB) that can be hydrolyzed into nucleobases. Addition of HCN across -CH=N- bonds of the polyimine creates pendant cyano groups that can undergo thermal or photochemical cyclization to form triazines, purines and pyrimidines linked together in a polynucleobase macromolecule **F**. Small molecule cyclization chemistry related to that shown has been explored in reactions of HCN tetramer with HCN to form adenine (Ferris & Orgel 1966, Glaser *et al.* 2007). This new model helps to explain the sequence of carbons and nitrogens in purines and pyrimidines.

Turning to extraterrestrial chemistry, there is a growing body of evidence for the presence of HCN polymers on bodies such as moons (Titan), meteorites, and comets. Titan's atmosphere consists primarily of a mixture of nitrogen, a few percent of methane, and traces of acetylene, HCN, CO, and CO_2. In 1980, the Voyager 1 mission passed by Titan and showed it was covered by a thick orange-colored aerosol smog. Matthews (1982) proposed this was due to the presence of HCN polymers in Titan's atmosphere. In experimental simulations of Titan atmospheric chemistry, discharge-induced reactions of the N_2 and CH_4 produce solids called tholins that are primarily non-volatile high molecular weight hydrocarbons and nitrogen heterocycles. The major gaseous product is HCN that can be trapped together with ammonia at low temperature and which polymerizes at higher temperatures (Imanaka 2004). Work in the labs of Bar-Nun *et al.* (1988) and Scattergood *et al.* (1992) showed that, under high energy conditions, acetylene polymerizes more readily than HCN. These observations lead us to propose a new model for Titan's atmospheric chemistry. Methane and acetylene polymerize in the upper atmosphere to give aerosol droplets of colorless hydrocarbons and polyacetylene. HCN does not readily polymerize in the gas phase, but instead is absorbed by the hydrocarbon droplets where it concentrates to the point that it can undergo ammonia or radiation initiated polymerization. As this occurs, the droplet obtains the orange color of HCN polymer. The composite hydrocarbon/HCN polymer droplets gradually rain or settle out onto the surface of Titan. The recent Cassini-Huygens mission to Titan showed evidence of ammonia and hydrogen cyanide from the Aerosol Collector Pyrolyzer-Mass Spectrometer on the Huygens lander (Israel *et al.* 2005). Waite *et al.* (2007) have recorded spectra from Cassini's ion beam and electron spectrometers that indicate the presence of negative ions with molecular weights up to 8,000 consistent with complex carbon-nitrogen precursors. This Titan atmospheric chemistry is likely relevant to early Earth.

As seen for HCN polymer, amino acids and nucleobases can be released by hydrolysis from the macromolecular material found in carbonaceous meteorites. There are many

similarities in the types and relative amounts of the compounds detected from meteorites and HCN polymer hydrolysates. Hydrolysis experiments with D_2O instead of H_2O
produce similar deuterium labeling patterns for glycine derived from HCN polymer and
from the Murchison meteorite, but not from glycine monomer (Matthews et al. 1977).

In 1986, the Giotto mission to Comet Halley showed that its nucleus was very dark in
color. Spectral analysis indicated the presence of C-H and C≡N species in the crust and
H_2O, HCN, and CN radicals in the coma (Kissel & Krueger 1987). Impact mass spectrometry indicated that some emanating particles consisted of primarily CHON or CHN
(Fomenkova 1997). Matthews & Ludicky (1986) proposed that this data was consistent
with the presence of HCN polymers on Halley and other comets. The Stardust mission
returned cometary matter to Earth in January 2006. The high O and N contents, lower
aromatic contents, and elevated $-CH_2-/CH_3-$ ratios are all qualitatively consistent with
what is expected from radiation processing of astrophysical ices and the polymerization
of simple species such as HCO, H_2CO, and HCN (Sandford et al. 2006).

Comets or meteorites bearing HCN polymers could have been an exogenous source
of these and other organic compounds. In addition, atmospheric reactions of methane
and nitrogen could have been an endogenous source. With both these possible inputs, it
seems likely that HCN polymers were present on the early Earth. These polymers are the
most likely starting point for the origin of protein/nucleic acid based life and therefore
are the critical link connecting cosmochemistry and biochemistry.

References

Bar-Nun, A., Kleinfeld, I., & Ganor, E. 1988, *Journal of Geophysical Research*, [Atmospheres],
    93, 8383.
Bernstein, M. P., Sandford, S. A., & Allamandola, L. J. 1999, *Sci. Am.*, 281, 42-49
Ehrenfreund, P., Rasmussen, S., Cleaves, J. & Chen, L. 2006, *Astrobiology*, 6, 490.
Ferris, J. P. & Orgel, L. E. 1966, *J. Amer. Chem. Soc.*, 88, 1074.
Fomenkova, M. N. 1997 in *From Stardust to Planetesimals*, Ed. Y.J. Pendleton and A. G. G.
    M. Tielens, Astronomical Society of the Pacific, San Francisco, 415.
Glaser, R., Hodgen, B., Farrelly, D., McKee, E. 2007, *Astrobiology*, 7, 455.
Imanaka, H. 2004, Laboratory Simulations of Titan's Organic Haze and Condensation Clouds
    (Ph.D. Thesis). University of Tokyo, Tokyo, Japan.
Israel, G. et al. 2005, *Nature*, 438, 796.
Kissel, J. & Krueger, F. R. 1987, *Nature* (London), 326, 755.
Mamajanov, I. & Herzfeld, J. 2008, submitted for publication.
Matthews, C. N. 1982, *Origins of Life and Evolution of the Biospheres*, 12, 281
Matthews, C. N. & Ludicky, R. A. 1986, Proceedings 20th ESLAB Symposium on the Explo-
    ration of Halley's Comet, ESA SP-250, 273.
Matthews, C. N. & Minard, R. D. 2006, *Faraday Discuss.*, 133, 393-401 and subsequent discus-
    sion.
Matthews, C. N. & Moser, R. E. 1966, Proc. National Acad. of Science. USA, 56, 1087.
Matthews, C. N. & Moser, R. E. 1967, *Nature*, 215, 1230.
Matthews, C., Nelson, J., Varma, P., & Minard, R. 1977, *Science*, 198, 622
Minard, R., Yang, W., Varma, P., Nelson, J., & Matthews, C. 1975, *Science*, 190, 387.
Miyakawa, S., Cleaves, H. J., Miller, S. 2002, *Origins of Life and Evolution of the Biospheres*,
    32, 209.
Oró, J. 1961, *Arc. Biochem. and Biophys.*, 94, 217-227
Oró, J. & Kamat, S. S. 1961, *Nature*, 190, 442.
Platts, S. N. 2004, http://www.pahworld.com accessed 3/31/2008.
Proust, J. L. 1808, *The Philosophical Magazine*, Tilloch, A., Ed., 32, 336.
Sandford, S. A. et al. 2006, *Science*, 314, 1720.
Scattergood, T. W., Lau, E. Y., Stone, B. M. 1992, *Icarus*, 99, 98.

Thaddeus, P. 2006, *Phil. Trans. R. Soc. B*, 1681

Waite, J. H., Young, D. T., Cravens, T. E., Coates, A. J., Crary, F. J., Magee, B., & Westlake, J. 2007 *Science*, 316, 870

Discussion

SANDFORD: We have done lots of experiments where we irradiate ices that are similar to what we see in dense clouds to see what complex organics are made. Most of these ices are very rich in H_2O, CH_3OH, and CO, so there is enough oxygen around that we don't see CN polymers of exactly the type you are talking about. However, one of the more abundant products we do see even in these O-rich mixtures is hexamethylenetetramine. I am curious what you think about how this might be related to what you've talked about.

MATTHEWS: Some papers have come out from the French workers on the question of hexamethylenetetramine versus HCN polymer (Fray, N., Bénilan, H., Cottin, H., *et al.* 2004, *Meteoritics and Planetary Science*, 39, 581). They say their experiments fit the HCN polymer very well, but not the hexamethylenetetramine. That's not a very good answer, but they have studied it.

CODY: We have attempted to do some NMR on HCN polymer and we've recognized, and you've confirmed in your talk, that these polymers have a lot of free radicals in them. I am intrigued as to whether you know, on the basis of electron paramagnetic resonance spectroscopy, if these are di-radicaloid type radicals?

MATTHEWS: The polymers are free radical in character. We usually use base catalysis, but radical catalysis is just as good. One of the papers (Budil, D. E., Roebber, J. L., Liebman, S. A., & Matthews, C. N. 2003, *Astrobiology*, 3, 323) shows you have carbon and nitrogen free radicals.

KHARE: In life as we know it, we need 20 amino acids. Out of those 20, two amino acids contain sulfur. Do you have any recipe to include these sulfur-containing amino acids as well?

MATTHEWS: They will be contained if H_2S is in the HCN polymer. If you have H_2S in the HCN mixture, it will modify the polyamino structure with the active CN group in the middle and give you S-containing side chains.

Cliff and Sandra Matthews enjoying a walk at the Victoria Peak during the Wednesday tour (photo by Bill Irvine).

Organic Matter in Space
Proceedings IAU Symposium No. 251, 2008
S. Kwok & S. Sandford, eds.

Laboratory simulation of the evolution of organic matter in dense interstellar clouds

Vito Mennella

Istituto Nazionale di Astrofisica, Osservatorio Astronomico di Capodimonte,
via Moiariello 16, 80131 Napoli, Italy
email: mennella@na.astro.it

Abstract. Laboratory simulation of interstellar grain processing is a unique tool to better understand the nature and evolution of cosmic dust. In recent years this approach has been crucial to outline a new model of evolution of the aliphatic component of organic matter in the interstellar medium. Here, the results of a recent laboratory research on processing of nano-sized carbon particles by H atoms under simulated dense medium conditions are discussed. The experiments show that the formation of C-H bonds in the aliphatic CH_2 and CH_3 functional groups does not take place, while the activation of a band at 3.47 μm, due to the C-H stretching vibration of tertiary sp^3 carbon atoms, is observed. These results indicate that the assumption about inhibition of aliphatic C-H bond formation in interstellar dense clouds is correct. Moreover, they suggest that carbon grains responsible for the interstellar aliphatic band at 3.4 μm in diffuse regions can contribute to the absorption observed at 3.47 μm in dense clouds.

Keywords. ISM: dust, extinction, methods: laboratory

1. Introduction

The presence of an aliphatic organic dust component in diffuse regions of the interstellar medium is indicated by the 3.4 μm absorption band with subfeatures at 3.38, 3.42, and 3.48 μm due to the C-H stretching modes in the methyl (CH_3) and methylene groups (CH_2) (Sandford *et al.* 1991, Pendleton *et al.* 1994). The carrier of the feature is a widespread component of diffuse interstellar dust. The 3.4 μm band is absent in the spectrum of dense cloud dust: the number of aliphatic C-H bonds is reduced by at least 55% with respect to the diffuse regions (Muñoz Caro *et al.* 2001). In dense clouds, the C-H stretching spectral region is generally characterized by the absorption bands at 3.25, 3.54 and 3.47 μm. Allamandola *et al.* (1992) first detected the band at 3.47 μm in four protostars. Subsequent observations have confirmed that the 3.47 μm band is a common feature of young stellar objects embedded in molecular clouds (Sellgren *et al.* 1994, Brooke *et al.* 1996, 1999). Chiar *et al.* (1996) reported the first detection of the band in quiescent medium along the line of sight of the field star Elias 16 in the Taurus cloud. The optical depth of the 3.47 μm band is better correlated with the optical depth of the 3 μm water ice band than with that of the 9.7 μm silicate absorption feature, suggesting that the absorption is not due to a refractory component (Chiar *et al.* 1996, Brooke *et al.* 1999). Allamandola *et al.* (1992) attributed the band to the C-H stretching vibration of tertiary sp^3 carbon atoms, and they suggested interstellar diamond grains as carrier of the feature. More recently, the band has been assigned to a vibrational mode due to the interaction of the nitrogen atom of NH_3 with an OH bond of the water molecule, forming an ammonia hydrate (Dartois *et al.* 2002).

The distinct spectral difference between dense and diffuse medium represents a strong constraint for any description of formation and evolution of interstellar organic matter

(Allamandola *et al.* 1993). It has long been enigmatic. In recent years, on the basis of specific laboratory simulation of interstellar processing, a model has been proposed to reconcile the dichotomy of the 3.4 μm band through an evolutionary transformation of interstellar organic matter. During cycling of materials between diffuse and dense interstellar regions the degree of processing and the properties of grains change and these changes determine the observed spectral variations. In the diffuse medium, bare carbon grains are processed by UV photons and cosmic rays, which destroy C-H bonds, and H atoms, which form aliphatic bonds. Competition between destruction and formation determines an equilibrium value for grain hydrogenation and we can observed the 3.4 μm band. In dense regions an ice layer, consisting primarily of H_2O, covers grains. In these conditions destruction of C-H bonds by cosmic rays and the internal UV field remains active, while their formation should stop, determining a gradual reduction of the intensity of the C-H stretching feature (Mennella *et al.* 1999, 2001, 2002, 2003, Muñoz Caro *et al.* 2001, Mennella 2006). The previous model relies on the results of specific laboratory studies of the interaction of UV photons, ions, and hydrogen atoms with hydrogen-free and hydrogenated carbon grains under simulated diffuse and dense medium conditions. From these experiments the C-H bond formation and destruction cross sections have been obtained. Knowledge of these quantities has allowed a quantitative description of the evolution of the aliphatic component.

All processes driving the evolution of interstellar organic matter have been studied in the laboratory, except for the intercation of H atoms with carbon grains covered with an ice cap.

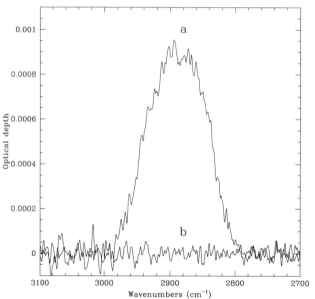

Figure 1. (a) Activation of the 2883 cm^{-1} (3.47 μm) band after H atom exposure to 1.0×10^{18} atoms cm^2 of nano sized carbon grains covered with a water ice layer at 12 K. (b) For comparison, the spectrum of a water ice layer, with a thickness of 27 nm, deposited on a Cs I window after an irradiation of 1.4×10^{18} atoms cm^2.

2. H atom irradiation of carbon grains in dense clouds

Only recently have simulations of the evolution of organic matter in the ISM been completed with laboratory research on H atom processing of nano-sized carbon particles under simulated dense medium conditions. The variations of the degree of hydrogenation of carbon particles covered with a water ice layer in response to H atom exposure have been studied (Mennella 2008).

Hydrogen-free carbon grain samples with a water ice layer have been exposed at 12 K to an atomic hydrogen beam obtained by microwave - excited discharge of molecular hydrogen. H atoms were cooled to a temperature of 80 K. The effects of H atom irradiation were studied with IR spectroscopy. Figure 1 shows the results obtained for carbon grains covered with a water ice layer of 22 nm. Exposure to 1×10^{18} H atoms cm^{-2} does not activate the aliphatic and aromatic C-H bonds, while, quite surprisingly, the activation of a band at 2883 cm^{-1} (3.47 μm) is observed. In a control experiment, H irradiation of a water ice layer deposited on a CsI substrate, shows that no feature appears in the C-H stretching region (see Figure 1), indicating that the absorption at 3.47 μm is activated in carbon grains by H atom exposure. Following Allamandola *et al.* (1992), the band at 2883 cm^{-1} (3.47 μm) has been attributed to the C-H vibration of tertiary sp^3 carbon atoms. The 3.47 μm band had been identified as a component of the 3.4 μm feature of hydrogenated amorphous carbon film using a line shape fitting analysis (Grishko & Duley 2000).

These experiments have shown that H irradiation of carbon grains under simulated dense cloud conditions does not activate the formation of aliphatic C-H bonds. This result confirms that the assumption about inhibition of formation of this type of bonds in dense clouds is correct. On the other hand, the activation of the 3.47 μm band is observed. The different behaviour of the two types of C-H bonds has been interpreted in terms of different activation energies for C-H bond formation in aliphatic and tertiary sp^3 carbon sites. H atoms, impinging on the sample, interact with the water layer at 12 K. Some of the H atoms are adsorbed in the water ice, while the remainder reach carbon grains. However, due to the interaction with the ice layer, their temperature is lowered well below the activation temperature of ~ 70 K for the formation of C-H bonds in the CH$_2$ and CH$_3$ groups (Mennella 2006).

On the other hand, the observed formation of C-H bonds in tertiary sp^3 carbon sites implies a negligible activation energy for these bonds.

Figure 2 shows the comparison of the profile of the band activated by H atoms in carbon grains with a water ice cap with that of interstellar sources. The interstellar spectra (from Brooke *et al.* 1999) refer to a quiescent region (Elias 16 in the Taurus cloud) and two young stellar objects (GL 2591 and W51). For two massive protostars (GL 989 and GL 2136), Dartois *et al.* (2002) have derived a profile of the 3.47 μm band broader than that obtained by other authors (e.g., Brooke *et al.* 1999). This result is a consequence of a different choice of the continuum used to obtain the optical depth of the feature (for more details, see Dartois *et al.* 2002). Note that also in the case of a broader interstellar feature the laboratory band remains compatible with it, accounting for only a part of the overall band profile. The spectral compatibility of an analog material is only a first step towards a solid assignment of the interstellar feature. Formation and destruction cross sections have to be estimated to see whether the proposed carrier is compatible, from the evolutionary point of view, with the conditions present in dense clouds.

The C-H stretching vibration of tertiary sp^3 carbon atoms, which was proposed as responsible for the 3.47 μm band by Allamandola *et al.* (1992), has been activated in carbon grains through experiments simulating dense cloud conditions. These authors proposed diamond grains as carrier of the feature, a grain population different from the carrier of the diffuse medium 3.4 μm band. The laboratory results suggest that there is no need to invoke a separate grain population. As a consequence of evolutionary

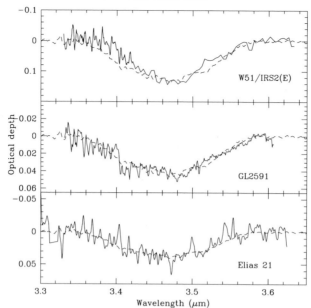

Figure 2. Comparison of the 3.47 μm band activated by H atom exposure in carbon grains covered with a water ice layer (thin dashed lines) with that of three dense clouds (solid lines) (Brooke *et al.* 1999).

transformations carbon grains responsible for the 3.4 μm band in the diffuse ISM can absorb at 3.47 μm in dense clouds.

Acknowledgements

This work has been supported by ASI, INAF and MIUR research contracts.

References

Allamandola, L. J., Sandford, S. A., Tielens, A. G. G. M., & Herbst, T. M. 1992, *ApJ*, 399, 134

Allamandola, L. J., Sandford, S. A., Tielens, A. G. G. M., & Herbst, T. M. 1993, *Science*, 260, 64

Brooke, T. Y., Sellgren, K., & Geballe T. R. 1999, *ApJ*, 517, 883

Brooke, T. Y., Sellgren, K. & Smith, R. G. 1996, *ApJ*, 459, 209

Chiar, J. E., Adamson, A. J. & Whittet, D. C. B. 1996, *ApJ*, 472, 665

Dartois, E., d'Hendecourt, L., Thi, W., Pontoppidan, K. M., & van Dishoeck E. F. 2002, *A&A*, 394, 1057

Grishko, V. I. & Duley, W. W. 2000, *ApJ* (Letter), 543, L85

Mennella, V. 2006, *ApJ* (Letter), 647, L49

Mennella, V. 2008, *ApJ* (Letter), submitted

Mennella, V., Baratta, G. A., Esposito, A., Ferini, G., & Pendleton, Y. J. 2003, *ApJ*, 587, 727

Mennella, V., Brucato, J. R., Colangeli, L., & Palumbo, P. 1999, *ApJ*, 524, L71

Mennella, V., Brucato, J. R., Colangeli, L., & Palumbo, P. 2002, *ApJ*, 569, 531

Mennella, V., Muñoz Caro, G., Ruiterkam, R., Schutte, W. A., Greenberg, J. M., Brucato, J. R., & Colangeli, L. 2001, *A&A*, 367, 355

Muñoz Caro, G., Ruiterkam, R., Schutte, W. A., Greenberg, J. M., & Mennella, V. 2001, *A&A*, 367, 347

Pendleton, Y. J., Sandford, S. A., Allamandola, L. J., Tielens, A. G. G. M., & Sellgren, K. 1994, *ApJ*, 437, 683

Sandford, S. A., Allamandola, L. J., Tielens, A. G. G. M., Sellgren, K., Tapia, M., & Pendleton, Y. 1991, *ApJ*, 371, 607

Sellgren, K., Smith, R. G., & Brooke, T. Y. 1994, *ApJ*, 433, 179

Discussion

BERNSTEIN: I understand that in your experiments the over layer of water suppresses the formation of the aliphatic C-H stretch at 3.4 microns. However, I would be cautious about strong statements that only the tertiary C-H can form in dense clouds. Earlier today we heard from Dr. Cuppen, who talked about H atom addition to carbon monoxide forming reduce compounds. I also showed earlier that the photolysis of PAHs in water ice, where hydrogen atoms are generated in situ, can add hydrogen atoms to aromatic systems.

MENNELLA: For my samples, I am evaluated the formation cross-section and there is a small activation energy for the aliphatic component. On the other hand, there is destruction of the aliphatic bonds by UV photons and cosmic rays. I do not see why if you produce the aliphatic bonds by photolysis of ices you do not see aliphatic feature in dense regions. You should explain to me why the aliphatic modes are lacking in dense regions while you can observe the 3.47 micron band. The results of the experiments I have presented indicate a possible origin for the difference between diffuse and dense medium spectral behaviour.

Vito Mennella talking about the evolution of organic material in dense clouds.

From left to right: Susana Kglesias-Groth, Arturo and Anna Manchado at the banquet.

Organic Matter in Space
Proceedings IAU Symposium No. 251, 2008
S. Kwok & S. Sandford, eds.

© 2008 International Astronomical Union
doi:10.1017/S1743921308022199

Formation of amino acid precursors with large molecular weight in dense clouds and their relevance to origins of bio-homochirality

Kensei Kobayashi[1], Takeo Kaneko[1], Yoshinori Takano[2], and Jun-ichi Takahashi[3]

[1]Graduate School of Engineering, Yokohama National University
79-5 Tokiwadai, Hodogaya-ku, Yokohama 240-8501, Japan
email: kkensei@ynu.ac.jp, t-kaneko@ynu.ac.jp

[2]IFREE, Japan Agency for Marine-Earth Science and Technology
2-15 Natsushimacho, Yokosuka 237-0061, Japan
email: takano@jamstec.go.jp

[3]NTT Microsystem Integration Laboratories
3-1 Morinosato-Wakamiya, Atsugi 243-0198. Japan
email: jitaka@aecl.ntt.co.jp

Abstract. A wide variety of organic compounds have been found in carbonaceous chondrites and comets, which suggests that extraterrestrial organic compounds could have been an important source of the first terrestrial biosphere. In the Greenberg model, these organic compounds in the small bodies were originally formed in interstellar dusts (ISD) in dense clouds by the action of cosmic rays and ultraviolet light. We irradiated a frozen mixture of methanol, ammonia and water with high-energy heavy ions from an accelerator ("HIMAC" in NIRS, Japan) to simulate the action of cosmic rays in dense clouds. Racemic mixtures of amino acids were detected after hydrolysis of the irradiation products. A mixture of carbon monoxide, ammonia and water also gave such complex amino acid precursors with large molecular weights. When such amino acid precursors were irradiated with circular polarized UV light from a synchrotron, enantiomeric excesses were detected. The yield of amino acids was not largely changed between, before, and after CPL-irradiation. The present results suggest that the seed of homochirality of terrestrial amino acids were originally formed in interstellar space.

Keywords. Astrobiology, cosmic rays, ISM: dust, ISM: molecules, ultraviolet: stars

1. Introduction

Miller (1953) reported that amino acids could be formed in spark discharges in a gas mixture of methane, ammonia, hydrogen and water. Since then, a great number of experiments have been reported to study possible abiotic formation of bioorganic compounds. In earlier studies, strongly reducing gas mixtures such as a mixture of methane, ammonia and water were used as starting materials ("simulated primitive Earth atmospheres"). Many kinds of amino acids were detected in the products when various kinds of energies including ultraviolet light (Sagan & Khare 1971) and heat (Harada & Fox 1964). It is suggested, however, that the primitive atmosphere of the Earth was not strongly reduced, but only slightly reduced: Carbon dioxide, carbon monoxide, nitrogen and water were among major constituents of the primitive Earth atmosphere (Kasting 1990). It is not easy to form bioorganic compounds from slightly reduced gases by such major energies as ultraviolet light, spark discharge and heat. It is still possible that amino acids and nucleic acid bases can be synthesized from slightly reduced gas mixtures by high-energy particles (Kobayashi *et al.* 1998), X-rays (Takahashi *et al.* 1999) or high-temperature plasma (Miyakawa *et al.* 2002), but the total production rate of bioorganic compounds from slightly-reduced atmospheres are much less than that from strongly-reduced atmospheres.

The other major problems in the origin of life include the riddle of the origin of homochirality of bioorganic compounds, especially amino acids. Proteinous amino acids (except glycine) have D- and L-enantiomers, and both enantiomers can be formed equally when they are abiotically synthesized. On the other hand, terrestrial organisms use only L-amino acids when they synthesize proteins biologically. The development of specific chirality in amino acids has remained one of the most important problems with regard to chemical evolution on the primitive Earth.

A wide variety of organic compounds have been found in materials in carbonaceous chondrites and comets, and the relevance between them and terrestrial origins of life is discussed. Here we discuss possible formation of organic compounds in extraterrestrial environments and show a persuasive scenario for the exogenous origins of homochirality of amino acids by using our recent results of laboratory simulation experiments.

2. Organic compounds in space

Carbonaceous chondrites, comets and interplanetary dust particles are three of the major candidates, since they contain considerable amount of organic compounds. Chyba & Sagan (1992) estimated the inventory rates of organic (or carbon-containing) compounds by meteorites, comets and interplanetary dust particles (IDPs), and concluded that approximately 6×10^7 kg of organic carbon was delivered by extraterrestrial bodies.

Organic compounds in carbonaceous chondrites have been comprehensively analyzed. A wide variety of organic compounds have been detected in extracts from carbonaceous chondrites, including amino acids (Kvenvolden *et al.* 1970) together with unidentified complex organic compounds. Organic globules were found in Tagish Lake meteorite, and isotopic ratio of the carbon suggested that these organic compounds were formed at quite low temperature (10-20 K) (Nakamura-Messenger *et al.* 2006).

Comets also bring a wide variety of organic compounds. Mass spectra of dusts of Comet Halley (Kissel & Krueger 1987) suggested that cometary dusts have various types of complex organic compounds including heterocyclic compounds. There were, however, no indications for amino acids in cometary dust. Cometary dust from Comet 81P/Wild 2 was captured by Stardust spacecraft. Preliminary analysis of the dusts showed that they have novel types of complex organic compounds that seemed to be more primitive than meteoritic organics (Sandford *et al.* 2006).

Greenberg proposed a scenario that these exogenous organic compounds were originally formed in interstellar dust (ISD) environments in dense clouds (Greenberg & Li 1997). Water, carbon monoxide, methanol and ammonia were detected among major constituents of ice mantles of ISDs (Greenberg & Mendoza-Gomez 1993), but more complex organics could not been observed. A number of studies have been done to see what kind of organic compounds can be formed in such icy environments.

3. Possible formation of amino acids in ISDs

Inner dense clouds are as cold as 10-20 K, and most molecules are frozen onto ISDs to form ice mantles. High-energy cosmic rays can go through the dense clouds, and ultraviolet light is generated when cosmic rays interact with molecules. Thus, cosmic rays and secondary ultraviolet light are possible energy sources for the reactions in ice mantles of ISDs. When a frozen mixture of molecules found in interstellar dusts such as carbon monoxide and ammonia was irradiated with high-energy protons or UV light, amino acids were detected in hydrolyzed products (Kobayashi *et al.* 1995, Muñoz Caro *et al.* 2002, Bernstein *et al.* 2002). It was difficult, however, to discuss the formation rate of amino acids in ISDs or to characterize the amino acid precursors before hydrolysis since the only limited amount of amino acid precursors were obtained in these experiments.

In order to discuss the energy yield of amino acid precursors, we chose a mixture of methanol, ammonia and water, found in ISD ice mantles as major constituents, and (ii) it is easy to make a large quantity of ice. Energies applied to the materials are high-energy heavy ions, γ-rays and ultraviolet light. The starting mixture was placed in a liquid nitrogen bath (77 K), or at ambient temperature so that the starting material was kept as solid ice or liquid, respectively. We performed heavy ion irradiation by using the "HIMAC" accelerator in National Institute of Radiological Sciences (Chiba, Japan). Such ions as helium (150 MeV amu^{-1}), carbon (290 MeV amu^{-1}), neon (400 MeV amu^{-1}) and argon (500 MeV amu^{-1}) were irradiated to the solid (77 K) or liquid (ambient) mixture of methanol, ammonia and water. Fifty gram each of the mixture was irradiated with heavy ions at the dose rate of 250 - 4800 Gy h^{-1}. The total dose was 700 - 15800 Gy, and the total energy deposit was $2.2 \times 10^{20} - 4.9 \times 10^{21}$ eV (Kobayashi *et al.* 2007). After individual irradiation experiments, an aliquot of the irradiation products was hydrolyzed with 6 M HCl at 383 K for 24 hours. Amino acids in the hydrolyzed and unhydrolyzed fraction were analyzed with an ion exchange high performance liquid chromatography (HPLC) system where a post-column derivatization with o-phthalaldehyde and N-acetyl-L-cystein was applied, and/or gas chromatography / mass spectrometry (GC/MS) after derivatization with ethyl chloroformate and heptafluorobuthanol.

Figure 1 shows mass spectra of glycine derivative: The glycine was obtained by carbon ion irradiation of the frozen mixture of ^{13}C-methanol, ammonia and water at 77 K. Besides glycine, proteinous amino acids such as alanine, aspartic acid and non-proteinous amino acids such as β-alanine, and α- and γ-aminobutyric acid were detected. In the unhydrolyzed fraction, only small amount of glycine was detected. It showed no free amino acids, but amino acid precursors were formed during irradiation.

The energy yields (G-values) of glycine were determined by irradiation of the mixture with high-energy carbon ions (290 MeV amu^{-1}). The G-values of glycine by irradiation with carbon ions at 77 K were 0.007, while that at ambient temperature was 0.014. The former is about 50% of the latter. The present results show that amino acid precursors can be formed in ice mantles of ISDs in dense clouds by cosmic rays and/or UV light effectively, even if the temperature of the source mixtures is so low that they are solid ices.

4. Characterization of complex organic compounds formed from a mixture of carbon monoxide, ammonia and water

It is proved that complex organic compounds including amino acid precursors can be formed in solid, liquid or gas phase mixtures (Kobayashi *et al.* 2004a). When a gas mixture of carbon monoxide, ammonia and water was irradiated with high-energy protons, complex organic compounds containing amino acid precursors could be formed quite effectively. The G-value of glycine (after hydrolysis) was as high as 0.23 (Kobayashi *et al.* 2007). We characterized the complex organic compounds (hereafter referred as CAW).

Discussion

MUMMA: I was very interested in your comments on mechanisms that might lead to homochirality in meteoritic amino acids. The problem I have with this is that, although they might do so in free space as you have suggested, those amino acids probably didn't form there. Indeed if you go to the CM, CI, and CR meteorites, the number of amino acids dramatically decrease, leading one to wonder whether any amino acids were present in organic material before entering the solar nebula. If you form them in the solar nebula, then the whole issue of circularly polarized light causing this homochirality or enantiomeric excesses, or energetic electrons which you have also suggested, seems to me to not be relevant. So I then pose the question that there must be some other processes that is active in the protoplanetary disk that creates this enantiomeric excess. Perhaps you could comment on this.

KOBAYASHI: Thank you very much for your comments. Yes, my idea is that complex organic acids are formed originally in dense clouds, but they are modified in other environments like diffuse clouds, protoplanetary nebula, and so on. This action could take place in any of these environments.

Organic Matter in Space
Proceedings IAU Symposium No. 251, 2008
S. Kwok & S. Sandford, eds.

© 2008 International Astronomical Union
doi:10.1017/S1743921308022205

High-power laser-plasma chemistry in planetary atmospheres

Svatopluk Civiš[1] and Libor Juha[2]

[1] J. Heyrovský Institute of Physical Chemistry ASCR, Dolejškova 3, 182 23 Prague 8,
Czech Republic
email: civis@jh-inst.cas.cz

[2] Institute of Physics ASCR, Na Slovance 2, 182 21 Prague 8, Czech Republic
email: juha@fzu.cz

Abstract. Large laser sparks created by a single shot of a high-power laser system were used for the laboratory simulation of the chemical consequences of high-energy-density events (lightning, high-velocity impact) in planetary atmospheres, e.g., the early Earth's atmosphere.

Keywords. Astrobiology, astrochemistry, plasmas, methods: laboratory, techniques: spectroscopic

It is currently a generally accepted opinion that Earth's early atmosphere was a mildly reduced mixture of molecular gases. This represents a crucial difficulty for the classical Miller's experiments. These experiments were carried out in a strongly reducing gaseous mixture. In a weakly reducing environment they give a poor yield of amino acids and other compounds of relevance to chemical evolution. In our study we intended to simulate impact shocks and lightning in the mixtures of molecular gases modeling Earth's early atmosphere with help of a focused beam from high-power laser systems (Civis *et al.* 2004, Babankova *et al.* 2006a and references cited therein). The main goal of this program is to diagnose the laser-produced plasma, determine its basic physical characteristics, and investigate their links to the chemical action of the laser spark in such a reaction system. Varying laser-plasma interaction conditions and composition of the gas mixtures of the chemical evolution during various stages of Earth's early atmosphere evolution was investigated in dependence on the kind and energetics of the initializing high-energy density event (i.e., impact shocks and atmospheric discharges).

A single laser pulse delivered from a Prague Asterix Laser System (a pulse duration of 400 ps and a wavelength of 1.315 μm) with energy content $\leqslant 100$ J was used for irradiation of the CO-N_2-H_2O gas mixtures (mildly reducing systems) and their components at atmospheric pressure. The high pulse energy and the relatively short pulse duration result in the observed large volume of the laser spark.

Chemical consequences of the laser-produced plasma generation in such a gaseous system were investigated by gas/liquid chromatography and high-resolution Fourier transform infrared absorption spectrophotometry with a Bruker IFS 120 in the spectral interval from 500-7000 cm^{-1} at a resolution of 0.0035 cm^{-1}. Several organic compounds were identified in the reaction mixture exposed to a few laser shots. The reaction mechanism of CO_2 formation was investigated using stable isotope-labeled water $H_2^{18}O$ (Figure 1).

Optical emission spectra (OES) of the large laser spark were measured by a MS255 spectrometer (Oriel) equipped with time-resolved ICCD detector (Andor) in the spectral range of 350-1000 nm. A significant difference has been found in the optical spectra of LIDB plasmas created in CO and N_2 containing mixtures in the static cell and gas puff (Babankova *et al.* 2006b). In the case of the jet, there is no examined emission from

Gilbert, W. 1986, *Nature*, 319, 618

Janas, T. & Yarus, M. 2003, *RNAPubl. RNA Soc.*, 9, 1353

Jan, E., Kinzy, T. G., & Sarnow, P. 2003, *Proc. Natl. Acad. Sci., USA 100*, 15, 410

Lyons, A. J. & Robertson, H. D. 2003, *J. Biol. Chem.*, 278, 26 844

Miller, S. L. 1998, in: Brack, A. (ed.), *The endogenous synthesis of organic compounds*, The Molecular Origins of Life, (Cambridge University Press, Cambridge), p. 59

Orgel, L. E. 2004, *Critical Rev. Biochem. Molec. Biol.*, 39, 99

Schmidt, J. G., Nielsen, P. E., & Orgel, L. E. 1997,*Nucleic Acids Res.*, 25, 4797

Jorge Enrique Bueno Prieto and Adriana Patricia Lozano Medellin during the welcome reception at Renaissance Harbour View Hotel

Organic Matter in Space
Proceedings IAU Symposium No. 251, 2008
S. Kwok & S. Sandford, eds.

© 2008 International Astronomical Union
doi:10.1017/S1743921308022229

Guanine synthesis from interstellar organic molecules as an example of prebiotical chemistry

Adriana Patricia Lozano Medellin

Universidad Nacional de Colombia, Faculty of Science, Department of Chemistry, Colombia
email: aplozanom@unal.edu.co

Abstract. The idea that some nitrogen bases, such as adenine and guanine, are easily formed and can be found in interstellar space is should not be rejected. In fact, they have been found in the soluble fraction of some ancient meteors. It is possible that the "seed" of life had its origin in space, and thanks to Earth conditions, a synergistic interaction ocurred and allowed life to spring forth on our planet.

Keywords. Guanine, synthesis, prebiotical, chemistry, DNA

1. Introduction

Guanine is one of the five main nucleobases found in the nucleic acids DNA and RNA, the others being adenine, cytosine, thymine, and uracil. With the formula $C_5H_5N_5O$, guanine is a derivative of purine, consisting of a fused pyrimidine-imidazole ring system with conjugated double bonds.

Two experiments conducted by Levy *et al.* showed that heating 10 M NH_4CN at 80 °C for 24 hours gave a yield of 0.0007%, while using 0.1 M NH_4CN frozen at -20 °C for 25 hours gave a 0.0035% yield. These results indicate guanine could arise in frozen regions of the primitive Earth.

2. Theoretical proposal

Is it possible to make guanine from organic molecules from interstellar space?

The live organisms require a molecular response RNA or DNA, it is vital to understand whether predecessors exist in interstellar space.

This synthesis is proposed as a reaction of hydrogen cyanide molecules with formaldehyde, going through mediators; malonitrile, acid malonic, nitro malonic acid, 2-amino malomic acid, and 3-oxoserinamide. The last, due to cyclization, forms 2-oxo 4,5 dihydro-1H- imidazole-4 carboxylic acid. This, because of a hydroxyl functional exchange for chloride and reaction with guanine, forms 1,2,3,9 tetrahydro 6H purinone which produces guanine through reduction. As a result, the importance of prebiotic chemistry to generate life is proven.

Acknowledgements

We thank the IAU 251 Committee for travel support and Airline AVIANCA, and Mary Ruth Garcia Conde, a teacher at the National University of Colombia, and thanks to Ernesto Silva of the Chemistry Department.

478   Adriana Patricia Lozano Medellin

References

Bada, J. L. 2004, *Earth Planet. Sci. Lett.*, 226, 1

Bennet, J., Shostak, S., & Jakosky, B. 2003, *Life in the Universe*, (Addison-Wesley, New York), p. 27

Echols, H. &, Goodman, M. F. 1991, *Ann. Rev. Biochem.*, 60, 477

Lahav, N. 1999, *Biogenesis, Theories of Life's Origin*, (Oxford University Press, New York), p. 11

Miller, S. L. 1998, in: Brack, A. (ed.), *The endogenous synthesis of organic compounds*,The Molecular Origins of Life, (Cambridge University Press, Cambridge), p. 59

Miller, S. L. 1987, *Symp. Quant. Biol.*, 52, 17

Sirover, M. A., & Loeb, L. A. 1976, *Science*, 194, 1434

Willians, R. J. 2002, *J. Inorg. Biochem.*, 88, 241

Zubay, G., & Mui, T. 2001, *Orig. Life Evol. Biosph.*, 31, 87

Anisia Tang acknowledging the expression of thanks by the conference participants during the banquet.

BANQUET SPEECH
Full Circle: Star Ferry to Stardust

Clifford N. Matthews

Good evening. I'd like to invite you to join me on a journey that could be entitled "Full Circle: Star Ferry to Stardust". "Star Ferry" represents Hong Kong, my home town, and especially its university - Hong Kong University - as I knew it during the years of World War II. "Stardust" refers to our gathering here to report on our research on possible organic chemistry in space.

In late 1941, some 700 students of the arts and sciences, engineering and medecine lived and worked and played together on this cozy hillside campus. Though well aware of the Japanese presence in China, few, if any, could imagine that their lives would be transformed shortly by the onset of war. On December 8, 1941, Japanese troops occupying south China crossed the border of the New Territories and, using guerrilla tactics, soon had control of the whole peninsula of Kowloon. By December 18, they invaded the main Hong Kong island itself, crossing the harbor at Lyemun and North Point. Defending the colony were British troops - the Middlesex Regiment; Canadian forces - the Winnipeg Grenadiers and the Royal Rifles of Canada; an Indian regiment, and last but not least, the Hong Kong Volunteer Defense Corps. I was one of these Volunteers, who had been rapidly mobilized in response to the emergency, at the same time that I was a student of science at the university. The invaders fought their way up Wongneichong Gap and down to Repulse Bay, where fierce fighting took place just where we are sitting tonight. My unit, the Number 3 Machine Gun Company of the Volunteers, was stationed that night in concrete "pillboxes" on the hillside of the Gap, and suffered severe casualties.

My brother and I had been sent as couriers further up the Gap to warn the Canadians, who had only recently arrived in Hong Kong and hardly knew the terrain, of the Japanese landing. There I met up with other Volunteers and together 8 of us continued down to battle stations at the Wanchai waterfront - near where the opening reception of the Symposium was held last Sunday evening. There, on Christmas Day, we faced advancing Japanese troops who were heading for Central Hong Kong. We were prepared to make a final stand, and I remember thinking that this was going to be the last day of my life. Of course, this thought does not worry you during wartime. But suddenly and unexpectedly, a British officer appeared ahead of us, waving a white flag. Hong Kong had surrendered! All the defending troops were ordered to assemble on the Murray Parade Ground, on Garden Road. We then crossed over to Kowloon by ferry and marched several miles up Nathan Road, somewhat bedraggled, until we reached Sham Shui Po Barracks on the waterfront. There we were to become prisoners of war for the duration.

What happened to the university during those challenging times? An account is given in the book entitled *Dispersal and Renewal: Hong Kong University during the War Years*. You will notice on the cover of the book (Figure 1) a photograph of the Main Building, where we have met every day for the conference sessions, slightly damaged by air bombardment. The campus remained mostly empty during the occupation. What about the students and faculty? There were essentially 3 situations. Those who had fought as volunteers became prisoners of war. Non-combatant British faculty were interned at the civilian camp at Stanley, which many of you visited yesterday. Most of the students, who were predominantly Chinese, were not interned and many escaped into China. You will see on the cover of the book a photograph of a group of students who successfully made the trip through occupied China into Free China on foot and by boat, looking happy and relieved to have arrived. There were many such groups. Accustomed to speaking Cantonese, their first job was to learn Mandarin! In China, most of them taught or studied at Chinese universities.

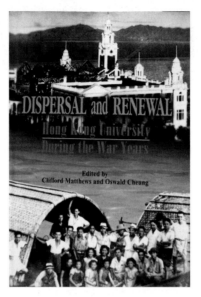

Figure 1.

Let's look at some individual examples of what Hong Kong University faculty and students experienced during the war. Lindsay Ride was a professor of physiology at the University and, during the war, commanded the medical unit of the Volunteers. He became a prisoner of war at Sham Shui Po, but not for long. With the help of his assistant Y.P. Lee, who brought a boat to the Sham Shui Po waterfront, he escaped through barbed wire and they made their way to Free China. There Ride founded the British Army Aid Group, an intelligence unit that operated in South China throughout the war. After the war, he rejoined the University, which had just been restarted by Vice Chancellor Duncan Sloss, and was elected the new vice Chancellor, a position he held for 15 very active years, during which he promoted the rapid growth of the university. Duncan Sloss himself was interned at Stanley. My mother and sister were also there, and remember well his inspiring intellectual presence. My sister even took a Shakespeare course with him.

In Figure 2 you see a reunion gathering in 1995 of former Hong Kong University students who were friends before the war. At the end of the war, a number of them were invited by the British Council to become graduate students at English universities. On the left in the back row, Oswald Cheung took a law degree in London, and returned to Hong Kong for a career in law and politics that earned him a knighthood. He was also the first Chinese to become head of the Hong Kong Jockey Club, where he used his influence to start the new University of Science and Technology in the New Territories.

Next to him is H.T. Huang, who had an unusual experience in China during the war. Teaching at Chengdu, he saw a notice one day announcing the arrival of Joseph Needham, a scientist from Cambridge University, whose mission was to assist Chinese universities

Figure 2.

in any way possible. H.T. became his driver and translator, and for a year and a half they visited Chinese universities, at the same time frequenting bookstores in back streets where they found many manuscripts on science in China. Returning to Cambridge after the war, Needham planned to publish a slim volume on science and civilization in China. Instead this became a lifetime endeavor; there are now over 20 books in print and 3 more to come. While working after the war at the National Science Foundation in the U.S., H.T. Huang remained an active collaborator on this profound scholarly project, authoring the section entitled "Fermentations and Food Science".

Also in the reunion group is Rayson Huang. Originally a science student at Hong Kong University (I was his lab partner in chemistry), Rayson went on, like H.T., to obtain his PhD. at Oxford working with Sir Robert Robertson. Returning to Asia, he became first a professor of chemistry at Kuala Lumpur and then the first Vice Chancellor of the new Nanyang University in Singapore. From there, he received an offer he could not refuse - to become Vice Chancellor of Hong Kong University, his alma mater. During his 14 years as Vice Chancellor, he met with Deng Xiao Ping, China's Premier, to discuss the imminent return of Hong Kong to China. Upon retiring, Rayson Huang continued to be active on the committee that drafted the Basic Law for Hong Kong, establishing the principle of "One Country: Two Systems".

His distinguished career as the first Chinese Vice Chancellor of the University shows the great change in Hong Kong after the war. Before the war, the Vice Chancellor of the University would have been brought in from England. After the war, local Chinese were able to take on many leadership positions, and Hong Kong became what it still is today - a multi-cultural center bridging East and West.

At our reunion, I recall us discussing the possibility that there should be a memorial of some kind to the members of the University community who lost their lives during the war. One idea was to erect a memorial fountain on campus. Another was to have a special room in the library devoted to wartime themes. What actually happened was the publication of the book I have already mentioned, which brings together reminiscences by many alumni and former faculty. I became the editor and my old school and university friend Oswald Cheung assisted, particularly with the graphics. On the dedication page of the book we named all those who lost their lives as a result of the war.

Of all those at the reunion, I was the only one who was not in China during the war, because I was a POW. At Sham Shui Po I was working alongside thousands of other prisoners at tasks such as building a new runway at Kai Tak Airport. After about a year, unexpectedly, a hundred Volunteers, including my brother and I, were chosen to be sent as POWs to Japan, to an island called Innoshima, in the Inland Sea about 40 miles south of Hiroshima. There we met a hundred other British POWs from Java and Singapore, and we all worked together in the Habu Ship-Building Yard. We lived in wooden barracks such as those pictured in this cartoon by a fellow prisoner (Figure 3). Thirty-two of us were in each room, sleeping on tatamis in bunk beds. Early each morning we would march down to the dock yard to take up our work positions. My group of 6 worked in the foundry, where we threw scrap metal into the furnace to make ingots - a nice job in the winter! Fortunately one of us spoke Japanese from previous experience in Japan, and this enabled us to communicate somewhat with our 4 Japanese supervisors, who we soon grew to respect and appreciate as individuals.

Figure 3.

Most dramatic was V-J day. When we reached the dockyard that morning, we were told not to enter, but to wait outside. Inside, the emperor was about to broadcast on the radio for the first time in history. When we heard that he had announced the Japanese surrender, we immediately returned to our camp, tore down the wooden fence that surrounded it, and roamed around the island for the first time, bringing back chickens and vegetables. We had our first chicken dinner in Japan that night.

I decided not to return to Hong Kong, since the University was not yet functioning, but to complete my studies in England. There, I joined Birkbeck College at London University, where I took classes at night and worked during the day as the chemistry laboratory superintendant. Most inspiring for me was the presence of the physics professor John Desmond Bernal, arguably the most eminent British scientist of that era. A recent biography has appeared entitled *J.D. Bernal: the Sage of Science.* (Figure 4). He was known as sage because of his wide interest in all aspects of science and its role in society. Before the war, he started X-ray crystallography on proteins; during the war he was a major advisor on the Normandy landing and later became scientific advisor to Lord Louis Mt. Batten in the Asian theater. After the war in London, his research included studies on the structure of water, cement, cosmic rays and viruses. He also started research on meteorites arising from his newfound interest in the origin of life, influenced by the early writings of the biochemists Alexander Oparin in Russia and J.B.S. Haldane in England.

In 1947 Bernal gave his famous Guthrie lecture on the origin of life. I was there in the audience at Imperial College, and was amazed to think that one could study such a profound subject through laboratory experiments. I was especially struck by Bernal's insistence on the importance of investigating the structure and origin of proteins, an idea which stuck with me throughout the following years, when I earned my PhD. at Yale followed by fundamental research in industry on hydrogen cyanide. This research brought

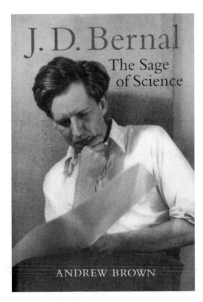

Figure 4.

together for me the connection between hydrogen cyanide and the origin of life, which I continued to pursue through further research at the University of Illinois at Chicago and more recently at Penn State in collaboration with Robert Minard. Tomorrow I will be lecturing on our model, which states that hydrogen cyanide polymers are the bridge between cosmochemistry and biochemistry.

The great interest in Bernal's lecture led him to write his definitive volume on the origin of life, a necessary background to the founding of ISSOL - the International Society for the Study of the Origin of Life - and its journal "The Origins of Life and Chemical Evolution". Bernal, Oparin and Haldane actually took part in early meetings of ISSOL. International gatherings, such as those of ISSOL and the IAU, not only bring scientists together here on Earth, but remind us that ultimately we, and all life, are recycled stardust.

Encore: Enjoying this banquet, I couldn't help but be reminded of the meager diet we had as POWs in Japan. All POWs are obsessed with food: they think food, they talk food, they dream food, and they even sing food, as I shall demonstrate by singing this song, entitled "Protein Deficiency", written by Ted Tandy, brother of the acress Jessica Tandy. Here goes:

There'd be nothing wrong with life on the Inland Sea,
If only I could have a poached egg for my tea.
The local lads stick to rice and fish, quite forgetting that an egg makes a tasty dish.
In the days before the war I'd have an egg with every meal,
but when eggplant supplanted egg my sufferings were real.

For breakfast I used to have ham and eggs; now I have yam and eggplant stew.
I never used to suffer from swollen legs; that's a thing that an egg-fed leg won't do.
For dinner I would have an underdone rump steak, with a nice poached egg on the top.
For tea I would boil em, for boiling doesn't spoil em,
And there perhaps you'd think that I would stop, but NO!

I love to think of suppers in the days before the war
When I'd have a dozen fried and then tell Ma to fry some more.
In the far future days when the Inland Sea is nothing more to us than a memory,
There'll be eggs in my marmalade and eggs in my tripe,
I'll drink them in my beer and I'll smoke them in my pipe.
Every Friday night I'll bath in eggs before I go to bed, and I'll have a kidney omelet in my coffin when I'm dead.

There'll always be an England, and what England means to me is
Steak and chips and two fried eggs and a nice hot cup o' tea.

Acknowledgements

I would like to thank Sun Kwok, Bill Irvine and Sandra Matthews for their considerable assistance, making it possible for me, legally blind, to take part in this stimulating symposium.

Author Index

Object Index

Subject Index

21 μm feature – 179, 213, 215
30 μm feature – 179
2175 Å feature – 11, 12, 15, 35, 36, 57, 67, 71, 73, 75, 77, 433

acenaphthene ($C_{12}H_{10}$) – 361
acetaldehyde (CH_3CHO) – 124
acetamide (CH_3CONH_2) – 20, 28
acetone (($CH_3)_2CO$) – 20
acetylene (C_2H_2) – 7, 9, 48, 82, 139, 180, 197, 198, 223, 322, 427, 445
adenine – 454, 475, 477
adsorption – 112
aerosols – 286, 321, 410
Alais meteorite – 413
alanine – 444, 454
aldehydes – 454
alkyl naphthalenes – 278
Allende – 278
AlO – 223
amides – 11, 468
amino acids – 8, 28, 147, 305, 349, 371, 443, 447, 454, 465, 467, 473
ammonia – 162, 453, 454, 467
amorphization – 436
amorphous carbon – 71, 384, 412, 413
amphiphiles – 305
anions – 157
annealing – 416, 426
anthracene ($C_{14}H_{10}$) – 358, 468
aqueous alteration – 295
aromatic infrared bands (AIB) – 176, 207, 367, 425
aspartic acid – 454
asteroids – 181, 273, 286, 291, 295, 299
astromineralogy – 176
asymptotic giant branch (AGB) stars – 35, 147, 166, 169, 171, 173, 175, 191, 197, 201, 217, 218, 296, 341, 343, 426, 431

benzene (C_6H_6) – 180, 223, 427
benzoperylene ($C_{22}H_{12}$) – 361
buckyonions – 71

C^+ – 65, 92, 93, 139, 140
C_3 – 49, 153, 155, 214, 395, 396

C_4 – 395
C_5 – 51, 395, 402
C_{14} – 398
C_{15} – 397
C_{17} – 397
C_{18} – 51, 398, 399
C_{19} – 397
C_{22} – 398
C_{24}^+ – 67, 68, 399
C_{60} – 57, 58, 59, 61, 183
C_{60}^+ – 5, 51, 400
C_{70} – 57
C_{80} – 58
C_{84} – 57
C_{240} – 57, 58
C_{320} – 58
C_{540} – 58
carbon chains – 4, 5, 395
carbon clusters – 417
carbon dioxide (CO_2) – 8, 11, 47, 92, 202, 223, 377, 443, 451
carbon monoxide (CO) – 8, 47, 49, 92, 112, 142, 145, 171, 221, 223, 252, 378, 443, 445, 457
carbon nano tubes – 43
carbon nanoparticles – 178, 383, 428
carbon nitrides – 413
carbon rings – 395
carbon stars – 197
carbonaceous chondrites – 11, 267, 277, 413, 465, 466, 468
carbonates – 4
carboxylic acids – 447
catalytic reduction – 403
cations – 208
c-C_3H_2 – 260
CCP – 151
CCS – 18
Centaurs – 285
C_2H – 91, 260
C_2H^- – 157
C_2H_3 – 353
C_2H_4 – 178, 322, also see ethylene
C_2H_6 – 9, 322, 445
C_3H – 396

493